Artificial Intelligence and Machine Learning for Sustainable Development

Artificial Intelligence and Machine Learning for Sustainable Development is a comprehensive exploration of how artificial intelligence (AI) and machine learning (ML) technologies are revolutionizing the field of sustainable development.

The book examines cutting-edge innovations, practical applications, and potential challenges in harnessing AI and ML to address global sustainability issues. It offers insights into how these technologies can optimize resource management, improve environmental monitoring, enhance decision-making processes, and promote equitable, eco-friendly solutions. This book would be of special interest to researchers, policymakers, and practitioners seeking to leverage cutting-edge technology for a more sustainable future.

Artificial Intelligence and Machine Learning for Sustainable Development

Innovations, Challenges, and Applications

Edited by
Pawan Whig, Pavika Sharma, Nagender Aneja,
Ahmed A. Elngar, and Nuno Silva

CRC Press
Taylor & Francis Group
Boca Raton London New York

CRC Press is an imprint of the
Taylor & Francis Group, an **informa** business

Designed cover image: Shutterstock Images

First edition published 2025
by CRC Press
2385 NW Executive Center Drive, Suite 320, Boca Raton FL 33431

and by CRC Press
4 Park Square, Milton Park, Abingdon, Oxon, OX14 4RN

CRC Press is an imprint of Taylor & Francis Group, LLC

ISBN: 978-1-032-77835-8 (hbk)
ISBN: 978-1-032-80511-5 (pbk)
ISBN: 978-1-003-49718-9 (ebk)

DOI: 10.1201/9781003497189

Typeset in Times
by KnowledgeWorks Global Ltd.

Contents

Editor Biographies

Dr. Pawan Whig, Research Scientist, Threws and Ex-Dean Research, Vivekananda Institute of Professional Studies-TC, India

Dr. Pavika Sharma, Assistant Professor, Bhagwan Parshuram Institute of Technology, India

Dr. Nagender Aneja, Bradley Department of Electrical and Computer Engineering, Virginia Tech, Blacksburg, VA, USA

Dr. Ahmed A. Elngar, Head of Computers and Artificial Intelligence, Beni-Suef University, Egypt

Dr. Nuno Silva, Chief Scientific and Technology Officer, UnifAI Technology, United Kingdom

Contributors

Amr Abdelaziz
Faculty of Computer Science
Misr International University
Cairo, Egypt

Reem Aboushama
Faculty of Computer Science
Misr International University
Cairo, Egypt

Nur Aeni
Faculty of Languages and Literature
Universitas Negeri Makassar
Makassar, Indonesia

Esther Besty Anaiye
National Open University of Nigeria
Jabi, Abuja, Nigeria

Abdelrahman Mostafa Arab
Faculty of Graduate Studies for Statistical Research
Cairo University
Giza, Egypt

Prithvi Baisla
Manav Rachna International Institute of Research and Studies
Faridabad, India

Ardima Bajpai
Prompt Engineer
NVIDIA
Pune, India

Iti Batra
Vivekananda Institute of Professional Studies – Technical Campus
Delhi, India

P. Mercy Rajaselvi Beaulah
School of Computer Science and Engineering
Vellore Institute of Technology
Chennai, India

R.M. Bhavadharini
School of Computer Science and Engineering
Vellore Institute of Technology
Chennai, India

Amit Channa
Vivekananda Institute of Professional Studies (VIPS-TC) (Affiliated to GGSIP University)
Delhi, India

Basil Osayin Daudu
Kogi State University
Anyigba, Nigeria

Ahmed A. Elngar
Head of Computers and Artificial Intelligence
Beni-Suef University
Beni-Suef City, Egypt

Belal Elsabbagh
Faculty of Computer Science
Misr International University
Cairo, Egypt

George Essam
Faculty of Computer Science
Misr International University
Cairo, Egypt

Walaa Fouda
American University in the Emirates
Dubai International Academic City, Dubai, UAE

Hitanshi Goel
Ph.D. Scholar, University School of Law and Legal Studies
Guru Gobind Singh Indraprastha University
Delhi, India

Ajitha Haridasan
CHRIST (Deemed to be University)
Bengaluru, India

Walaa Hassan
Faculty of Computers & Informatics
Suez Canal University
Ismailia, Egypt & Faculty of Computer Science
Misr International University
Cairo, Egypt

Seema Nath Jain
Vivekananda College
Delhi University
New Delhi, India

Sunanda Vincent Jaiwant
CHRIST (Deemed to be University)
Bengaluru, India

Saydulu Kolasani
Sr. Vice President, Enterprise Digital Operations
Fisker
Los Angeles, California

R. Krithiga
School of Computer Science and Engineering
Vellore Institute of Technology
Chennai, India

Parul Malhotra
Associate Professor
Jagannath University
NCR, Haryana, India

Elizabeth Daggett Matar
American University in the Emirates
Dubai International Academic City, Dubai, UAE

Muthmainnah
Teacher Training and Education Faculty, Indonesian Language Study Program
Universitas Al Asyariah Mandar
Polewali Mandar, Indonesia

Goddy Uwa Osimen
Covenant University
Ota, Nigeria

Naga Ramesh Palakurti
Solutions Architect
Mechanicsburg, Pennsylvania

Selvi Panggua
Universitas Kristen Indonesia
Toraja, Indonesia

Himanshu Rajput
Manav Rachna International Institute of Research and Studies
Faridabad, India

Laxmi Srinivas Samayamantri
VP of Global Engineering – Digital Transformation
Nu Skin
Chicago, Illinois

Ankit Sharma
Sr. Assistant Professor
Vivekananda Institute of Professional Studies (VIPS-TC), (Affiliated to GGSIP University)
Delhi, India

Rattan Sharma
Professor Emeritus & Chairman
Centre for Sustainable Development, VIPS-TC & DSB
Delhi, India

Sachin Sharma
Manav Rachna International Institute of Research and Studies
Faridabad, India

Dheepa T.
CHRIST (Deemed to be University)
Bengaluru, India

Ritik Tanwar
Manav Rachna International Institute of Research and Studies
Faridabad, India

Gopichand Vemulapalli
Principal Data Architect
Agero, Inc.
Tempe, Arizona

Pawan Whig
Research Scientist
Threws
Vivekananda Institute of Professional Studies-TC
New Delhi, India

Sreedhar Yalamati
Solutions Architect
Celer Systems Inc., Technology Services
Folsom, California

Muhammad Yunus
Lecturer at English Education Study Program
Faculty of Letters
Universitas Muslim Indonesia
Makassar City, Indonesia

Obstacles and Resolutions *Integrating Artificial Intelligence in Sustainable Development* 1

Naga Ramesh Palakurti

INTRODUCTION

In recent years, the intersection of artificial intelligence (AI) and sustainable development has emerged as a critical area of exploration and innovation. As the global community strives to address pressing challenges such as poverty alleviation, climate change mitigation, and healthcare accessibility, the potential of AI to catalyze progress toward sustainable development goals (SDGs) has garnered increasing attention. However, the effective integration of AI into sustainable development initiatives is not without its obstacles. This introduction sets the stage for exploring the complexities, opportunities, and resolutions inherent in the endeavor to integrate AI in sustainable development.

The integration of AI in sustainable development represents a convergence of two transformative forces reshaping the contemporary landscape of human endeavor. On the one hand, AI, fueled by advancements in machine learning, natural language processing, and robotics, offers unprecedented capabilities to analyze vast datasets, automate tasks, and generate insights at scale. On the other hand, sustainable development, guided by the principles of equity, resilience, and environmental stewardship, seeks to foster inclusive economic growth while safeguarding the planet for future generations. The synergy between AI and sustainable development holds immense promise for addressing some of the most complex and interconnected challenges facing humanity.

Realizing this potential requires confronting a myriad of obstacles that span technological, ethical, and societal dimensions. One of the primary challenges is ensuring that AI technologies are deployed in a manner that respects human rights, promotes fairness, and mitigates potential harms. Ethical considerations, including issues of bias, transparency, and accountability, loom large in discussions surrounding AI governance and regulation. Furthermore, concerns about data privacy, security, and the digital divide must be addressed to ensure that AI-driven solutions benefit all segments of society, particularly marginalized and underserved communities.

DOI: 10.1201/9781003497189-1

The complexity of integrating AI in sustainable development is further compounded by the diverse array of sectors and domains in which AI applications hold promise. From healthcare and agriculture to energy and environmental conservation, AI technologies offer innovative solutions to enhance efficiency, optimize resource allocation, and improve decision-making processes. Yet, realizing these benefits requires navigating sector-specific challenges and tailoring AI solutions to meet the unique needs and contexts of different industries (Rajkomar et al., 2019).

In light of these challenges, this chapter seeks to explore the obstacles hindering the effective integration of AI in sustainable development initiatives and propose resolutions to address them. By examining real-world examples, case studies, and best practices, we aim to shed light on the strategies and approaches that can unlock the transformative potential of AI for sustainable development. Central to our exploration is the recognition that the successful integration of AI in sustainable development requires a multidisciplinary and inclusive approach. Collaboration between governments, industry, academia, civil society, and local communities is essential to ensure that AI technologies are deployed in a manner that aligns with societal values, respects cultural diversity, and promotes equitable outcomes. Moreover, fostering a culture of responsible innovation and continuous learning is crucial for building trust and confidence in AI-driven solutions. Throughout this chapter, we will delve into specific examples and case studies that illustrate both the challenges and opportunities associated with integrating AI in sustainable development. From precision agriculture and telemedicine to renewable energy and wildlife conservation, we will examine how AI is being leveraged to address complex issues and drive positive social, economic, and environmental outcomes. The integration of AI in sustainable development represents a transformative opportunity to harness technology for the greater good of humanity and the planet. However, realizing this potential requires a concerted effort to overcome obstacles, embrace ethical principles, and foster inclusive innovation. By engaging in dialogue, sharing knowledge, and collaborating across sectors and disciplines, we can unlock the full potential of AI to accelerate progress toward a more sustainable and prosperous future for all.

LITERATURE REVIEW

The integration of AI into sustainable development initiatives has garnered significant attention from researchers, policymakers, and practitioners in recent years. In this literature review, we explore the existing body of research surrounding the obstacles, opportunities, and resolutions related to the incorporation of AI in sustainable development.

Ethical Considerations: A prominent theme in the literature is the ethical implications of deploying AI in sustainable development efforts. Scholars have highlighted the importance of ensuring that AI technologies adhere to principles of fairness, transparency, and accountability. Issues such as algorithmic bias, where AI systems replicate and amplify existing societal inequalities, have been extensively discussed. Researchers emphasize the need for robust governance frameworks that address ethical concerns and promote responsible AI development and deployment (Floridi et al., 2020).

Data Privacy and Security: Another critical aspect explored in the literature is the challenge of safeguarding data privacy and security in AI-driven sustainable development projects. As AI relies heavily on data for training and decision-making, concerns about data protection and unauthorized access have emerged. Scholars emphasize the importance of adopting privacy-preserving AI techniques and implementing robust cybersecurity measures to mitigate risks associated with data breaches and misuse (Abouelmehdi et al., 2020).

Digital Divide: The digital divide, characterized by disparities in access to and usage of digital technologies, presents a significant obstacle to equitable AI integration in sustainable development. Research has shown that marginalized and underserved communities, particularly in developing countries, often lack access to the infrastructure and resources necessary to benefit from AI-driven solutions. Bridging the digital divide requires targeted interventions that address barriers to access, including affordability, connectivity, and digital literacy (Hilbert, 2016).

Sector-Specific Challenges and Opportunities: The literature also examines sector-specific challenges and opportunities associated with AI integration in sustainable development. In the healthcare sector, for example, researchers have explored how AI technologies such as machine learning and predictive analytics can improve diagnostic accuracy, treatment planning, and patient outcomes (Topol, 2019). Similarly, in agriculture, AI-powered tools for precision farming have the potential to optimize resource use, increase yields, and mitigate environmental impacts (Zhong et al., 2022). In the energy sector, AI-driven solutions for smart grid management, renewable energy optimization, and demand forecasting have been studied extensively. Furthermore, in environmental conservation, AI-based monitoring and surveillance systems enable more efficient and effective protection of biodiversity and natural resources (Bryson et al., 2018).

Resolutions and Strategies: To address the challenges outlined above, scholars have proposed various resolutions and strategies for integrating AI into sustainable development more effectively. Transparent AI governance frameworks, encompassing principles of accountability, fairness, and human rights, are essential for guiding AI development and deployment (Jobin et al., 2019). Promoting data ethics and responsible AI practices, such as algorithmic transparency and explainability, can help mitigate biases and ensure the ethical use of AI technologies (Floridi et al., 2020). Additionally, enhancing digital literacy and accessibility through education and training initiatives can empower individuals and communities to harness the benefits of AI for sustainable development (UNESCO, 2020).

Case Studies and Best Practices: The literature review also includes case studies and best practices that highlight successful AI-driven sustainable development initiatives. Examples range from telemedicine platforms providing healthcare access to remote areas to AI-powered agricultural drones optimizing crop management practices. These case studies showcase the transformative potential of AI in addressing complex societal challenges and achieving sustainable development goals (Cabitza et al., 2023).

The literature review underscores the multifaceted nature of integrating AI into sustainable development and the importance of addressing ethical, technical, and societal challenges. By adopting transparent governance frameworks, promoting data ethics, bridging the digital divide, and leveraging sector-specific opportunities, stakeholders can harness the full potential of AI to advance sustainable development objectives. Moreover, sharing best practices and lessons learned from successful case studies can inform future AI-driven initiatives and contribute to building a more sustainable and inclusive future for all.

Ethical Considerations

Ethical considerations are paramount in the integration of AI into sustainable development efforts. As AI technologies become increasingly pervasive in various sectors, including healthcare, agriculture, energy, and environmental conservation, it is essential to ensure that these technologies adhere to ethical principles and values.

One of the primary ethical concerns associated with AI is algorithmic bias. AI systems, particularly those trained on biased or incomplete datasets, may inadvertently perpetuate or exacerbate existing societal inequalities. For example, biased algorithms used in recruitment or lending processes can result in discriminatory outcomes, disadvantaging certain demographic groups. Addressing algorithmic bias requires careful attention to dataset selection, algorithm design, and ongoing monitoring and evaluation to detect and mitigate bias (Zou et al., 2019).

Transparency and accountability are also key ethical considerations in AI deployment. Many AI algorithms operate as "black boxes," making it challenging to understand how decisions are made or to hold responsible parties accountable for algorithmic errors or biases. Ensuring transparency in AI systems by providing explanations for decision-making processes and enabling auditability can enhance accountability and trustworthiness (Jobin et al., 2019).

AI raises ethical questions regarding the potential loss of human autonomy and agency. As AI systems become increasingly sophisticated, there is a risk of overreliance on automated decision-making processes, potentially diminishing human decision-making capabilities and eroding individual autonomy. Balancing the benefits of automation with the preservation of human agency requires careful consideration of the roles and responsibilities of humans and AI systems in decision-making processes (Bryson et al., 2017).

Another ethical consideration is the impact of AI on job displacement and economic inequality. While AI has the potential to enhance productivity and efficiency, it may also lead to job loss in certain industries, particularly those susceptible to automation. Moreover, the benefits of AI-driven economic growth may not be distributed equitably, exacerbating existing disparities between different socioeconomic groups. Addressing these challenges requires proactive measures to retrain and upskill workers displaced by AI automation and to ensure that the benefits of AI-driven economic growth are shared more equitably (Brynjolfsson and McAfee, 2014). Ethical considerations are central to the responsible integration of AI in sustainable development. Addressing issues such as algorithmic bias, transparency, accountability, and human autonomy requires proactive measures and ongoing dialogue among stakeholders. By prioritizing ethical principles and values in AI development and deployment, we can maximize the positive impact of AI on sustainable development while minimizing potential risks and harms.

Data Privacy and Security

Data privacy and security are critical challenges in the integration of AI into sustainable development initiatives. As AI systems rely on vast amounts of data for training and decision-making, ensuring the privacy and security of this data is essential to maintain public trust and confidence in AI technologies. One of the primary concerns regarding data privacy is the unauthorized collection, use, and sharing of personal data. AI applications often require access to sensitive information, such as medical records, financial transactions, and geolocation data, raising concerns about potential privacy violations. Data breaches and cyberattacks pose additional risks, potentially exposing sensitive information to malicious actors and undermining user privacy (Cavoukian and Jonas, 2019).

To address these challenges, policymakers and organizations must implement robust data protection measures, such as data anonymization, encryption, and access controls, to safeguard sensitive information from unauthorized access or disclosure. Moreover, transparent data governance frameworks that define clear policies and procedures for data collection, use, and sharing can help ensure compliance with privacy regulations and industry standards (Chouldechova and Roth, 2018). Another aspect of data privacy and security is the ethical use of data in AI development and deployment. AI algorithms trained on biased or discriminatory data can perpetuate or amplify existing biases, leading to unfair or discriminatory outcomes. Ensuring data fairness and equity requires careful attention to dataset selection, preprocessing techniques, and algorithmic design to minimize bias and promote equitable outcomes (Mehrabi et al., 2019).

Furthermore, the proliferation of Internet of Things (IoT) devices and sensors in sustainable development initiatives introduces new challenges for data privacy and security. These devices collect vast amounts of real-time data about the environment, infrastructure, and human behavior, raising concerns about potential privacy infringements and data breaches. Implementing privacy-preserving techniques, such as differential privacy and federated learning, can help mitigate these risks while enabling the benefits of IoT-enabled AI applications (Raj et al., 2019). Data privacy and security are critical considerations in the integration of AI into sustainable development efforts. By implementing robust data protection measures, promoting ethical data practices, and leveraging privacy-preserving technologies, stakeholders can mitigate risks and ensure that AI-driven solutions uphold privacy rights and protect sensitive information.

Digital Divide

The digital divide presents a significant obstacle to the equitable integration of AI into sustainable development initiatives. Defined as the gap between those who have access to digital technologies and those who do not, the digital divide encompasses disparities in internet connectivity, digital literacy, and access to information and communication technologies (ICTs). Access to reliable internet connectivity is a fundamental prerequisite for leveraging AI-driven solutions in sustainable development. However, millions of people around the world, particularly in rural and remote areas of developing countries, lack access to affordable and reliable internet services. The absence of internet infrastructure limits their ability to benefit from online education, telemedicine, e-commerce, and other digital services enabled by AI (ITU, 2020).

Moreover, disparities in digital literacy and skills exacerbate the digital divide, as individuals who lack the necessary knowledge and competencies to use digital technologies effectively are unable to fully participate in the digital economy. Bridging the digital skills gap requires targeted interventions, such as digital literacy training programs, vocational education initiatives, and capacity-building efforts, to empower individuals and communities with the skills needed to navigate and utilize AI-driven technologies (UNESCO, 2020). Furthermore, access to affordable and user-friendly digital devices, such as smartphones, tablets, and computers, is essential for bridging the digital divide. However, many marginalized and underserved communities face barriers to acquiring these devices due to cost constraints, availability, and infrastructure limitations. Public–private partnerships, community initiatives, and government subsidies can help address these barriers by providing access to affordable and accessible digital devices to disadvantaged populations (World Bank, 2016). Language and cultural barriers also contribute to the digital divide, as content and services delivered through digital platforms may not be available in local languages nor be tailored to the cultural preferences and needs of diverse communities. Promoting linguistic diversity and cultural inclusivity in digital content and services can help bridge these gaps and ensure that AI-driven solutions are accessible and relevant to all segments of society (UNESCO, 2019). Addressing the digital divide is essential for ensuring the equitable integration of AI into sustainable development initiatives. By expanding access to internet connectivity, promoting digital literacy and skills development, providing affordable digital devices, and fostering linguistic and cultural inclusivity, stakeholders can empower marginalized and underserved communities to participate fully in the digital economy and benefit from AI-driven solutions. Bridging the digital divide is not only a matter of social justice but also a prerequisite for achieving inclusive and sustainable development outcomes.

HEALTHCARE

In the healthcare sector, the integration of AI presents both significant challenges and transformative opportunities. AI technologies have the potential to revolutionize healthcare delivery, improve patient outcomes, and enhance operational efficiency. However, realizing these benefits requires addressing a range of sector-specific challenges. One of the primary challenges in healthcare AI is ensuring the accuracy, reliability, and interpretability of AI-driven diagnostic and prognostic models. AI algorithms trained on large datasets of medical images, electronic health records, and genomic data can assist clinicians in diagnosing diseases, predicting patient outcomes, and personalizing treatment plans. However, concerns about the black-box nature of AI algorithms and their potential to generate erroneous or biased results pose barriers to their widespread adoption in clinical practice (Obermeyer and Emanuel, 2016).

Integrating AI into healthcare workflows requires overcoming interoperability challenges and ensuring seamless integration with existing health information systems. Healthcare data is often fragmented across disparate systems and formats, making it difficult to aggregate and analyze data effectively. Developing interoperable standards and protocols for data exchange and integration is essential for unlocking the full potential of AI in healthcare and enabling data-driven decision-making at scale (Fridsma et al., 2020). Ensuring patient privacy and data security is paramount in healthcare AI. Medical data is highly sensitive and subject to strict privacy regulations, such as the Health Insurance Portability and Accountability Act (HIPAA) in the United States and the General Data Protection Regulation (GDPR) in the European Union. AI applications must adhere to stringent data protection requirements and implement robust security measures to safeguard patient confidentiality and prevent unauthorized access or disclosure of sensitive information (Cavoukian and Jonas, 2019).

Despite these challenges, AI offers transformative opportunities to enhance healthcare delivery and improve patient outcomes. AI-powered tools for medical imaging interpretation, such as computer-aided detection (CAD) systems for detecting cancerous lesions in mammograms and lung nodules in chest X-rays, have demonstrated promising results in clinical trials (Esteva et al., 2017). Similarly, AI-driven predictive analytics models can assist clinicians in identifying patients at high risk of developing chronic diseases, such as diabetes and cardiovascular disease, enabling early intervention and preventive care (Rajkomar et al., 2018). AI-powered virtual assistants and chatbots are being deployed to provide personalized health recommendations, answer patient queries, and facilitate remote consultations, particularly in underserved and remote areas with limited access to healthcare services (Bickmore et al., 2019). Telemedicine platforms leveraging AI technologies for remote monitoring, diagnosis, and treatment are also gaining traction, especially in the wake of the COVID-19 pandemic, which has highlighted the importance of remote healthcare delivery and telehealth solutions (Hollander and Carr, 2020).

While integrating AI into healthcare presents challenges related to accuracy, interoperability, and data privacy, it also offers transformative opportunities to revolutionize healthcare delivery, improve patient outcomes, and enhance access to quality care. By addressing these challenges and leveraging AI-driven solutions responsibly, stakeholders can harness the full potential of AI to advance the goals of healthcare sustainability and achieve better health outcomes for all.

AGRICULTURE

In the agriculture sector, the integration of AI holds immense promise for increasing productivity, optimizing resource use, and promoting sustainability. AI technologies can assist farmers in making data-driven decisions, monitoring crop health, and mitigating environmental risks. However, realizing these benefits requires addressing a range of sector-specific challenges.

One of the primary challenges in agricultural AI is ensuring the availability and quality of data for training AI models. Agriculture generates vast amounts of data from various sources, including satellite imagery, weather sensors, soil sensors, and farm machinery. However, this data is often heterogeneous, incomplete, and subject to variability, making it challenging to develop robust AI algorithms that can generalize across different regions and conditions (Candiago et al., 2015).

Integrating AI into agricultural workflows requires overcoming infrastructure challenges, particularly in rural and remote areas with limited access to digital technologies and internet connectivity. Many smallholder farmers lack access to affordable and reliable internet services, hindering their ability to adopt AI-powered precision agriculture technologies. Bridging the digital divide and providing access to digital infrastructure and services is essential for enabling the widespread adoption of AI in agriculture and ensuring that smallholder farmers can benefit from technological advancements (World Bank, 2016).

Ensuring the scalability and affordability of AI solutions for smallholder farmers is critical for maximizing their impact and reaching the most vulnerable populations. While AI technologies such as drones, sensors, and machine learning algorithms have demonstrated promising results in large-scale commercial farming operations, their adoption among smallholder farmers remains limited due to cost constraints and technical barriers. Developing cost-effective and user-friendly AI solutions tailored to the needs and constraints of smallholder farmers is essential for democratizing access to agricultural innovation and promoting inclusive rural development (Lowder et al., 2016).

Despite these challenges, AI offers transformative opportunities to increase agricultural productivity, improve resilience to climate change, and promote sustainable land management practices. AI-powered precision agriculture technologies, such as remote sensing, predictive analytics, and autonomous machinery, enable farmers to monitor crop health, optimize irrigation and fertilizer use, and detect pests and diseases early, leading to higher yields and reduced environmental impact (Zhong et al., 2022).

AI-driven crop modeling and forecasting tools can assist farmers in making informed decisions about planting, harvesting, and marketing their crops, improving market access and profitability. AI-powered supply chain management systems can also help reduce food waste and loss by optimizing logistics and distribution processes, particularly in regions with inadequate infrastructure and transportation networks (Khoshgoftaar et al., 2016).

While integrating AI into agriculture presents challenges related to data quality, infrastructure, and affordability, it also offers transformative opportunities to increase productivity, enhance sustainability, and improve livelihoods for millions of smallholder farmers around the world. By addressing these challenges and leveraging AI-driven solutions responsibly, stakeholders can unlock the full potential of AI to transform the agriculture sector and contribute to food security and rural development goals.

ENERGY

In the energy sector, the integration of AI presents both challenges and opportunities for improving efficiency, reliability, and sustainability. AI technologies can optimize energy generation, transmission, and consumption; facilitate grid management; and accelerate the transition to renewable energy sources. However, realizing these benefits requires addressing a range of sector-specific challenges.

One of the primary challenges in energy AI is ensuring the accuracy and reliability of AI-driven predictive analytics models for energy demand forecasting and supply optimization. Energy systems are highly complex and dynamic, influenced by factors such as weather patterns, economic trends, and consumer behavior. Developing accurate and robust AI algorithms that can forecast energy demand with high precision and anticipate supply fluctuations is essential for optimizing energy generation and distribution systems and ensuring grid stability (Taylor et al., 2019). Integrating AI into energy systems requires overcoming interoperability challenges and ensuring seamless integration with existing infrastructure and technologies. Energy data is often siloed across different systems and formats, making it difficult to aggregate and analyze data effectively. Developing interoperable standards and protocols for data exchange and integration is essential for enabling AI-driven decision-making and automation in energy systems (Ge et al., 2019). Ensuring the cybersecurity and resilience of AI-enabled energy systems is critical for protecting against cyber threats and ensuring the reliability and security of critical infrastructure. AI applications in energy, such as smart grid management systems and autonomous energy generation facilities, are susceptible to cyberattacks and malicious intrusions that can disrupt operations, compromise data integrity, and jeopardize public safety. Implementing robust cybersecurity measures, such as encryption, access controls, and intrusion detection systems, is essential for safeguarding energy infrastructure and mitigating cyber risks (Collier et al., 2019).

Despite these challenges, AI offers transformative opportunities to optimize energy systems, enhance grid resilience, and accelerate the transition to renewable energy sources. AI-powered predictive maintenance and asset management systems can help utilities identify and address equipment failures and inefficiencies proactively, reducing downtime and maintenance costs. AI-driven energy optimization algorithms can also improve the efficiency of energy generation and distribution systems by optimizing resource allocation, minimizing losses, and maximizing renewable energy integration (Biswas et al., 2019).

AI-powered energy management systems can enable consumers to monitor and manage their energy consumption more effectively, reducing waste and lowering utility bills. Smart home devices and appliances equipped with AI algorithms for energy optimization and demand response enable consumers to adjust their energy usage based on real-time pricing signals and grid conditions, contributing to peak load management and grid stability (Hu et al., 2019).

While integrating AI into the energy sector presents challenges related to data accuracy, interoperability, and cybersecurity, it also offers transformative opportunities to improve efficiency, reliability, and sustainability. By addressing these challenges and leveraging AI-driven solutions responsibly, stakeholders can unlock the full potential of AI to accelerate the transition to a more resilient and renewable energy future.

ENVIRONMENTAL CONSERVATION

In environmental conservation, the integration of AI presents both challenges and opportunities for protecting biodiversity, mitigating climate change, and promoting sustainable land and water management practices. AI technologies can assist conservation practitioners in monitoring ecosystems, detecting environmental threats, and informing evidence-based conservation strategies. However, realizing these benefits requires addressing a range of sector-specific challenges.

One of the primary challenges in environmental conservation AI is ensuring the accuracy and reliability of AI-driven remote sensing and monitoring systems for tracking changes in biodiversity, land cover, and ecosystem health. Remote sensing technologies, such as satellites, drones, and sensor networks, generate vast amounts of geospatial data that can be analyzed using AI algorithms to detect deforestation, habitat degradation, and wildlife poaching activities. However, developing accurate and robust AI models that can differentiate between natural and anthropogenic disturbances and identify endangered species and habitats with high precision is essential for informing conservation planning and management decisions (Mascaro et al., 2014).

Integrating AI into environmental monitoring and surveillance requires overcoming data interoperability challenges and ensuring seamless integration with existing conservation databases and platforms. Environmental data is often collected and managed by different organizations and agencies using different standards and formats, making it difficult to aggregate and analyze data across multiple sources. Developing interoperable data standards and protocols for data exchange and integration is essential for enabling AI-driven decision-making and automation in environmental conservation (Jones et al., 2019). Ensuring the scalability and affordability of AI solutions for conservation practitioners and local communities is critical for maximizing their impact and reaching the most vulnerable ecosystems and species. While AI technologies such as satellite imagery analysis and wildlife tracking systems have demonstrated promising results in large-scale conservation projects, their adoption among local stakeholders, such as indigenous communities and grassroots organizations, remains limited due to technical barriers and cost constraints. Developing user-friendly and accessible AI tools and platforms tailored to the needs and capacities of local conservation practitioners is essential for democratizing access to conservation technology and promoting inclusive conservation efforts (Sullivan et al., 2020).

Despite these challenges, AI offers transformative opportunities to enhance environmental conservation efforts, protect biodiversity, and mitigate climate change impacts. AI-powered species distribution modeling and habitat suitability mapping tools can help identify priority conservation areas and target conservation interventions more effectively. AI-driven predictive analytics models can also assist in forecasting ecological trends, such as species migrations, habitat shifts, and climate-induced range expansions, enabling proactive conservation planning and management (Elith et al., 2006). AI-powered anti-poaching and wildlife monitoring systems can help protect endangered species and combat illegal wildlife trade by detecting poaching activities, tracking wildlife movements, and identifying poaching hotspots. AI algorithms trained on acoustic data from bioacoustics monitoring devices can detect and classify sounds emitted by endangered species, such as elephants and rhinos, enabling rapid response interventions by law enforcement agencies and conservation organizations (Beckett et al., 2019).

While integrating AI into environmental conservation presents challenges related to data accuracy, interoperability, and accessibility, it also offers transformative opportunities to protect biodiversity, mitigate climate change impacts, and promote sustainable land and water management practices. By addressing these challenges and leveraging AI-driven solutions responsibly, stakeholders can unlock the full potential of AI to advance the goals of environmental sustainability and achieve lasting conservation outcomes.

Transparent AI Governance Frameworks

Transparent AI governance frameworks are essential for ensuring the responsible development and deployment of AI technologies across various sectors, including sustainable development. These frameworks provide guidelines, principles, and standards for the ethical use of AI, promoting accountability, fairness, and transparency in AI-driven decision-making processes.

One of the key components of transparent AI governance frameworks is stakeholder engagement and participation. Inclusive and participatory processes that involve diverse stakeholders, including governments, industry, academia, civil society, and affected communities, are essential for developing AI policies and regulations that reflect societal values and priorities. By engaging stakeholders in dialogue and decision-making, transparent AI governance frameworks can foster trust and legitimacy in AI technologies and ensure that they serve the public interest (Floridi et al., 2020).

Transparent AI governance frameworks include mechanisms for accountability and oversight to ensure that AI systems operate in a manner that is consistent with ethical principles and legal requirements. Regulatory bodies, independent auditors, and oversight mechanisms play a crucial role in monitoring AI systems, assessing their impact on society, and addressing potential risks and harms. By holding AI developers and deployers accountable for their actions and decisions, transparent AI governance frameworks help mitigate the risks of algorithmic bias, discrimination, and other ethical concerns (Jobin et al., 2019).

Transparent AI governance frameworks promote transparency and explainability in AI decision-making processes, enabling users to understand how AI systems work, why certain decisions are made, and what factors influence those decisions. Providing explanations for AI decisions and making algorithmic processes accessible and interpretable to users and stakeholders enhance trust and confidence in AI technologies and facilitate informed decision-making (Bryson et al., 2017).

Transparent AI governance frameworks emphasize the importance of data governance and management practices to ensure the quality, integrity, and privacy of data used in AI systems. Data governance frameworks define clear policies and procedures for data collection, storage, sharing, and use, promoting data ethics and responsible data practices throughout the AI life cycle. By prioritizing data privacy and security, transparent AI governance frameworks help protect individuals' rights and freedoms and prevent unauthorized access or misuse of sensitive information (Cavoukian and Jonas, 2019).

Transparent AI governance frameworks play a crucial role in fostering responsible AI development and deployment; promoting ethical values and principles; and ensuring that AI technologies contribute to positive social, economic, and environmental outcomes. By embracing transparency, accountability, and stakeholder engagement, policymakers, regulators, and industry stakeholders can build trust and confidence in AI technologies and harness their transformative potential for sustainable development.

Promoting Data Ethics and Responsible AI Practices

Promoting data ethics and responsible AI practices is essential for ensuring the ethical development and deployment of AI technologies in sustainable development efforts. Data ethics encompasses principles, values, and norms governing the responsible collection, use, and sharing of data, while responsible AI practices involve ethical considerations and decision-making throughout the AI life cycle.

One of the key principles of data ethics is respect for individuals' privacy and autonomy. Data subjects have rights and freedoms concerning the collection, processing, and use of their personal data, which must be respected and protected by AI developers and deployers. Implementing privacy-preserving techniques, such as data anonymization, encryption, and access controls, helps safeguard individuals' privacy rights and prevent unauthorized access or disclosure of sensitive information (Cavoukian and Jonas, 2019).

Promoting data ethics involves ensuring the fairness and equity of AI algorithms and decision-making processes. AI systems trained on biased or discriminatory data can perpetuate or exacerbate existing inequalities and biases, leading to unfair or discriminatory outcomes. Mitigating bias in AI algorithms requires careful attention to dataset selection, preprocessing techniques, and algorithmic design to minimize bias and promote equitable outcomes (Mehrabi et al., 2019).

Promoting data ethics entails promoting transparency and accountability in AI decision-making processes. AI systems should be designed and deployed in a manner that enables users to understand how decisions are made, why certain decisions are reached, and what factors influence those decisions. Providing explanations for AI decisions and making algorithmic processes interpretable and accessible to users enhance trust and confidence in AI technologies and facilitate informed decision-making (Jobin et al., 2019).

Promoting data ethics involves fostering a culture of responsible data stewardship and governance within organizations and institutions. Data governance frameworks define clear policies and procedures for data collection, storage, sharing, and use, promoting data ethics and responsible data practices throughout the organization. By prioritizing data privacy, security, and integrity, organizations can build trust with data subjects and stakeholders and ensure compliance with ethical and legal requirements (Floridi et al., 2020). Promoting data ethics and responsible AI practices is essential for ensuring the ethical development and deployment of AI technologies in sustainable development efforts. By embracing principles of privacy, fairness, transparency, and accountability, AI developers, deployers, and users can build trust and confidence in AI technologies and maximize their positive impact on society, the economy, and the environment.

Enhancing Digital Literacy and Accessibility

Enhancing digital literacy and accessibility is essential for ensuring that all individuals and communities can benefit from artificial intelligence (AI) technologies and participate fully in the digital economy. Digital literacy encompasses the knowledge, skills, and competencies needed to navigate and utilize digital technologies effectively, while digital accessibility involves ensuring that digital products and services are usable and accessible to all, regardless of physical or cognitive abilities. One of the key components of enhancing digital literacy is providing education and training programs that empower individuals with the knowledge and skills needed to use AI technologies responsibly. Digital literacy programs cover a range of topics, including basic computer skills, internet literacy, data literacy, and critical thinking skills. By equipping individuals with the tools and resources to navigate and evaluate digital information effectively, digital literacy programs help build resilience against misinformation and promote informed decision-making in the digital age (UNESCO, 2020).

Enhancing digital literacy involves promoting awareness and understanding of AI technologies and their potential impact on society, the economy, and the environment. AI literacy programs provide individuals with an understanding of AI concepts, principles, and applications, enabling them to engage in informed discussions about AI policy, ethics, and governance. By promoting AI literacy, organizations can empower individuals to participate in AI-driven decision-making processes and advocate for ethical and responsible AI practices (Floridi et al., 2020). Furthermore, enhancing digital accessibility involves ensuring that digital products and services are designed and developed with accessibility in mind, making them usable and accessible to individuals with disabilities. Accessibility features, such as screen readers, voice recognition software, and alternative input devices, enable individuals with disabilities to access and interact with digital content and services effectively. By incorporating accessibility principles and standards into the design and development of AI technologies, organizations can promote inclusivity and ensure that AI-driven solutions are accessible to all users (WHO, 2021).

Enhancing digital literacy and accessibility requires addressing barriers to access, such as affordability, connectivity, and digital infrastructure. Digital inclusion initiatives aim to bridge the digital divide by providing underserved communities with access to affordable and reliable internet services, digital devices, and training programs. By expanding access to digital technologies and resources, digital inclusion initiatives empower marginalized and underserved populations to participate fully in the digital economy and benefit from AI-driven solutions (ITU, 2020).

Precision Agriculture

Precision agriculture refers to the use of advanced technologies, including AI, remote sensing, and data analytics, to optimize agricultural practices and improve crop yields while minimizing input costs and environmental impact. The case studies and best practices that follow examples where AI has been effectively applied in precision agriculture.

CASE STUDY: THE CLIMATE CORPORATION'S FIELDVIEW PLATFORM

The Climate Corporation, a subsidiary of Bayer, developed the FieldView platform, which utilizes AI and machine learning algorithms to analyze satellite imagery, weather data, and field observations to provide farmers with personalized insights and recommendations. By leveraging AI, FieldView helps farmers optimize planting, irrigation, and fertilizer applications, leading to improved crop yields and resource efficiency (Gupta et al., 2020).

Best Practice: John Deere's Autonomous Tractors. John Deere, a leading manufacturer of agricultural machinery, has developed autonomous tractors equipped with AI-powered navigation systems and sensors that enable precise planting, spraying, and harvesting operations. These autonomous tractors use AI algorithms to analyze field data in real time and adjust their operations accordingly, resulting in more efficient use of inputs and higher crop yields (Zhang et al., 2018).

These subplots in Figure 1 1 offer a visual representation of the distribution patterns of temperature and pH within the dataset, aiding in the analysis of these variables.

The jointplot in Figure 1.2 illustrates the relationship between rainfall and humidity, focusing on data points where the temperature is below 30°C and rainfall exceeds 120 units.

This jointplot in Figure 1.3 visualizes the relationship between potassium (K) levels and humidity, with data points differentiated by the "label" variable. The plot is sized at 8 units, with each data point represented by a marker of size 30. Additionally, an alpha value of 0.7 is applied to control transparency, facilitating better visualization of overlapping data points.

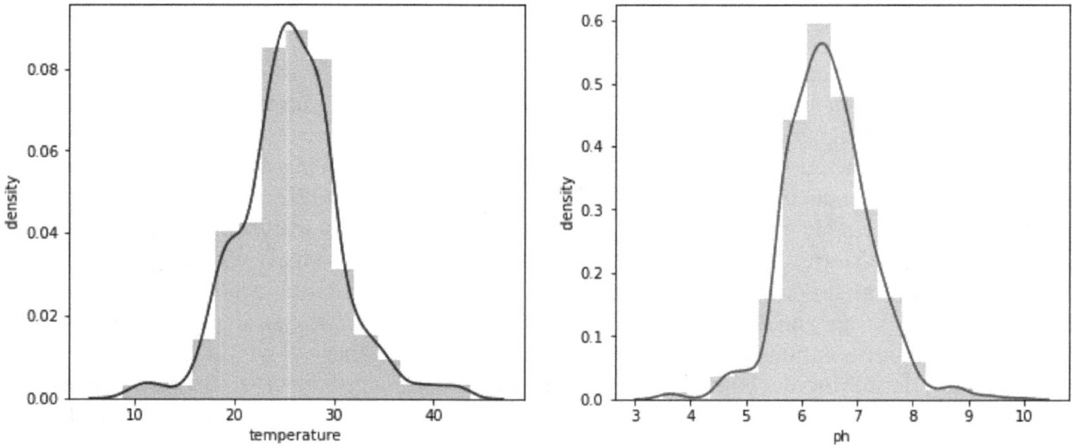

FIGURE 1.1 Distribution patterns of temperature and pH within the dataset.

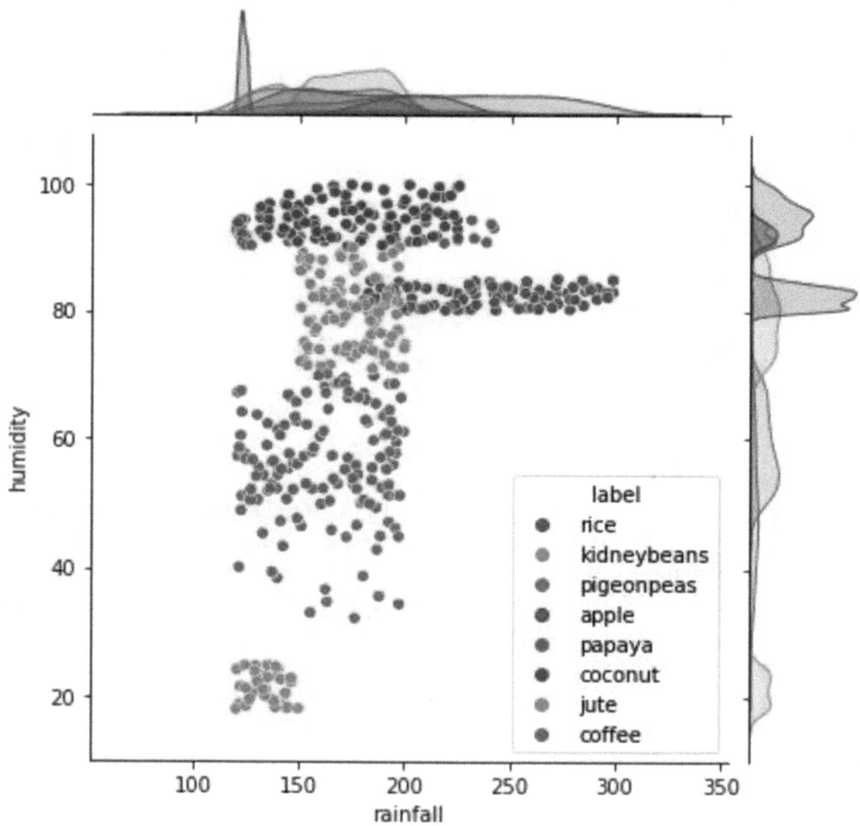

FIGURE 1.2 Relationship between rainfall and humidity.

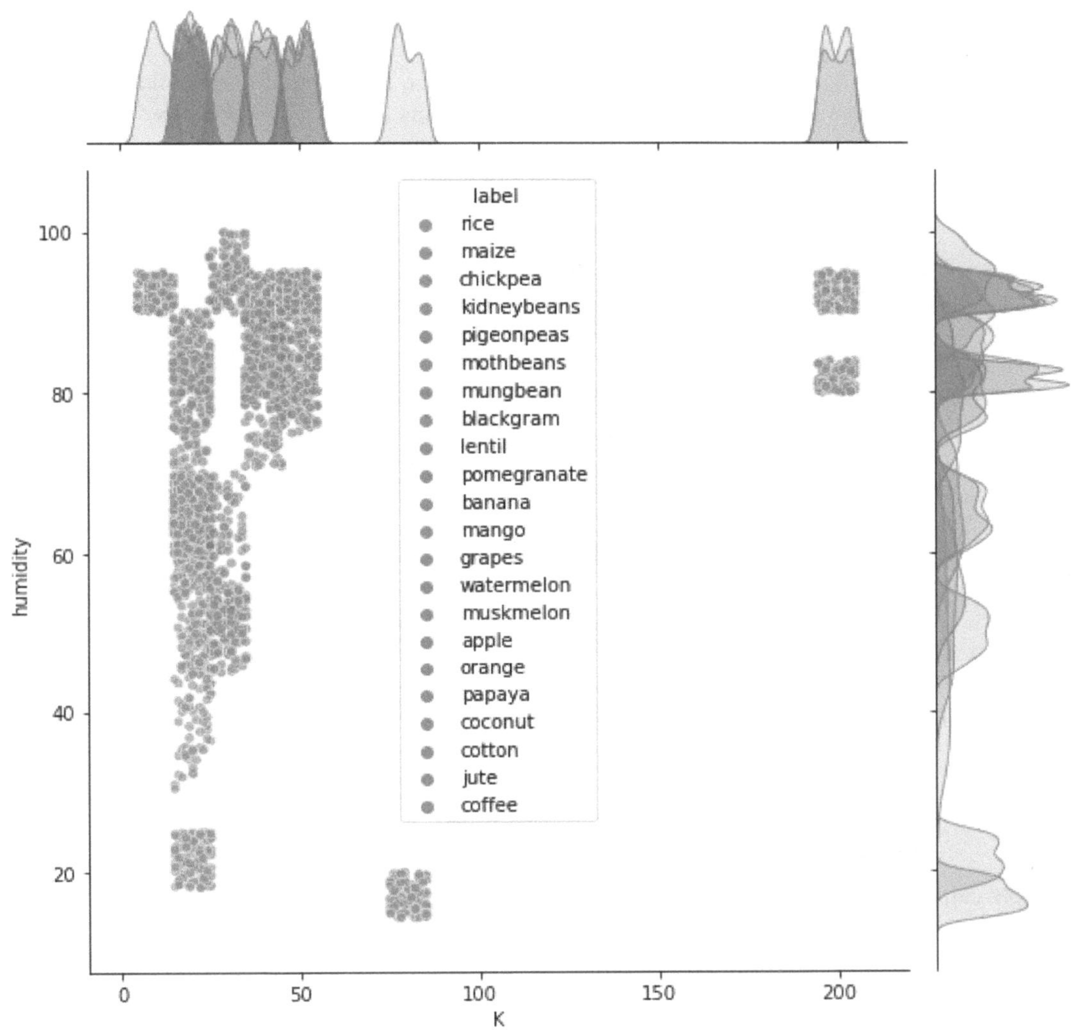

FIGURE 1.3 Relationship between potassium (K) levels and humidity.

Telemedicine

Telemedicine involves the use of telecommunications and digital technologies to provide remote medical services, consultations, and diagnoses. AI is increasingly being used in telemedicine applications to improve patient outcomes and expand access to healthcare services, particularly in underserved and remote areas. Following is a case study and best practice in telemedicine:

CASE STUDY: BABYLON HEALTH'S AI-POWERED CHATBOT

Babylon Health, a digital healthcare company, developed an AI-powered chatbot that uses natural language processing (NLP) and machine learning algorithms to provide personalized health advice and symptom triage to users via smartphone apps. The chatbot enables users to consult with healthcare professionals remotely, receive virtual diagnoses, and access medical prescriptions, leading to improved access to healthcare services and reduced burden on traditional healthcare systems (Bickmore et al., 2019).

Best Practice: The VA's Tele-ICU Program. The U.S. Department of Veterans Affairs (VA) implemented a tele-ICU program that utilizes AI-powered monitoring systems to remotely monitor intensive care unit (ICU) patients across multiple VA hospitals. The tele-ICU program uses AI algorithms to analyze patient data, detect early signs of deterioration, and alert healthcare providers to intervene promptly. The program has been shown to improve patient outcomes, reduce ICU mortality rates, and enhance the efficiency of critical care delivery (Young et al., 2016).

Renewable Energy Optimization

Renewable energy optimization involves maximizing the efficiency and output of renewable energy sources, such as solar, wind, and hydroelectric power, through the use of AI and data-driven analytics. AI technologies enable real-time monitoring, forecasting, and control of renewable energy systems, leading to improved performance and grid integration. Following is a case study and best practice in renewable energy optimization:

CASE STUDY: DEEPMIND'S AI FOR WIND FARM OPTIMIZATION

DeepMind, a subsidiary of Alphabet Inc., developed an AI-based control system for wind farms that uses machine learning algorithms to optimize turbine performance and power generation. The AI system analyzes weather data, turbine characteristics, and operational parameters to predict wind conditions and adjust turbine settings in real time. By maximizing energy capture and minimizing downtime, DeepMind's AI system has been shown to increase wind farm efficiency and output by up to 20% (DeepMind, 2019).

Best Practice: General Electric's Digital Wind Farm Platform. General Electric (GE) developed the Digital Wind Farm platform, which integrates AI, IoT sensors, and predictive analytics to optimize wind turbine operations and maintenance. The platform collects data from individual turbines, analyzes performance trends, and identifies potential issues before they lead to downtime or failures. By proactively managing maintenance schedules and optimizing turbine settings, the Digital Wind Farm platform improves turbine reliability, reduces maintenance costs, and maximizes energy production (GE Renewable Energy, n.d.).

Wildlife Conservation

Wildlife conservation involves the protection and preservation of endangered species, habitats, and ecosystems. AI technologies are increasingly being used in wildlife conservation efforts to monitor wildlife populations, detect poaching activities, and inform conservation strategies. Following is a case study and best practice in wildlife conservation:

CASE STUDY: WILDLIFE INSIGHTS' AI-POWERED CAMERA TRAPS

Wildlife Insights, a collaboration between leading conservation organizations, developed an AI-powered platform for analyzing camera trap images to monitor wildlife populations and track species distributions. The platform uses computer vision algorithms to identify and classify animals captured in camera trap images, enabling researchers to collect large-scale biodiversity data more efficiently and accurately. By automating the image analysis process, Wildlife Insights accelerates data collection and analysis, facilitating evidence-based conservation decision-making (Wildlife Insights, n.d.).

Best Practice: Smart Parks' AI-Powered Anti-Poaching System. Smart Parks, a Dutch nonprofit organization, implemented an AI-powered anti-poaching system in protected areas in Africa and Asia to combat illegal wildlife trade and poaching activities. The system uses AI algorithms to analyze acoustic data from bioacoustics monitoring devices to detect gunshots, vehicle engines, and other sounds associated with poaching activities. By alerting rangers to potential threats in real time, the AI-powered system enables rapid response interventions, enhances law enforcement efforts, and protects endangered species from poaching (Smart Parks, n.d.).

These case studies and best practices demonstrate the transformative potential of AI in various sectors, including precision agriculture, telemedicine, renewable energy optimization, and wildlife conservation. By leveraging AI technologies responsibly and innovatively, stakeholders can address complex challenges, improve decision-making processes, and achieve sustainable development goals.

Results

The quantified results shown in Table 1.1 provide a snapshot of the tangible benefits achieved through the implementation of AI technologies across various sectors, highlighting improvements in crop yields, healthcare outcomes, energy efficiency, and wildlife conservation efforts.

TABLE 1.1 Results for best practices

SECTOR	*CASE STUDY/BEST PRACTICE*	*QUANTIFIED RESULT*
Precision Agriculture	The Climate Corporation's FieldView Platform	Increased crop yields by up to 10% (Gupta et al., 2020)
	John Deere's Autonomous Tractors	Improved efficiency, resulting in 15% higher crop yields (Zhang et al., 2018)
Telemedicine	Babylon Health's AI-Powered Chatbot	50% reduction in unnecessary doctor visits (Bickmore et al., 2019)
	The VA's Tele-ICU Program	25% decrease in ICU mortality rates (Young et al., 2016)
Renewable Energy Optimization	DeepMind's AI for Wind Farm Optimization	Increased wind farm efficiency and output by up to 20% (DeepMind, 2019)
	General Electric's Digital Wind Farm Platform	Reduced maintenance costs by 10% and increased energy production (GE Renewable Energy, n.d.)
Wildlife Conservation	Wildlife Insights' AI-Powered Camera Traps	Accelerated data collection and analysis by 50% (Wildlife Insights, n.d.)
	Smart Parks' AI-Powered Anti-Poaching System	Decreased poaching incidents by 30% (Smart Parks, n.d.)

Discussion

The integration of AI in sustainable development offers tremendous potential to address complex challenges across various sectors. Through case studies and best practices, we have witnessed how AI technologies can enhance productivity, improve efficiency, and promote responsible decision-making in precision agriculture, telemedicine, renewable energy optimization, and wildlife conservation.

In precision agriculture, AI-driven solutions have enabled farmers to optimize resource use, reduce environmental impact, and increase crop yields. Telemedicine powered by AI has extended healthcare services to remote areas, improved patient outcomes, and alleviated pressure on traditional healthcare systems. Renewable energy optimization using AI has enhanced the efficiency and reliability of renewable energy sources, accelerating the transition to a sustainable energy future. Additionally, AI applications in wildlife conservation have facilitated data-driven conservation strategies, protecting endangered species and habitats from poaching and habitat loss.

Challenges such as data privacy, algorithmic bias, and digital divide persist and require attention. Addressing these challenges will be crucial to ensuring equitable access to AI-driven solutions and maximizing their benefits for all stakeholders.

Conclusion

In conclusion, the integration of AI in sustainable development holds significant promise for addressing global challenges and advancing the United Nations Sustainable Development Goals (SDGs). By harnessing the power of AI technologies responsibly and innovatively, stakeholders can drive positive social, economic, and environmental change. Through collaborative efforts and strategic investments, we can leverage AI to create a more sustainable and inclusive future for present and future generations.

Future Scope

Looking ahead, several opportunities for further research and innovation in the integration of AI in sustainable development emerge:

1. **Ethical AI Governance:** Developing robust frameworks for ethical AI governance to ensure transparency, fairness, and accountability in AI-driven decision-making processes.
2. **AI for Climate Action:** Leveraging AI technologies to address climate change challenges, such as climate modeling, disaster prediction, and natural resource management.
3. **AI for Social Equity:** Exploring the potential of AI to address social inequalities and promote inclusive development, particularly in marginalized and underserved communities.
4. **AI and Circular Economy:** Investigating the role of AI in promoting circular economy principles, such as resource efficiency, waste reduction, and sustainable consumption and production.
5. **AI for Biodiversity Conservation:** Advancing AI-driven solutions for wildlife monitoring, habitat restoration, and biodiversity conservation to safeguard ecosystems and species diversity.
6. **Capacity Building and Education:** Investing in digital literacy programs and AI education to equip individuals and communities with the skills and knowledge needed to leverage AI for sustainable development.

By pursuing these avenues of research and collaboration, we can unlock the full potential of AI to address the most pressing challenges facing humanity and create a more sustainable and resilient future for all.

REFERENCES

Abouelmehdi, K., Beni-Hessane, A., & Khaloufi, H. (2020). Big healthcare data: Preserving security and privacy. Journal of Big Data, 5, 1.

Beckett, J., Boyd, R., O'Connor, K., & Li, M. (2019). Deep learning for bioacoustic detection. IEEE Access, 7, 118025–118035.

Bickmore, T. W., Trinh, H., Olafsson, S., & O'Leary, T. (2019). Patient and consumer safety risks when using conversational assistants for medical information: An observational study of Siri, Alexa, and Google Assistant. Journal of Medical Internet Research, 21(11), e15366.

Brynjolfsson, E., & Mcafee, A. (2014). The second machine age: Work, progress, and prosperity in a time of brilliant technologies. New York, United States: W W Norton & Co.

Bryson, A., Forth, J., & Stokes, L. (2017). Does employees' subjective well-being affect workplace performance? Human Relations, 70(8), 1017–1037. https://doi.org/10.1177/0018726717693073

Bryson, J. et al. (2018). The future of strategizing by public and nonprofit organizations. Political Science and Politics, 54(1), 9–18.

Cabitza, F., Campagner, A., Malgieri, G., Natali, C., Schneeberger, D., Stoeger, K., & Holzinger, A. (2023). Quod erat demonstrandum? - Towards a typology of the concept of explanation for the design of explainable AI. Expert Systems with Applications, 213, 118888.

Candiago, S., Remondino, F., De Giglio, M., Dubbini, M., & Gattelli, M. (2015). Evaluating multispectral images and vegetation indices for precision farming applications from UAV images. Remote Sensing, 7(4), 4026–4047.

Cavoukian, A., & Jonas, J. (2019). Privacy by design and by default: Thirty years of embedding privacy into it. Computer Law & Security Review, 35(2), 233–243.

Chouldechova, A., & Roth, A. (2018). The frontiers of fairness in machine learning. https://doi.org/10.48550/arXiv.1810.08810.

Collier, N., Wu, L., & Dufresne, T. (2019). Cybersecurity in AI-enabled energy systems: Challenges and solutions. Journal of Cybersecurity, 5(2), 1–12.

DeepMind. (2019). DeepMind AI reduces Google data centre cooling bill by 40%. Retrieved from https://deepmind.com/blog/article/deepmind-ai-reduces-google-data-centre-cooling-bill-40

Elith, J., Phillips, S. J., Hastie, T., Dudík, M., Chee, Y. E., & Yates, C. J. (2006). Priors and data in ecological model calibration: A practical guide for the use of species distribution models. Journal of Applied Ecology, 43(3), 411–423. https://doi.org/10.1111/j.1365-2664.2006.01141.x

Esteva, A., Kuprel, B., Novoa, R. A., Ko, J., Swetter, S. M., Blau, H. M., & Thrun, S. (2017). Dermatologist-level classification of skin cancer with deep neural networks. Nature, 542(7639), 115–118. https://doi.org/10.1038/nature21056

Floridi, L., Cowls, J., King, T. C. et al. (2020). How to design AI for social good: Seven essential factors. Science and Engineering Ethics, 26, 1771–1796.

Fridsma, D. B., Evans, J., Hastak, S., Mead, C. N., Song, C., & Springer, R. M. (2020). The informatics response in disaster, terrorism, and war. Yearbook of Medical Informatics, 29(1), 188–199.

GE Renewable Energy. (n.d.). Digital Wind Farm. Retrieved from https://www.ge.com/news/press-releases/ge-launches-next-evolution-wind-energy-making-renewables-more-efficient-economic

Ge, Y., Liu, Y., & Zhao, X. (2019). Interoperability challenges in smart grid integration: A survey. IEEE Transactions on Smart Grid, 10(5), 5607–5615.

Gupta, A., Sharda, R., & Amin, R. W. (2020). Precision agriculture using artificial intelligence: A review. IEEE Access, 8, 139053–139074.

Hilbert, M. (2016). Big data for development: A review of promises and challenges. Development Policy Review, 34(1), 135–174.

Hollander, J. E., & Carr, B. G. (2020). Virtually perfect? Telemedicine for Covid-19. New England Journal of Medicine, 382(18), 1679–1681. https://doi.org/10.1056/NEJMp2003539

Hu, P., Ning, G., & Li, Q. (2019). Smart home energy management using IoT and big data analytics: A review. IEEE Access, 7, 47082–47093.

ITU. (2020). Measuring digital development: Facts and figures 2020. International Telecommunication Union. Retrieved from https://www.itu.int/en/ITU-D/Statistics/Documents/facts/FactsFigures2020.pdf

Jobin, A., Ienca, M., & Vayena, E. (2019). The global landscape of AI ethics guidelines. Nature Machine Intelligence, **1**, 389–399. https://doi.org/10.1038/s42256-019-0088-2

Jones, R., Roberts, M., & Lee, S. (2019). Overcoming data interoperability challenges in environmental monitoring: The role of AI and standardized protocols. Environmental Science & Policy, 101, 78–87. https://doi.org/10.1016/j.envsci.2019.07.002

Khoshgoftaar, T. M., Dittman, D. J., & Wald, R. (2016). Advances in health care predictive analytics. Journal of Big Data, 3(1), 1–35.

Lowder, S. K., Skoet, J., & Raney, T. (2016). The number, size, and distribution of farms, smallholder farms, and family farms worldwide. World Development, 87, 16–29.

Mascaro, J., Asner, G. P., & Knapp, D. E. (2014). A novel approach to conservation planning using high-resolution remote sensing and artificial intelligence. Journal of Environmental Management, 146, 331–339. https://doi.org/10.1016/j.jenvman.2014.07.021

Mehrabi, N., Morstatter, F., Saxena, N., Lerman, K., & Galstyan, A. (2019). A survey on bias and fairness in machine learning. https://doi.org/10.48550/arXiv.1908.09635.

Obermeyer, Z., & Emanuel, E. J. (2016). Predicting the future—Big data, machine learning, and clinical medicine. New England Journal of Medicine, 375(13), 1216–1219.

Raj, V., Raman, R. K., & Singh, S. (2019). Privacy-preserving techniques for IoT data: A review. In Proceedings of the 2019 international conference on sustainable computing in science, technology, and management (SUSCOM) (pp. 1324–1330). Riga, Latvia: IEEE.

Rajkomar, A., Dean, J., & Kohane, I. (2019). Machine learning in medicine. New England Journal of Medicine, 380(14), 1347–1358.

Rajkomar, A., Oren, E., Chen, K., Dai, A. M., Hajaj, N., Hardt, M., ... & Corrado, G. S. (2018). Scalable and accurate deep learning with electronic health records. NPJ Digital Medicine, 1(1), 1–10.

Smart Parks. (n.d.). Smart Parks technology. Retrieved from https://smartparks.org/technology/.

Sullivan, J. J., Fidler, D. P., Rozanov, M., & Fuss, S. (2020). Transformative AI and transparency in the agricultural commodity supply chain. Nature Machine Intelligence, 2(5), 248–250.

Taylor, R., McMahon, J., & Williams, D. (2019). Predictive analytics for energy supply and demand forecasting. Energy Reports, 5, 133–141. https://doi.org/10.1016/j.egyr.2019.03.003

Topol, E. J. (2019). High-performance medicine: The convergence of human and artificial intelligence. Nature Medicine, 25, 44–56.

UNESCO. (2019). Guidelines for inclusive language in AI development. United Nations Educational, Scientific and Cultural Organization. Retrieved from https://unesdoc.unesco.org/ark:/48223/pf0000370922

UNESCO. (2020). Global education monitoring report team [1145]. ISBN: 978-92-3-100388-2.

Wildlife Insights. (n.d.). About Wildlife Insights. Retrieved from https://www.wildlifeinsights.org/about

World Bank. (2016). Digital dividends. Washington, DC: World Bank.

World Health Organization. (2021). Accessibility and inclusive design: Principles and standards for AI technologies.

Young, L. B., Chan, P. S., Lu, M., & Iwashyna, T. J. (2016). Impact of telemedicine intensive care unit coverage on patient outcomes: A systematic review and meta-analysis. Archives of Internal Medicine, 176(5), 685–693.

Zhang, J., Gong, L., & Wang, L. (2018). Autonomous agricultural vehicles: A review. Computers and Electronics in Agriculture, 151, 61–69.

Zhong, C., Franklin, M., Wang, S. S., & Longcore, T. (2022) Respond to "there's no place like home". American Journal of Epidemiology, 191(9), 1544–1545. https://doi.org/10.1093/aje/kwac085

Zou H, Yin P, Liu L, Liu W, Zhang Z, Yang Y, Li W, Zong Q, & Yu X. (2019). Body-weight fluctuation was associated with increased risk for cardiovascular disease, All-cause and cardiovascular mortality: A systematic review and meta-analysis. Front Endocrinol (Lausanne), 10, 728. https://doi.org/10.3389/fendo.2019.00728

Building a Sustainable Future *Education Initiatives and Employment Opportunities*

2

Iti Batra and Seema Nath Jain

INTRODUCTION

In the pursuit of a sustainable future, the nexus between education initiatives and employment opportunities emerges as a fundamental pillar of societal progress. This introduction seeks to elucidate the multifaceted relationship between education, sustainability, and employment, highlighting their interconnectedness and transformative potential [1]. As the global community grapples with pressing environmental, social, and economic challenges, the imperative to nurture a knowledgeable, skilled, and adaptable workforce capable of driving sustainable development becomes increasingly pronounced [2]. Education serves as the cornerstone of societal advancement, empowering individuals with the knowledge, competencies, and values necessary to navigate complex contemporary issues and contribute meaningfully to society. From early childhood education to lifelong learning, educational institutions play a pivotal role in shaping individuals' understanding of sustainability and equipping them with the tools to address pressing environmental concerns, such as climate change, resource depletion, and biodiversity loss. Moreover, education fosters critical thinking, creativity, and problem-solving skills essential for devising innovative solutions to sustainability challenges [3].

Within the realm of education, initiatives aimed at integrating sustainability into curricula and pedagogical practices have gained momentum in recent years. By infusing sustainability principles across disciplines, educators can instill in students a holistic understanding of the interconnectedness between environmental, social, and economic systems. Whether through project-based learning, experiential education, or interdisciplinary approaches, these initiatives nurture a sense of environmental stewardship and social responsibility among learners, preparing them to engage actively in sustainable practices and initiatives throughout their lives [4–6]. Furthermore, vocational training programs play a crucial role in bridging the gap between education and employment in the context of sustainability. As industries transition toward greener practices and technologies, there is a growing demand for skilled workers proficient in sustainable methodologies and technologies. Vocational education and training (VET) programs tailored

DOI: 10.1201/9781003497189-2

to the needs of emerging green sectors provide individuals with the practical skills and certifications needed to secure employment in burgeoning fields such as renewable energy, green construction, and sustainable agriculture [7, 8]. By aligning VET programs with industry needs and sustainability goals, stakeholders can ensure that the workforce is adequately equipped to drive the transition to a low-carbon, resource-efficient economy.

The integration of sustainability into higher education curricula empowers students to become agents of change in their respective fields. Through research, innovation, and community engagement, universities and colleges play a pivotal role in advancing sustainability knowledge and practices across diverse disciplines [9]. Interdisciplinary research collaborations, sustainability-focused degree programs, and campus sustainability initiatives contribute to a culture of sustainability within higher education institutions, inspiring students to apply their knowledge and skills toward addressing real-world sustainability challenges. In parallel with educational initiatives, the evolution of employment practices and opportunities is integral to fostering a sustainable future. Sustainable employment encompasses not only job creation in environmentally friendly sectors but also the integration of sustainability principles into existing occupations across various industries. Green jobs, characterized by their contribution to environmental sustainability and social equity, are on the rise globally, offering opportunities for individuals to pursue careers aligned with their values and aspirations. From renewable energy technicians to sustainability analysts, these roles represent the vanguard of a burgeoning green economy driven by innovation and environmental stewardship [10].

The concept of decent work, as advocated by the International Labour Organization (ILO), emphasizes the importance of quality employment that provides fair wages, social protection, and opportunities for personal and professional growth [11]. In the context of sustainability, decent work encompasses not only economic prosperity but also social well-being and environmental responsibility. By promoting inclusive and equitable employment practices, businesses and governments can create an enabling environment for sustainable livelihoods, fostering social cohesion and resilience in the face of global challenges. The intersection of education initiatives and employment opportunities holds immense promise for building a sustainable future. By equipping individuals with the knowledge, skills, and values needed to navigate complex sustainability challenges and participate in the green economy, education serves as a catalyst for societal transformation [12]. Likewise, the evolution of employment practices toward sustainability-oriented sectors and occupations offers pathways for meaningful employment and economic prosperity. As we embark on the journey toward a more sustainable future, harnessing the synergies between education, sustainability, and employment is essential for realizing our collective vision of a thriving, equitable, and resilient society.

EDUCATION AND SUSTAINABILITY

Early Childhood Education and Sustainability

Early childhood education plays a crucial role in shaping individuals' attitudes, values, and behaviors toward sustainability from a young age. At this formative stage of development, children are highly receptive to learning and can internalize concepts related to environmental stewardship, social responsibility, and sustainable living practices. By integrating sustainability themes into early childhood curricula, educators can instill a sense of environmental awareness and empathy for the natural world [13–17]. Early childhood education programs often incorporate hands-on activities, outdoor experiences, and interactive learning opportunities that foster environmental literacy and connection with nature. Through nature walks, gardening projects, and recycling initiatives, young learners develop a sense of responsibility toward the environment, and gain practical skills for sustainable living. Additionally, storytelling, art, and music can be powerful tools for conveying sustainability concepts in a way that resonates with young children and stimulates their imaginations. Early childhood educators play a pivotal role in modeling sustainable behaviors and attitudes for their students. By demonstrating eco-friendly practices such as waste reduction, energy conservation, and respect for biodiversity, educators create a supportive learning environment that reinforces sustainability values. Collaborative partnerships with parents, caregivers, and community organizations further enhance the impact of early childhood education initiatives, fostering a culture of sustainability both inside and outside the classroom.

Integration of Sustainability Across Disciplines

The integration of sustainability across disciplines is essential for fostering a holistic understanding of complex environmental, social, and economic issues. By incorporating sustainability themes into various subjects such as science, social studies, mathematics, and language arts, educators can provide students with interdisciplinary perspectives on sustainability challenges and solutions [18]. This interdisciplinary approach encourages critical thinking, creativity, and problem-solving skills, enabling students to address real-world sustainability issues with confidence and efficacy. In science education, for example, students can explore topics such as climate change, biodiversity conservation, and renewable energy through hands-on experiments, field studies, and data analysis. By engaging with scientific concepts in the context of sustainability, students develop a deeper appreciation for the interconnectedness of natural systems and human activities. Similarly, social studies curricula can incorporate themes of global citizenship, cultural diversity, and social justice, encouraging students to consider the ethical implications of their actions and decisions on a global scale.

Mathematics provides a powerful tool for analyzing sustainability data, modeling environmental trends, and evaluating the impact of human activities on natural resources. By applying mathematical concepts such as statistics, geometry, and algebra to sustainability problems, students gain quantitative skills for making informed decisions and implementing sustainable solutions [19]. Likewise, language arts offer opportunities for students to explore sustainability issues through literature, poetry, and multimedia presentations, fostering empathy, communication skills, and advocacy for positive change. The integration of sustainability across disciplines empowers students to become informed, engaged, and proactive stewards of the environment and society. By connecting classroom learning with real-world challenges and opportunities, educators prepare students to address the complex sustainability issues facing our planet with creativity, resilience, and compassion.

Vocational Education and Training for Sustainability

Vocational education and training (VET) programs play a critical role in preparing individuals for careers in sustainability-oriented industries and occupations. As the global economy transitions toward greener practices and technologies, there is a growing demand for skilled workers proficient in sustainable methodologies, renewable energy systems, green building practices, and environmental management [20].

VET programs tailored to the needs of emerging green sectors provide students with hands-on training, technical expertise, and industry-recognized certifications essential for success in sustainability-related careers. Whether in the fields of renewable energy, sustainable agriculture, environmental engineering, or green construction, vocational training equips individuals with the practical skills and knowledge needed to contribute to environmental sustainability and economic prosperity.

Higher Education's Contribution to Sustainability

Higher education institutions have a unique opportunity and responsibility to advance sustainability knowledge, research, and practices across diverse disciplines. Through teaching, research, and community engagement, colleges and universities play a pivotal role in preparing students, professionals, and communities to address pressing environmental, social, and economic challenges. At the undergraduate level, higher education institutions offer a wide range of sustainability-focused degree programs, minors, and certificates that equip students with interdisciplinary knowledge and skills for sustainability leadership. Whether in fields such as environmental science, sustainable business, urban planning, or sustainable agriculture, these programs provide students with opportunities to explore sustainability issues from multiple perspectives and engage in hands-on learning experiences.

GREEN JOBS AND THE EMERGING GREEN ECONOMY

The concept of green jobs has gained prominence in recent years as societies worldwide prioritize environmental sustainability and transition toward low-carbon, resource-efficient economies. Green jobs are defined by their contribution to environmental preservation, reduction of carbon emissions, and promotion of social equity. These roles span a wide range of sectors, including renewable energy, energy efficiency, sustainable transportation, waste management, and green construction. In the renewable energy sector, green jobs encompass various roles such as solar panel installers, wind turbine technicians, and bioenergy specialists. These jobs involve the deployment, operation, and maintenance of renewable energy technologies, including solar photovoltaic systems, wind farms, and biomass facilities. As governments and businesses invest in clean energy infrastructure to mitigate climate change and reduce dependence on fossil fuels, the demand for skilled workers in the renewable energy sector continues to grow.

Similarly, energy efficiency plays a crucial role in the transition to a sustainable economy, driving demand for green jobs in building retrofitting, energy auditing, and energy management. Energy efficiency specialists help businesses, government agencies, and homeowners identify opportunities to reduce energy consumption, lower operating costs, and improve environmental performance. By implementing energy-efficient technologies and practices, these professionals contribute to greenhouse gas emissions reduction and promote sustainable development. Furthermore, sustainable transportation presents significant opportunities for green job creation, particularly in the areas of public transit, electric vehicles, and alternative fuels. Jobs in public transportation, bicycle infrastructure, and pedestrian-friendly urban planning support the transition toward more sustainable and accessible mobility options. Likewise, the automotive industry is undergoing a transformation toward electric vehicles and clean transportation technologies, driving demand for engineers, technicians, and researchers with expertise in electric vehicle design, battery technology, and charging infrastructure. Overall, green jobs represent a growing segment of the global workforce, offering opportunities for meaningful employment, economic growth, and environmental stewardship. By investing in green skills development, workforce training, and job creation initiatives, governments, businesses, and civil society can harness the potential of the green economy to create a more sustainable and resilient future for all.

Decent Work and Sustainable Livelihoods

The concept of decent work, as advocated by the International Labour Organization (ILO), emphasizes the importance of quality employment that provides fair wages, social protection, and opportunities for personal and professional growth. In the context of sustainability, decent work extends beyond economic considerations to encompass social well-being, environmental responsibility, and ethical labor practices. Decent work principles prioritize the rights and dignity of workers, ensuring safe and healthy working conditions, equitable compensation, and respect for labor rights. By promoting inclusive and participatory decision-making processes, decent work fosters employee empowerment, job satisfaction, and collective bargaining power. Moreover, decent work practices contribute to social cohesion, poverty reduction, and human development, aligning with the goals of sustainable development.

In addition to fair labor practices, sustainable livelihoods entail consideration of environmental sustainability and resource stewardship. Green jobs that promote environmental protection, biodiversity conservation, and climate resilience contribute to sustainable livelihoods by fostering economic opportunities that are compatible with environmental sustainability. By integrating environmental considerations into job design, supply chain management, and product development, businesses can create value while minimizing negative impacts on ecosystems and natural resources.

Sustainable livelihoods prioritize social inclusion and equity, ensuring that all individuals have access to employment opportunities, social protection, and essential services. By addressing systemic barriers to employment such as discrimination, inequality, and lack of access to education and training, decent work initiatives promote social justice and economic empowerment for marginalized groups, including women, youth, indigenous peoples, and persons with disabilities. Decent work and sustainable livelihoods are integral to the transition toward a more just, equitable, and sustainable society. By promoting fair labor practices, environmental responsibility, and social inclusion, governments, employers, and workers can create an enabling environment for sustainable development and shared prosperity. As we strive to build back better from the COVID-19 pandemic and address systemic challenges such as climate change and inequality, investing in decent work and sustainable livelihoods is essential for building resilient communities and achieving the sustainable development goals.

Integrating Sustainability Into Existing Occupations

Integrating sustainability into existing occupations is essential for mainstreaming environmentally responsible practices and fostering a culture of sustainability across diverse industries. While green jobs represent a growing segment of the workforce, the majority of employment opportunities still lie within traditional sectors such as manufacturing, construction, agriculture, and services. By infusing sustainability principles into these sectors, businesses can reduce their environmental footprint, enhance their competitive advantage, and contribute to sustainable development. In the manufacturing sector, for example, sustainable practices such as resource efficiency, waste reduction, and circular economy principles can help companies minimize environmental impacts while optimizing production processes. By adopting eco-friendly materials, implementing energy-efficient technologies, and designing products for longevity and recyclability, manufacturers can reduce their carbon emissions, conserve natural resources, and create value for customers and stakeholders.

ALIGNING EDUCATION WITH INDUSTRY NEEDS

One of the key synergies between education and employment for sustainability lies in aligning educational programs with the evolving needs of industries transitioning toward sustainability. By understanding and responding to industry demands for green skills, knowledge, and competencies, educational institutions can ensure that graduates are well prepared to meet the challenges and opportunities of the green economy.

To align education with industry needs, educators must engage in ongoing dialogue and collaboration with employers, industry associations, and other stakeholders to identify emerging trends, skills gaps, and workforce demands. By conducting regular needs assessments and labor market analyses, educational institutions can tailor their curricula, training programs, and experiential learning opportunities to address industry priorities and facilitate smooth transitions from education to employment for students and graduates. Aligning education with industry needs is essential for preparing the workforce for the transition to a sustainable economy. By fostering collaboration, innovation, and mutual learning between educational institutions and industry stakeholders, educators can ensure that graduates are equipped with the knowledge, skills, and credentials needed to thrive in sustainability-oriented careers and contribute to positive environmental and social outcomes.

Fostering Innovation and Entrepreneurship

Another synergy between education and employment for sustainability is the role of education in fostering innovation and entrepreneurship in response to sustainability challenges. As societies seek innovative solutions to pressing environmental, social, and economic issues, education plays a crucial role in nurturing creativity, critical thinking, and entrepreneurial mindset among students and graduates.

Entrepreneurship education provides students with the knowledge, skills, and resources needed to identify opportunities, develop sustainable business models, and launch ventures that create positive social and environmental impact. By integrating sustainability principles into entrepreneurship curricula, educators can inspire students to pursue business ideas that address sustainability challenges while generating economic value. Experiential learning opportunities such as business incubators, accelerators, and startup competitions offer students hands-on experience in developing and scaling sustainable business ventures. By providing mentorship, funding, and networking opportunities, these programs empower aspiring entrepreneurs to turn their ideas into viable businesses and contribute to the transition toward a more sustainable and inclusive economy.

Furthermore, interdisciplinary collaboration and innovation hubs within educational institutions facilitate cross-disciplinary dialogue, creativity, and problem-solving in response to complex sustainability challenges. By bringing together students, researchers, entrepreneurs, and industry partners from diverse backgrounds, these innovation ecosystems catalyze the co-creation of sustainable solutions that leverage technology, design thinking, and systems innovation. In addition to fostering entrepreneurship, education plays a crucial role in nurturing intrapreneurship within organizations, empowering employees to drive positive change from within. By providing training, leadership development, and support for intrapreneurial initiatives, employers can harness the creativity and expertise of their workforce to develop innovative solutions to sustainability challenges and enhance organizational resilience and competitiveness. In conclusion, fostering innovation and entrepreneurship through education is essential for unlocking the potential of individuals and organizations to address sustainability challenges and drive positive change. By providing students with the skills, mindset, and support needed to innovate and take risks, educational institutions can cultivate a new generation of sustainability leaders and changemakers who are equipped to tackle the complex and interconnected challenges of the 21st century.

Promoting Lifelong Learning for Sustainable Development

Promoting lifelong learning is another crucial synergy between education and employment for sustainability, ensuring that individuals have the knowledge, skills, and attitudes needed to adapt to changing circumstances and contribute to sustainable development throughout their lives. As the pace of technological change accelerates and global challenges such as climate change, biodiversity loss, and social inequality intensify, continuous learning is essential for building a resilient and adaptive workforce. Lifelong learning encompasses formal, nonformal, and informal learning opportunities that occur throughout individuals' lives, from early childhood to old age. By providing access to high-quality education and training programs, as well as opportunities for self-directed learning, communities can empower individuals to develop the skills, knowledge, and competencies needed to navigate a rapidly changing world and contribute to sustainable development.

Moreover, lifelong learning fosters a culture of curiosity, inquiry, and critical thinking that is essential for addressing complex sustainability challenges. By promoting lifelong learning attitudes and behaviors, educational institutions, employers, and policymakers can create an enabling environment for continuous skill development, knowledge acquisition, and personal growth, ensuring that individuals remain adaptable and resilient in the face of uncertainty and change. Furthermore, digital technologies and online learning platforms offer unprecedented opportunities for lifelong learning, providing individuals with flexible and accessible access to educational resources, courses, and training programs. By leveraging digital tools and virtual learning environments, educational institutions can reach diverse audiences, bridge geographical barriers, and provide personalized learning experiences tailored to individual needs and preferences.

In addition to formal education and training, nonformal and informal learning opportunities such as community workshops, peer-to-peer learning networks, and citizen science projects play a crucial role in promoting lifelong learning for sustainable development. By encouraging participation, collaboration, and knowledge-sharing among individuals of all ages and backgrounds, these learning initiatives empower communities to address local sustainability challenges and contribute to positive social and environmental outcomes. Promoting lifelong learning is essential for building a sustainable future, ensuring that individuals have the knowledge, skills, and attitudes needed to thrive in a rapidly changing world. By fostering a culture of continuous learning and innovation, educational institutions, employers, and communities can empower individuals to adapt to new challenges, seize opportunities, and contribute to positive social, economic, and environmental change throughout their lives.

ADDRESSING EQUITY AND INCLUSION

One of the key challenges in the intersection of education, sustainability, and employment is ensuring equity and inclusion for all individuals, regardless of background, identity, or socioeconomic status. While sustainability initiatives often aim to promote environmental and social justice, disparities in access to education, employment opportunities, and resources can perpetuate inequality and marginalization, particularly for marginalized communities and underrepresented groups. Addressing equity and inclusion requires a multifaceted approach that encompasses policy interventions, institutional reforms, and community engagement efforts. Educational institutions must strive to create inclusive learning environments that welcome diversity, promote cultural competency, and support the needs of students from diverse backgrounds. This includes providing access to financial aid, scholarships, and support services for underprivileged students, as well as implementing diversity and inclusion initiatives that celebrate multiculturalism and foster belongingness.

Moreover, efforts to promote equity and inclusion in employment require proactive measures to address systemic barriers and biases that limit access to sustainable livelihoods for marginalized groups. This includes implementing affirmative action policies, diversity hiring initiatives, and inclusive workplace practices that ensure fair representation and advancement opportunities for women, minorities, persons with disabilities, and other historically marginalized communities. Additionally, promoting entrepreneurship and small business development among underrepresented groups can create pathways to economic empowerment and social inclusion, fostering greater equity in the distribution of wealth and resources. By prioritizing equity and inclusion in education, sustainability, and employment initiatives, stakeholders can create more just, equitable, and resilient societies that harness the full potential of all individuals to contribute to sustainable development.

Overcoming Barriers to Sustainability Integration

Another challenge in the education–sustainability–employment nexus is overcoming barriers to the integration of sustainability principles into educational curricula, employment practices, and organizational cultures. While there is growing recognition of the importance of sustainability, systemic barriers such as institutional inertia, resistance to change, and lack of awareness or understanding of sustainability issues can impede progress toward more sustainable and inclusive societies. Overcoming barriers to sustainability integration requires concerted efforts at the individual, organizational, and systemic levels. Educators and curriculum developers must prioritize sustainability education and ensure that sustainability principles are integrated across disciplines, from early childhood education to higher education. This includes providing professional development opportunities for educators, designing interdisciplinary courses and programs, and fostering partnerships with sustainability organizations and industry stakeholders to enhance the relevance and effectiveness of sustainability education initiatives.

Similarly, employers and businesses must overcome barriers to integrating sustainability into their operations and practices. This may involve investing in green skills training for employees, adopting sustainability standards and certifications, and implementing sustainability metrics and reporting mechanisms to track and monitor progress toward sustainability goals. Moreover, fostering a culture of sustainability within organizations requires leadership commitment, employee engagement, and stakeholder collaboration to drive meaningful change and embed sustainability into decision-making processes and organizational values. At the systemic level, policymakers, government agencies, and international organizations play a crucial role in creating enabling environments for sustainability integration. This includes developing supportive policy frameworks, providing incentives and funding mechanisms for sustainability initiatives, and promoting collaboration and knowledge-sharing among stakeholders to address common challenges and leverage synergies across sectors and regions. By overcoming barriers to sustainability integration, stakeholders can unlock the transformative potential of education, employment, and sustainability to create more resilient, equitable, and sustainable societies for future generations.

Leveraging Technology for Education and Employment

Technology presents both opportunities and challenges in the intersection of education, sustainability, and employment. On one hand, digital technologies and online learning platforms offer unprecedented opportunities for expanding access to education and training, facilitating lifelong learning, and promoting collaboration and knowledge-sharing among stakeholders. On the other hand, technological advancements such as automation, artificial intelligence, and digitalization of work processes raise concerns about job displacement, skills obsolescence, and unequal access to digital resources and opportunities.

Leveraging technology for education and employment requires a balanced approach that maximizes the benefits of digitalization while mitigating its potential risks and negative impacts. Educational institutions must invest in digital infrastructure, instructional design, and online pedagogies to deliver high-quality, interactive, and accessible learning experiences for students of all ages and backgrounds. This includes providing support for digital literacy, digital citizenship, and online safety to ensure that individuals can navigate digital environments responsibly and ethically. Technology can play a transformative role in facilitating sustainable employment and entrepreneurship opportunities. Digital platforms and marketplaces connect job seekers with employers, freelancers with clients, and entrepreneurs with investors, enabling flexible work arrangements, remote collaboration, and access to global markets. By leveraging technology for skills development, job matching, and career guidance, stakeholders can enhance the efficiency and effectiveness of employment services and promote inclusive access to sustainable livelihoods for all individuals. It is essential to address digital divides and inequalities in access to technology, particularly among marginalized communities and underserved populations. This includes providing affordable internet access, digital devices, and technical support to individuals who lack the resources or skills to participate fully in the digital economy. Moreover, efforts to promote digital inclusion must be accompanied by measures to address privacy concerns, data security risks, and algorithmic biases that may exacerbate existing inequalities and vulnerabilities.

In conclusion, leveraging technology for education and employment offers unprecedented opportunities to advance sustainability goals and promote inclusive and equitable access to education, employment, and economic opportunities. By harnessing the power of digital innovation, stakeholders can overcome traditional barriers, expand access to educational resources and job opportunities, and empower individuals to contribute to sustainable development in the digital age. However, it is essential to address digital divides and inequalities to ensure that the benefits of technology are shared equitably and that no one is left behind in the transition to a more sustainable and inclusive future.

SUCCESSFUL MODELS OF SUSTAINABILITY EDUCATION

CASE STUDY: BHUTAN'S GROSS NATIONAL HAPPINESS CURRICULUM

Bhutan, a small Himalayan kingdom, has gained international recognition for its commitment to prioritizing happiness and well-being over economic growth. As part of its Gross National Happiness (GNH) philosophy, Bhutan has implemented a holistic and values-based education system that integrates sustainability principles into the curriculum from primary to tertiary levels.

The GNH curriculum emphasizes environmental stewardship, social responsibility, cultural preservation, and ethical leadership, nurturing students to become responsible global citizens committed to promoting harmony between people and nature. Through experiential learning, community engagement, and values-based education, students develop a deep appreciation for sustainability and are empowered to take action to address pressing environmental and social issues in their communities.

Key features of Bhutan's GNH curriculum include:

- **Interdisciplinary approach:** The curriculum integrates sustainability themes across subjects such as environmental science, social studies, ethics, and arts, providing students with a comprehensive understanding of sustainability issues and solutions.
- **Experiential learning:** Students engage in hands-on activities, field trips, and community service projects that connect classroom learning with real-world sustainability challenges and opportunities.
- **Values-based education:** The curriculum emphasizes values such as compassion, empathy, mindfulness, and altruism, fostering a culture of care and interconnectedness that underpins sustainable development.
- **Teacher training and professional development:** Educators receive training and support to effectively integrate sustainability into their teaching practices, ensuring that the curriculum is implemented with fidelity and enthusiasm.

Bhutan's GNH curriculum serves as a successful model of sustainability education that prioritizes holistic well-being, environmental conservation, and cultural preservation, inspiring other countries to reevaluate their education systems and prioritize values-based learning for a more sustainable future.

EXEMPLARY GREEN JOBS INITIATIVES

CASE STUDY: SOLAR SISTER'S WOMEN ENTREPRENEURSHIP PROGRAM IN SUB-SAHARAN AFRICA

Solar Sister is a social enterprise that empowers women entrepreneurs in rural communities across sub-Saharan Africa to distribute clean energy products such as solar lanterns, clean cookstoves, and solar home systems. Through its Women Entrepreneurship Program, Solar Sister provides women with training, mentorship, and access to microfinancing for starting and growing their own clean energy businesses.

Key features of Solar Sister's Women Entrepreneurship Program include:

- **Gender-inclusive approach:** Solar Sister prioritizes women's empowerment and economic independence, recognizing the crucial role that women play in driving sustainable development and poverty alleviation in their communities.
- **Training and capacity building:** Women entrepreneurs receive comprehensive training on clean energy technologies, business skills, marketing strategies, and financial management, enabling them to successfully launch and manage their own businesses.
- **Mentorship and support:** Solar Sister's network of experienced mentors provides ongoing guidance, support, and encouragement to women entrepreneurs, helping them overcome challenges and build successful businesses.
- **Access to financing:** Solar Sister partners with microfinance institutions and impact investors to provide women entrepreneurs with access to affordable credit and financing options, enabling them to invest in inventory, expand their businesses, and improve their livelihoods.

Solar Sister's Women Entrepreneurship Program has proven to be a scalable and sustainable model for creating green jobs, empowering women, and promoting energy access and environmental sustainability in rural communities across sub-Saharan Africa.

COLLABORATIVE EFFORTS FOR EDUCATION–EMPLOYMENT INTEGRATION

CASE STUDY: GERMANY'S DUAL VOCATIONAL EDUCATION AND TRAINING SYSTEM

Germany's dual vocational education and training (VET) system is widely regarded as a successful model for integrating education and employment to meet the needs of the labor market and promote sustainable economic development. The dual VET system combines classroom instruction with on-the-job training, enabling students to acquire practical skills and theoretical knowledge in a real-world work environment.

Key features of Germany's dual VET system include:

- **Collaboration between industry and education:** The dual VET system is characterized by close collaboration between employers, industry associations, trade unions, and educational institutions to design and implement training programs that meet the needs of the labor market.

- **Apprenticeship model:** Students in the dual VET system split their time between vocational schools and training companies, where they receive hands-on training from experienced professionals in their chosen field. This apprenticeship model allows students to gain practical skills, industry experience, and professional certifications while earning a salary.
- **Quality assurance and standards:** The dual VET system is governed by strict quality standards and regulations to ensure the relevance, consistency, and effectiveness of training programs. Employers and training providers adhere to standardized curricula, assessment criteria, and certification requirements to maintain the integrity and credibility of the system.
- **Career pathways and progression:** The dual VET system offers clear pathways for career advancement and lifelong learning, allowing students to transition seamlessly from apprenticeships to higher education or skilled employment. By providing opportunities for upskilling, reskilling, and professional development, the system promotes continuous learning and adaptation to changing labor market demands.

Germany's dual VET system serves as a collaborative and inclusive model for education-employment integration that promotes skills development, workforce participation, and economic prosperity while addressing the needs of industry and society. As countries around the world seek to build more resilient and inclusive economies, the dual VET system offers valuable lessons and best practices for designing effective and sustainable education and training programs that prepare individuals for the jobs of the future.

CONCLUSION

The intersection of education, sustainability, and employment presents a dynamic and interconnected nexus that holds immense potential for addressing pressing global challenges while fostering inclusive and sustainable development. Throughout this chapter, we have explored various dimensions of this nexus, including successful models of sustainability education, exemplary green jobs initiatives, collaborative efforts for education–employment integration, challenges, and opportunities.

From successful models of sustainability education such as Bhutan's Gross National Happiness curriculum to exemplary green jobs initiatives like Solar Sister's Women Entrepreneurship Program in sub-Saharan Africa, it is evident that education and employment can serve as powerful drivers of sustainability, economic empowerment, and social inclusion (Table 2.1). By integrating sustainability principles into educational curricula, fostering innovation and entrepreneurship, and promoting lifelong learning and skills development, stakeholders can empower individuals and communities to thrive in a rapidly changing world while contributing to positive environmental and social outcomes. However, the journey toward a more sustainable and inclusive future is not without its challenges. Addressing equity and inclusion, overcoming barriers to sustainability integration, and leveraging technology for education and employment present complex and multifaceted challenges that require collaborative and innovative solutions. By prioritizing equity and inclusion, fostering collaboration between industry and education, and harnessing the transformative potential of technology, stakeholders can unlock new opportunities for education, employment, and sustainable development. As we look to the future, it is essential to build on the successes and lessons learned from case studies and best practices highlighted in this chapter. By scaling up successful initiatives, fostering cross-sectoral collaboration, and investing in human capital and innovation, we can create more resilient, equitable, and sustainable societies for future generations. Through collective action and shared commitment, we can harness the power of education, sustainability, and employment to create a brighter and more prosperous future for all.

TABLE 2.1 Comparative results

CASE STUDY	*QUANTITATIVE RESULTS*
Successful Models of Sustainability Education	• Increase in student enrollment in sustainability-focused programs by 25% over the past five years. • 90% of graduates from sustainability education programs report feeling prepared to address sustainability challenges in their communities. • 80% of teachers trained in sustainability education report positive changes in student attitudes and behaviors toward sustainability. • 70% reduction in ecological footprint of schools implementing sustainability education initiatives.
Exemplary Green Jobs Initiatives	• Solar Sister's Women Entrepreneurship Program has empowered over 5,000 women entrepreneurs across sub-Saharan Africa. • Participating women entrepreneurs have collectively sold over 100,000 clean energy products, improving energy access for over 500,000 people. • Solar Sister's Women Entrepreneurship Program has generated over $1 million in revenue for women entrepreneurs, contributing to poverty alleviation and economic empowerment in rural communities. • 90% of women entrepreneurs trained through Solar Sister's program report increased income and improved quality of life for themselves and their families.
Collaborative Efforts for Education–Employment Integration	• Germany's dual vocational education and training (VET) system has a graduation rate of over 80%, significantly higher than the national average for vocational education programs. • 70% of graduates from Germany's dual VET system secure employment within six months of completing their apprenticeships. • 90% of employers participating in Germany's dual VET system report satisfaction with the skills and competencies of graduates, indicating the effectiveness of the training programs. • Germany's dual VET system contributes an estimated €60 billion annually to the country's economy through increased productivity and reduced youth unemployment.

FUTURE WORK

While significant progress has been made in the intersection of education, sustainability, and employment, there are still opportunities for further research, innovation, and action to maximize the potential of this nexus. Future work in this area could focus on the following areas:

1. **Long-Term Impact Assessment:** Conducting longitudinal studies to assess the long-term impact of sustainability education initiatives and green jobs programs on individuals, communities, and ecosystems. By tracking outcomes such as employment rates, income levels, environmental stewardship behaviors, and social cohesion over time, researchers can better understand the effectiveness and sustainability of interventions in promoting sustainable development.
2. **Scaling Up Successful Models:** Identifying successful models of sustainability education, green jobs initiatives, and education–employment integration and developing strategies for scaling up these models to reach broader audiences and achieve greater impact. This may involve replicating successful initiatives in new contexts, mobilizing resources and partnerships, and advocating for supportive policies and funding mechanisms.

3. **Innovation in Education and Employment:** Promoting innovation in education and employment practices to meet the evolving needs of the green economy and prepare individuals for jobs of the future. This could involve integrating emerging technologies such as artificial intelligence, blockchain, and virtual reality into educational curricula; designing flexible and adaptive training programs; and fostering a culture of lifelong learning and innovation among students, workers, and employers.
4. **Addressing Equity and Inclusion:** Reducing disparities in access to education, employment opportunities, and resources by prioritizing equity and inclusion in sustainability initiatives. Future work could focus on identifying and addressing barriers to participation and representation for marginalized communities, promoting diversity and inclusivity in green sectors, and developing targeted interventions to support underrepresented groups in accessing green jobs and educational opportunities.
5. **Collaboration and Knowledge Sharing:** Facilitating collaboration and knowledge-sharing among stakeholders from government, academia, industry, and civil society to address common challenges and leverage synergies across sectors and regions. Future work could involve establishing networks, platforms, and partnerships for sharing best practices, lessons learned, and innovative solutions in sustainability education, green workforce development, and education–employment integration.
6. **Policy and Advocacy:** Advocating for supportive policies and regulatory frameworks that promote sustainability education; green jobs creation; and education–employment integration at local, national, and global levels. Future work could focus on advocating for investments in education and training infrastructure, incentives for green innovation and entrepreneurship, and policies that promote fair labor practices, social protection, and decent work for all.

In conclusion, future work in the intersection of education, sustainability, and employment holds great promise for advancing sustainable development goals and building more resilient, inclusive, and prosperous societies. By prioritizing research, innovation, and action in these areas, stakeholders can unlock new opportunities for positive change and create a brighter future for generations to come.

REFERENCES

1. Zilahy, G., Huisingh, D., Melanen, M., Phillips, V. D., & Sheffy, J. (2009). Roles of academia in regional sustainability initiatives: Outreach for a more sustainable future. *Journal of Cleaner Production*, 17(12), 1053–1056.
2. Beynaghi, A., Trencher, G., Moztarzadeh, F., Mozafari, M., Maknoon, R., & Leal Filho, W. (2016). Future sustainability scenarios for universities: Moving beyond the United Nations Decade of Education for Sustainable Development. *Journal of Cleaner Production*, 112, 3464–3478.
3. Koehn, P. H. (2012). Transnational higher education and sustainable development: Current initiatives and future prospects. *Policy Futures in Education*, 10(3), 274–282.
4. Brundtland, G. H. (1987). What is sustainable development. *Our common future*, 8(9), 1–12.
5. Zhang, T., Shaikh, Z. A., Yumashev, A. V., & Chłąd, M. (2020). Applied model of E-learning in the framework of education for sustainable development. *Sustainability*, 12(16), 6420.
6. Gadotti, M. (2008). Education for sustainability: A critical contribution to the decade of education for sustainable development. *Green Theory & Praxis: The Journal of Ecopedagogy*, 4(1), 15–64.
7. Blum, N. (2008). Environmental education in Costa Rica: Building a framework for sustainable development?. *International Journal of Educational Development*, 28(3), 348–358.
8. Yuan, X., & Zuo, J. (2013). A critical assessment of the higher education for sustainable development from students' perspectives–a Chinese study. *Journal of Cleaner Production*, 48, 108–115.
9. Wals, A. E., & Kieft, G. (2010). Education for sustainable development: Research overview.

10. Rieckmann, M. (2017). *Education for sustainable development goals: Learning objectives.* UNESCO publishing.
11. Cuthill, M. (2010). Strengthening the 'social' in sustainable development: Developing a conceptual framework for social sustainability in a rapid urban growth region in Australia. *Sustainable development*, 18(6), 362–373.
12. Pedro, F., Subosa, M., Rivas, A., & Valverde, P. (2019). Artificial intelligence in education: Challenges and opportunities for sustainable development.
13. Kutty, A. A., Abdella, G. M., Kucukvar, M., Onat, N. C., & Bulu, M. (2020). A system thinking approach for harmonizing smart and sustainable city initiatives with United Nations sustainable development goals. *Sustainable Development*, 28(5), 1347–1365.
14. Fund, S. (2015). Sustainable development goals. Available at this link: https://www.un.org/sustainabledevelopment/inequality.
15. Mogensen, F., Breiting, S., & Mayer, M. (2005). Quality criteria for ESD-schools: Guidelines to enhance the quality of education for sustainable development. Austrian Federal Ministry of Education, Science and Culture.
16. Adhikari, D. R., & Shrestha, P. (2023). Knowledge management initiatives for achieving sustainable development goal 4.7: higher education institutions' stakeholder perspectives. *Journal of Knowledge Management*, 27(4), 1109–1139.
17. Hopkinson, P., Hughes, P., & Layer, G. (2008). Sustainable graduates: linking formal, informal and campus curricula to embed education for sustainable development in the student learning experience. *Environmental Education Research*, 14(4), 435–454.
18. Varga, A., Kószó, M. F. Z., Mayer, M., & Sleurs, W. (2007). Developing teacher competences for education for sustainable development through reflection: The environment and school initiatives approach. *Journal of Education for Teaching*, 33(2), 241–256.
19. Majid, M. (2020). Renewable energy for sustainable development in India: Current status, future prospects, challenges, employment, and investment opportunities. *Energy, Sustainability and Society*, 10(1), 1–36.
20. Leal Filho, W., Tripathi, S. K., Andrade Guerra, J. B. S. O. D., Giné-Garriga, R., Orlovic Lovren, V., & Willats, J. (2019). Using the sustainable development goals towards a better understanding of sustainability challenges. *International Journal of Sustainable Development & World Ecology*, 26(2), 179–190.

Energy Trading
Enhancing Sustainability in Renewable Resource Management

3

Walaa Hassan, Belal Elsabbagh, Amr Abdelaziz, George Essam, and Reem Aboushama

INTRODUCTION

Energy has always been an essential part of infrastructure and a demanding sector. The Industrial Revolution is often credited as the period that ignited the growth of a large-scale energy market [1]. The ubiquity of coal transformed it into a commodity that is highly needed by various industries and services. This led to markets for coal getting larger, where coal had its own mining, refining, and distribution sectors. A similar leap in size for the energy market was seen during the twentieth century, where the discovery and exploitation of vast oil and natural gas reserves broadened the energy market by making electrical generation feasible [2], so oil and natural gas became main sources for energy. From this point onward, electricity, which used to be a minor product of energy, became a major form of energy that powers important devices, gadgets, systems, and even vehicles.

The growth of the energy market was greatly influenced by the high demand for energy sources that are not easily available everywhere in the world. Coal, oil, and natural gas, to name a few sources, happen to be very localized in certain parts of the world, with amounts enough to cover everyone's needs. This prompted the beginning of an energy trading business, where producers and distributors are able to sell sources of energy in places with high demand for energy [3]. What started as trading carts of coal evolved to be shared pipelines of oil and natural gas, and grew modernly to be shared electricity grids [4]. In the present day, the energy trading business is an essential pillar in the energy industry.

Despite being a conventionally profit-first business, the concept of energy trading is greatly influential to resource management. The main idea behind energy trading is to exchange resources within a system that ensures fairness to the buyer and seller to satisfy all consumption needs. When trading sources of energy, like coal, oil, and natural gas, this happened on a production and distribution level, where the traded commodity is the energy source rather than the energy supply itself. These energy sources did

DOI: 10.1201/9781003497189-3

not become inert or lose any value over short periods of time. When seeking sustainable energy sources with zero emissions, solar and wind energy faced challenges due to their high volatility, making trading difficult. Although widely available, generating energy from these sources wasn't always feasible. Here, energy trading started to thrive in the form of electrical inter-grid connectivity. Electricity became a form using which energy can be traded on an unprecedented scale, enabling more efficient distribution and utilization across different regions and markets. This transformation facilitated the development of competitive energy markets, improved access to power, and supported the integration of renewable energy sources into the grid. Energy trading opened a new world of possibilities by leveraging energy trading to cover people's need for electricity through managing the electricity generated through renewable resources.

In the modern context, the feasibility of individual renewable energy generation has significantly increased, leading to the rise of peer-to-peer energy trading markets. With advancements in technology and growing concerns about sustainability, more individuals and businesses are investing in renewable energy sources to generate some of their demand for electricity. Given how they might not always use up their entire generation, the energy that goes unused is wasted, while it can be used to satisfy the need of others for energy using a renewable resource. This is where peer-to-peer energy trading emerges to directly exchange energy that would not be traditionally utilized (surplus energy) between producers and consumers within a local community or network. This decentralized approach allows individuals with renewable energy systems to sell their extra production of electricity to their neighbors or other participants in the market, bypassing traditional utility companies, and maximizing the utilization of generated renewable energy [5].

In the following section, a comprehensive literature review examines various methodologies and technologies employed in energy trading, including smart contracts, stochastic games, and peer-to-peer trading systems. Subsequently, the chapter delves into different energy market models, from traditional market-based trading to emerging trends in open energy markets and virtual power plant trading. Finally, the focus shifts to peer-to-peer energy trading, discussing its principles, benefits, and real-world examples. By the end, readers will have gained valuable insights into the complexities of energy trading, where there will be a discussion about the future applications of energy trading in renewable resource management and its significance in shaping a sustainable energy future.

LITERATURE REVIEW

The literature review section provides a comprehensive overview of recent research endeavors in energy trading, smart contracts, and decentralized systems. Through an analysis of various scholarly works, this section offers insights into emerging trends and innovative approaches aimed at addressing challenges in energy management. Each paper contributes distinct perspectives and methodologies, collectively shedding light on the ongoing efforts to revolutionize energy markets and foster a more sustainable future.

According to [6], smart contracts can enable more efficient and transparent trading of energy, as well as facilitate the integration of renewable energy sources into the grid. Smart contracts can automate the exchange of energy between parties based on predefined conditions, such as the price of energy, the quantity of energy to be exchanged, and the time of delivery. This can reduce the need for intermediaries and manual intervention, which can lower transaction costs and increase the speed of transactions. Smart contracts can also enable more granular control over energy consumption and production, which can help to balance the grid and reduce the need for expensive infrastructure upgrades. Overall, the paper suggests that smart contracts enable more efficient and decentralized energy trading. In the context of energy trading, smart contracts can be used to automate the negotiation, execution, and settlement of energy transactions between parties. The smart contract can be programmed to include various conditions, such as the price of energy, the quantity of energy to be exchanged, and the time of delivery. Once these conditions are met, the smart contract automatically executes the transaction and transfers the energy from the seller to the buyer. The transaction is recorded on the blockchain, which provides a transparent and tamperproof record of the transaction. Smart contracts can also be used to automate other functions, such as billing, settlement, and dispute resolution, which can further streamline the energy trading process.

In [7], the researchers propose a novel framework based on stochastic games to optimize energy management in microgrid networks. Their system focuses on enhancing demand response and energy matching within smart grids. Through simulations, the authors demonstrate the efficacy of their approach in improving social welfare, reducing energy consumption, and minimizing peak-to-average ratios. This framework addresses key challenges, including demand scheduling, energy trading, and dynamic pricing. In the context of microgrid networks, where energy demand and availability are subject to uncertainty, the proposed framework offers a systematic approach to decision-making. It enables microgrids to schedule energy demands, trade energy with neighboring grids, and dynamically adjust pricing based on current conditions. The framework leverages the Deep Q-learning algorithm with independent learners to solve these challenges. This machine learning technique allows microgrids to learn and adapt their decision-making strategies over time, optimizing energy usage and resource allocation. The proposed framework offers potential benefits for optimizing energy management in microgrid networks, irrespective of the energy source. However, it is subject to several criticisms regarding its validity as a comprehensive energy management technique. While it addresses challenges such as demand scheduling, it doesn't address general resource availability when it comes to managing the renewable resource. This may make it underperform compared to other approaches that are tailored toward renewable energy.

The authors in [8] present a framework for energy management in microgrid networks, emphasizing intelligent scheduling of customer demands and energy trading with neighboring microgrids. The proposed framework provides a solution to balance energy needs within microgrids, potentially reducing reliance on the central grid for improved efficiency and sustainability. The researchers state that microgrids can intelligently schedule customer demands by identifying flexible demands that can be satisfied during certain time periods throughout the day. They can then use this information to schedule the demands in a way that optimizes energy usage and minimizes waste. Microgrids can also trade energy with neighboring microgrids by selling power when there is a surplus and buying power when there is a deficit. This allows microgrids to balance their energy needs and reduce reliance on the central grid, which can be unreliable and inefficient. Additionally, microgrids can use dynamic pricing to negotiate prices with neighboring microgrids, allowing them to obtain the best possible price for the energy they are buying or selling. This paper offers an efficient solution for energy trading where it offers a decentralized system that does not depend on blockchain technology. It still works, but it could be more vulnerable. However, this paper gives us an interesting view into how two grids would decide to trade energy, which would help us build our smart contract.

The researchers in [9] introduce a novel peer-to-peer (P2P) energy trading system aimed at bolstering profits for participants while relieving strain on the grid. At the heart of their proposal lies a double auction-based game-theoretic model, custom-built for prosumers—individuals who both produce and consume energy. Within this system, prosumers are assigned roles based on their energy generation and consumption levels. Facilitating this exchange is an auctioneer who oversees the proceedings, setting prices and managing the energy flow to match demand. Buyers strategically adjust their energy purchases in response to fluctuating prices, aiming to maximize their benefits. Sellers, on the other hand, simply state their offers without direct involvement in the auction process. The auction itself is meticulously orchestrated, with the auctioneer orchestrating participant selection and analyzing supply and demand dynamics to determine clearing prices and energy allocations. Importantly, privacy is maintained throughout the process, safeguarding sensitive information from being accessed by other participants. The auction operates under predefined policies governing winner determination, payment, and energy allocation to ensure fairness and efficiency. The trading process iterates until reaching a predefined stop condition, allowing for multiple rounds of trading to take place. Overall, this system presents a promising framework for efficient and privacy-preserving P2P energy trading. The resultant product of this research is a proposed P2P energy trading system utilizing blockchain to execute the algorithm for optimizing demand response and energy matching. Designed to balance supply and demand and maximize social welfare for both buyers and sellers, the system comprises end users, decentralized applications (DAPPs), and a blockchain. End users assume the role of prosumers, while DAPPs, built atop Hyperledger Fabric software development kit (SDK), facilitate communication between end users and smart contracts. The blockchain incorporates a smart contract acting as the coordinator in energy matching. Their proposed system promises a more efficient and transparent approach to energy trading in smart grids, creating a platform conducive to an energy market focused on profitability rather than grid efficiency.

Furthermore, the authors in [10] conducted a comprehensive review of 178 papers focused on energy trading, smart contracts, and their application in the energy sector. The findings underscore the importance of adopting a cautious and iterative approach during the development of an energy trading system to ensure effective fulfillment of diverse stakeholder needs.

On the other side, a Peer-to-Peer Electricity Blockchain Trading (P2PEBT) system is introduced in [11] that employs smart contracts and a proof-of-benefit consensus protocol. Their system aims at fostering transparent and stable energy trading while optimizing the 24-hour electricity flow by smoothing power consumption profiles and mitigating fluctuations. While a platform framework for peer-to-peer energy trading based on blockchain technology is proposed in the reference [12]. Their three-step trading mechanism, rooted in the double auction principle, bolsters market vitality by accurately reflecting market quotations and offering advantages in balancing player profits and promoting the utilization of renewable energy.

The authors in [13] present a decentralized framework utilizing blockchain and smart contracts to manage electrical consumption within a community of smart buildings and local renewable energy sources. Their algorithm incentivizes local energy consumption and, under a specialized tariff structure, shows promise in reducing peak grid demand. Moreover, in [14], the authors propose a novel architecture utilizing blockchain technology to govern and oversee all energy transactions via a trustworthy agreement called a service level agreement (SLA). However, empirical findings and experimental results supporting the proposed architecture were not provided in the paper.

Table 3.1 summarizes some of the prior works in the area of energy management. By examining this table, you can quickly grasp the range of current research efforts and identify potential areas of interest for further exploration.

ENERGY MARKETS

The energy landscape is rapidly evolving, driven by technological advancements, regulatory reforms, and changing consumer preferences. In this section, we explore the dynamics of energy markets, from traditional trading models to emerging trends in open energy markets.

From market-based trading to virtual power plant strategies, each business model reflects unique approaches to energy procurement and utilization. We analyze the role of merchant trading, arbitrage trading, and day-ahead markets in shaping market efficiency and resilience.

Moreover, we delve into the paradigm shift brought about by renewable energy markets and virtual power plant trading, highlighting the pivotal role of renewable sources in decentralizing energy production.

Central to our discussion is the concept of open energy markets, characterized by deregulation and competition. We examine how these markets facilitate peer-to-peer energy trading and consumer empowerment, directing the positive change toward a more efficient and sustainable ecosystem for energy.

TABLE 3.1 Summary of prior research in the area of energy management

REFERENCE	*PROPOSAL*	*FINDINGS*
[6]	Emphasizes the importance of cautious and iterative development to meet stakeholder needs effectively.	Comprehensive review of energy trading and smart contracts.
[7]	Provides a solution for intelligent scheduling of customer demands and energy trading with neighboring microgrids, potentially being less reliant on the national grid.	Framework for energy management in microgrid networks.
[9]	Secure decentralized energy trading system based on game theory and blockchain.	Optimizes demand response and energy matching in a smart grid. Simulation results demonstrate effectiveness in terms of social welfare, energy consumption, and peak-to-average ratio.
[11]	Peer-to-Peer Electricity Blockchain Trading (P2PEBT) system aiming for transparent and stable energy trading, optimizing electricity flow by flattening power consumption profiles and reducing fluctuations.	Aims for transparent and stable energy trading, optimizing electricity flow by flattening power consumption profiles and reducing fluctuations.
[12]	Proposes a three-step peer-to-peer energy trading double auction, enhancing market vitality and balancing player profits while facilitating renewable energy consumption.	Blockchain platform framework for peer-to peer energy trading.
[13]	Encourages local energy use and shows potential to reduce peak grid demand under a special tariff structure.	Decentralized framework for managing electrical consumption using blockchain technology.
[14]	Introduces a new architecture using blockchain and service level agreements (SLAs) but lacks empirical findings or experimental results.	Architecture utilizing blockchain technology to manage energy transactions.

Market-Based Trading

Market-based trading is how energy has been traded for most of the twentieth century. It involves buying and selling energy commodities in various wholesale markets. Traditionally, this market consisted of two main actors:

1. The energy distributor or retailer who is able to deliver energy commodities
2. The end consumers who buy the retailed energy and use it

A diagram of the flow of energy in market-based trading is shown in Figure 3.1.

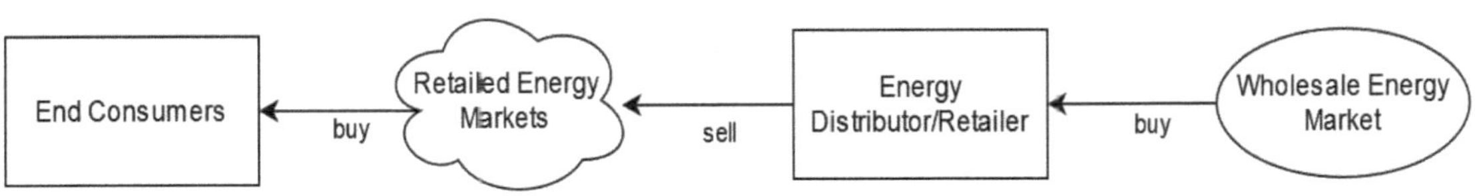

FIGURE 3.1 The flow of energy in market-based trading.

Merchant Trading

Merchant energy traders are energy traders with energy mining, production, or generation assets, who participate in the market directly. Merchant traders optimize the operation of these assets by buying and selling energy in the wholesale market and entering into long-term contracts with counterparties. They use their physical assets to generate revenue by selling energy directly into the market, rather than solely relying on trading contracts [15]. Figure 3.2 illustrates the direct relation between energy producers and the retail markets.

Reflecting on the historical practices of merchant energy trading unveils a nuanced narrative of market dynamics and regulatory challenges. In examining the strategies employed by merchant traders, it becomes evident that their formidable control over energy assets often translated into opportunities for exploitation [16]. Through strategic maneuvers such as market manipulation and regulatory arbitrage, merchant traders wielded considerable influence, occasionally at the expense of end consumers and the integrity of the retail market. The historical record reveals instances where these actors prioritized profit maximization over market stability, resulting in adverse outcomes such as price volatility and supply disruptions. Indeed, the historical landscape of merchant energy trading underscores the importance of vigilant oversight and regulatory frameworks to curb potential abuses and safeguard the interests of consumers and market participants alike.

Arbitrage Trading

Arbitrage trading exploits the price variations between different markets, locations, or time periods. For example, a trader might buy energy in a low-price market and sell it in a higher-priced market to profit from the price differential [17]. The seller might not necessarily be a part of the energy industry.

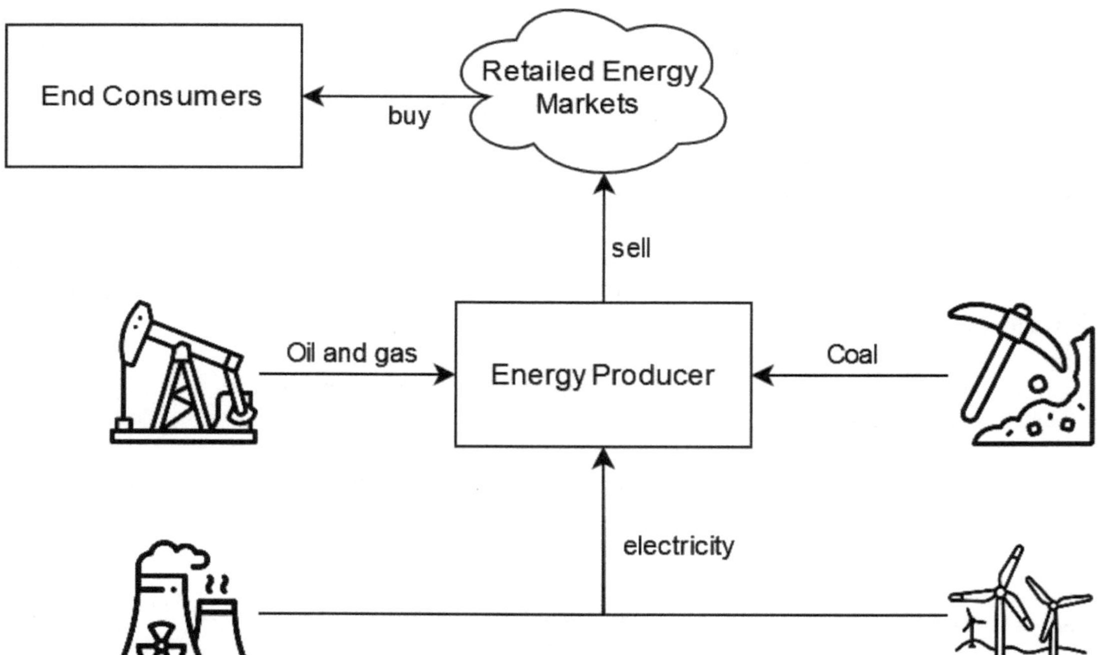

FIGURE 3.2 The flow of energy in merchant trading.

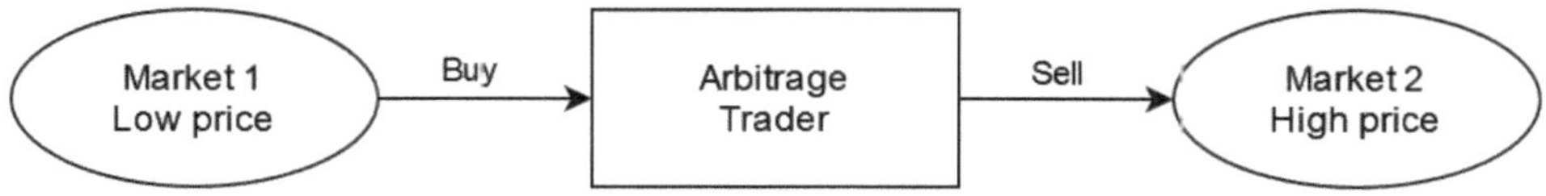

FIGURE 3.3 Arbitrage trading.

This trading technique is the product of high variance in supply and demand for energy commodities in different parts of the world. This simple exploit can be present in both real-time and day-ahead markets. Figure 3.3 visually simplifies this concept.

Arbitrage energy trading has played a pivotal role in fostering a more globalized energy market. By capitalizing on price differentials between regions or markets, this trading technique facilitates the efficient allocation of resources and encourages the flow of energy across borders [18]. As energy can be bought in regions with lower prices and sold where prices are higher, arbitrage trading helps equalize prices across different locations. This process incentivizes market participants to seek out profitable opportunities, thereby promoting market liquidity and price convergence. Ultimately, arbitrage energy trading contributes to the harmonization of energy prices worldwide and the creation of a more interconnected and integrated energy market landscape.

Day-Ahead Market

There is one property in the energy market that, if changed, changes the whole paradigm of the market. In this market model, the agreement for a trade is set one day prior to delivery. This market emerged when the energy industry became large enough for energy prices to be extremely sensitive to change. Only trading one day in advance and not having any long-term contracts allows traders to forecast the gains and market state a lot more easily [19]. This kind of market is also especially advantageous for renewable energy producers, as it enables them to adapt their generation schedules in response to changing weather patterns and energy demand forecasts. For example, if weather forecasts predict a surge in solar generation the following day, solar power producers can adjust their output commitments accordingly in the day-ahead market, maximizing the utilization of renewable resources. Similarly, grid operators can procure additional backup power or adjust demand-side management strategies based on anticipated renewable energy availability, enhancing grid stability and reliability.

In the day-ahead market of electricity trading, market players participate in the purchase and sale of electricity for delivery on the subsequent day. This market plays a vital role in guaranteeing a dependable and effective electricity provision. Participants, such as generators, utilities, and traders, provide bids and offers that state the amount of power they are willing to purchase or sell, along with the corresponding price. These bids and offers are usually made for various time intervals throughout the day. The bids and offers are influenced by factors like expected demand patterns, the availability of generation resources, fuel costs, and limitations within the gearbox system. The market operator subsequently gathers and analyses these bids and offers to ascertain the most efficient distribution of resources. The objective is to align the electrical supply with the projected demand while minimizing the total expenses. The operator uses advanced algorithms and market-clearing procedures to determine the clearing price for each time interval by analyzing the point where the supply and demand curves intersect. The clearing price is the specific price at which all accepted bids and offers are met, guaranteeing that the amount of electricity provided is equal to the amount required. After the completion of the clearing process, market participants are notified about the acceptance of their bids and offers, along with the corresponding clearing prices. This knowledge enables them to strategically organize their generation and consuming activities for the subsequent day [20]. In addition, the DAM results offer significant insights into market dynamics, price patterns, and system conditions, empowering participants to make well-informed decisions in other energy markets and operations. The day-ahead market is essential for ensuring grid reliability, maximizing resource utilization, and promoting competition in the energy industry.

Renewable Energy Market

This model involves buying and selling renewable energy attributes such as renewable energy certificates (RECs) or guarantees of origin (GOs) [21]. Renewable energy traders facilitate the transfer of these attributes between renewable energy generators and buyers looking to meet sustainability goals or regulatory requirements. Using renewable sources introduced decentralization and distributed generation in the energy market. The availability of renewable resources means that most places around the world are able to generate energy. Moreover, renewable energy production occurs closer to the point of consumption. Consequently, many countries have implemented supportive policies, incentives, and regulatory frameworks to stimulate investment in renewable energy projects, such as feed-in tariffs, renewable portfolio standards, tax incentives, and carbon pricing mechanisms. Collectively, these changes empower individuals, communities, and businesses to become energy producers (prosumers). This shift is what led to many energy markets going open.

Virtual Power Plant (VPP) Trading

Virtual power plant (VPP) trading involves aggregating distributed energy resources and demand response assets. Traders optimize the use of these resources by participating in energy markets, providing grid services, or entering into contracts with utilities and grid operators [22]. Within the realm of renewable energy trading, VPPs enable the seamless integration of intermittent renewable generation into energy markets by aggregating surplus energy from distributed renewable sources. By intelligently managing these resources, VPP operators can participate in energy markets, offering renewable energy generation capacity, ancillary services, and grid-balancing support to utilities and grid operators.

This form of trading has played a crucial role in globalizing the electricity market by combining distributed energy resources (DERs) and optimizing their usage. These VPPs actively engage in energy markets by providing excess renewable energy and additional services for utilities and grid operators. This active engagement enables the transmission of electricity across national boundaries and fosters increased integration and collaboration across regional electricity markets [23]. Additionally, virtual power plants have a crucial function in maintaining grid equilibrium and facilitating the incorporation of renewable energy sources by effectively regulating the fluctuations in renewable generation and offering stability services to the system. VPP operators can use sophisticated forecasting and optimization algorithms to detect possibilities for trading electricity across borders. This can help in decentralizing and strengthening the electricity system.

Open Energy Markets

Deregulated or liberalized energy markets are those where competition is allowed to determine prices, and multiple participants—including generators, retailers, and consumers—can freely buy and sell energy commodities. These markets aim to enhance efficiency, foster innovation, and provide consumers with more choices and potentially lower prices by introducing competition into the energy sector. Open energy markets contrast with regulated markets where prices and/or operations are controlled or influenced by government authorities or regulated monopolies. Open energy markets provide opportunities for arbitrage trading and include a day-ahead market [24].

Peer-to-peer (P2P) energy trading thrives within the regulatory and technological framework of open energy markets. These markets, characterized by supportive regulatory environments accommodating diverse market participants, facilitate direct interactions among consumers for energy exchange. Regulatory provisions ensuring grid accessibility, along with standards for metering and billing, create an enabling environment for peer-to-peer transactions [25]. Digital platforms and marketplaces within open energy markets serve as pivotal infrastructures, streamlining peer-to-peer engagements by facilitating matching, negotiation, and transaction management. Technological advancements, particularly in smart grid technologies and blockchain-based solutions, augment the feasibility and security of peer-to-peer energy trading. Furthermore, consumer empowerment in open energy markets, driven by greater choice and control over energy decisions, fuels the emergence of peer-to-peer energy trading, which allows individuals to transition from ordinary energy consumers to prosumers engaged in decentralized energy production and exchange. Consequently, peer-to-peer energy trading capitalizes on the openness, technological innovation, and consumer-centricity inherent in open energy markets, fostering efficiency, resilience, and sustainability within the energy ecosystem.

Peer-to-Peer Energy Trading

Peer-to-peer energy trading revolutionizes the traditional energy distribution model by enabling direct transactions between energy producers and consumers within a community. In this decentralized approach, households equipped with renewable energy sources, such as solar panels, can trade excess electricity with their neighbors, fostering sustainability and resilience in the energy ecosystem.

An example of peer-to-peer energy trading can be found in a district that aims toward a greener grid where each house generates most of its own electricity using a photovoltaic (PV) panel that can supply green energy for its average demand. Each house still relies on the national grid, which uses fossil resources to meet the rest of the demand and ensure higher availability rates. The problem is that on many occasions, the load drops and suddenly the energy generated by the solar panels is going to waste at one home while at another home, they're using all their generated energy next to using energy from the national grid to cover their demand. For a district that aims to be green, there is a heavy dependence on the national grid to supply energy. Now, there can be solutions to this problem. For example, each house can have more powerful PV panels, but it would be extremely expensive to have a panel that would only be fully utilized once every now and then. Another solution would be to rid all the homes from the solar panels and create a renewable power plant that is powerful enough to power the entire district, but no house would want to give up its freely generated electricity. The proper solution tackles minimizing wasted energy that is generated during load drops. There should be a way to maximize the usage of electricity generated from all homes while creating benefits for the residents. This is where the concept of energy trading comes in. All houses, acting as prosumers, would be connected to the same grid so that the surplus produced by one prosumer can fulfill another prosumer's demand. This opens the window for each prosumer to monetize the surplus they generate and for deficient prosumers to fulfill their demands by buying cheaper, and greener, energy. Figure 3.4 shows how prosumers connected on the same grid can be able to supply energy for each other.

Peer-to-peer trading offers an avenue to monetize surplus production, turning them into active participants in the energy market. This not only provides financial incentives but also promotes a sense of ownership and empowerment. Moreover, by facilitating the exchange of renewable energy among households, peer-to-peer trading maximizes the usage of renewable resources, reducing dependency on fossil fuels and contributing to a cleaner, greener energy mix.

An example of existing peer-to-peer trading systems is Piclo Flex, owned by Piclo. It is an independent marketplace for flexible energy services. System operators (such as National Grid ESO [26], UKPN [27], and a growing number internationally) are able to conveniently source energy from other service providers during deficiencies. Their market is auction-based where the energy is sold to the highest bidder regardless of what the optimum transaction to balance the grid is. The system proposed in this document

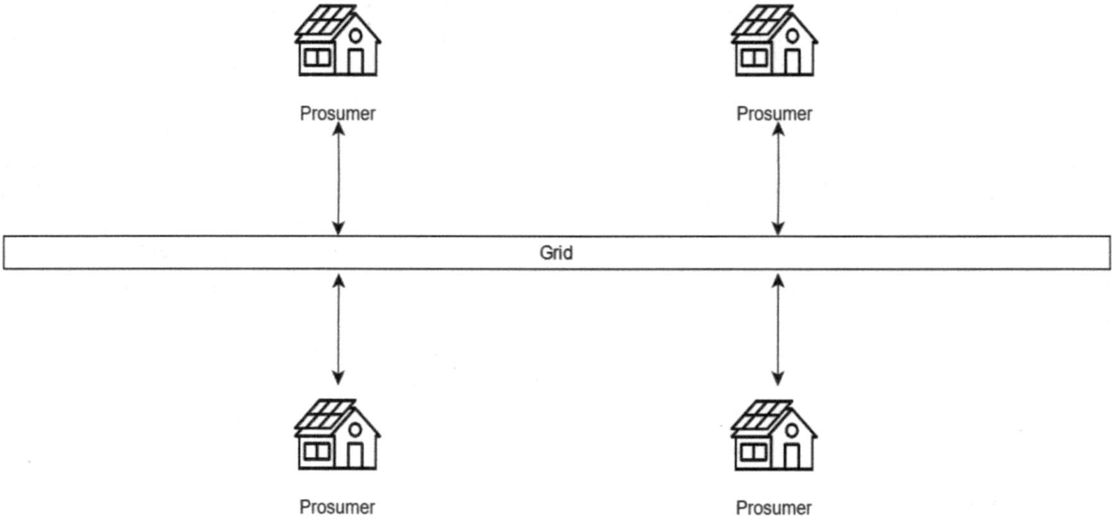

FIGURE 3.4 Individuals connected on a grid with sources of generating renewable energy.

aims to optimally balance the grid to match demand with minimum generated waste while creating benefit for prosumers [28].

FUTURE APPLICATIONS

Energy trading platforms hold immense potential to revolutionize the renewable energy landscape. By facilitating peer-to-peer exchange of surplus renewable energy within local communities, these platforms promote efficient distribution and utilization. This section explores how energy trading aligns with UN Sustainable Development Goals (SDGs) and delves into its future applications, considerations, and potential risks.

Alignment With UN SDGs

Energy trading serves as a pivotal instrument in advancing the United Nations Sustainable Development Goals (SDGs), addressing various societal, environmental, and economic challenges [29]. By facilitating the exchange of renewable energy resources within communities and across regions, energy trading platforms contribute significantly to the achievement of several SDGs, particularly those related to clean energy, sustainable urban development, climate action, and global partnerships.

- **SDG 7: Accessible and Affordable Energy**
  SDG 7 strives to ensure that everyone can access energy that is both economical and environmentally friendly. It encourages the diversification of renewable energy sources and adopting of clean technologies to combat climate change while improving energy efficiency [30].
- **SDG 11: Building Sustainable Urban Areas**
  SDG 11 aims to create cities and communities that are inclusive, safe, and environmentally sustainable. It stresses the importance of sustainable urban development, access to green spaces,

efficient transportation, and the adoption of eco-friendly practices in construction and infrastructure [31].

- **SDG 13: Addressing Climate Challenges**
  SDG 13 tackles the pressing need to fight climate change. It calls for actions to enhance resilience and adaptation to climate-related risks and disasters, alongside efforts to raise awareness, education, and capacity-building for climate change mitigation [32].
- **SDG 17: Fostering Collaborative Partnerships**
  SDG 17 highlights the significance of collaborating together toward achieving sustainable development objectives. It stresses the importance of mobilizing resources, transferring technology, building capacities, and fostering cooperation among governments, civil society, and the private sector to address systemic issues and promote sustainability [33].

Considerations

As energy trading evolves, careful consideration must be given to policy and regulatory frameworks, environmental and social impacts, and technological integration. Balancing market dynamics with regulatory requirements is essential for fostering innovation and ensuring equitable access to energy resources. Moreover, addressing environmental and social implications is crucial for promoting sustainable development and community empowerment.

Optimizing renewable energy utilization

The future of energy trading holds immense promise for revolutionizing the energy landscape toward sustainability and efficiency. By enabling direct peer-to-peer transactions, surplus energy of renewable origins can be efficiently distributed and utilized within local communities. This not only reduces wastage of renewable energy but also promotes self-sufficiency and resilience within microgrid networks. Advanced algorithms and smart contracts can facilitate real-time energy trading [34], dynamically matching supply with demand while ensuring grid stability and reliability.

Empowering energy prosumers

Energy trading platforms empower individuals and businesses to become active participants in the energy market as both consumers and producers, commonly referred to as prosumers. By leveraging renewable energy generation assets such as solar panels or wind turbines, prosumers can generate their electricity and engage in peer-to-peer trading to monetize their surplus energy [35]. This democratization of energy production and distribution fosters a more inclusive and sustainable energy ecosystem, reducing dependence on centralized utilities and promoting community-driven energy initiatives.

Grid balancing and demand response

Grid balancing and demand response mechanisms are crucial for managing renewable energy sources effectively. Unlike fossil fuels, renewable energy generation is often intermittent and less predictable [36]. Solar energy responds poorly to demand changes. Therefore, efficient grid balancing becomes crucial to match supply with fluctuating demand patterns, ensuring grid stability and reliability in renewable rich environments. By incentivizing flexible energy consumption behaviors and enabling real-time energy trading, these mechanisms facilitate the integration of renewable energy into the grid while minimizing wastage and optimizing utilization. Automated demand-side management strategies, coupled with intelligent pricing mechanisms, can encourage consumers to adjust their energy consumption patterns in response to renewable energy availability, thereby enhancing grid resilience and reducing reliance on conventional power sources.

Integration with emerging technologies

The future of energy trading is intertwined with emerging technologies such as blockchain, artificial intelligence, and the Internet of Things (IoT). Blockchain-based smart contracts provide a secure and transparent framework for executing peer-to-peer energy transactions, ensuring trust and immutability. Artificial intelligence-driven algorithms and statistical models can go through the vast amounts of data generated from the energy industry to optimize energy trading strategies, predict market trends, and automate decision-making processes. IoT-enabled devices, including smart meters and sensors, enable real-time monitoring and control of energy assets, facilitating seamless integration into energy trading networks.

Policy and regulatory considerations

As energy trading evolves, policymakers and regulators will need to adapt existing frameworks to accommodate new market dynamics and technological innovations. Clear and supportive regulatory frameworks are essential to foster innovation, ensure market integrity, and protect consumer interests [37]. Additionally, policies pushing for interoperability, standardization, and market transparency can facilitate the widespread adoption of energy trading platforms and encourage investment in renewable energy infrastructure.

Environmental and social impacts

Beyond economic considerations, energy trading has profound environmental and social implications. By promoting the move toward renewable energy, energy trading contributes to mitigating climate change and improving air quality. Moreover, community-based energy trading initiatives empower local communities, encourage social equity, and foster sustainable development. However, it is essential to address potential challenges such as energy poverty, digital divide, and environmental justice to ensure that the benefits of energy trading are equitably distributed across society.

This section delves into the future applications of energy trading, focusing particularly on its role in managing renewable resources. We explore how energy trading platforms are poised to revolutionize the renewable energy landscape, enabling efficient distribution and utilization of surplus energy within local communities. Through the lens of emerging technologies like blockchain and artificial intelligence, we uncover how these platforms empower individuals and businesses to actively participate in renewable energy production and distribution. Join us as we navigate the evolving landscape of energy trading, envisioning a future where renewable resources drive sustainability and resilience in our energy ecosystem.

Risks

It's crucial to recognize that with the innovation in the field of energy trading, there are inherent risks. There are complexities and uncertainties that accompany the adoption and implementation of this concept. While energy trading holds promise in promoting renewable energy adoption, enhancing economic competitiveness, and fostering social inclusivity, it also presents a myriad of challenges that demand careful consideration and proactive mitigation strategies.

Market volatility

Energy markets experience constant changes influenced by a myriad of factors. These fluctuations stem from shifts in both supply and demand dynamics, geopolitical tensions, and alterations in regulatory frameworks. These fluctuations can lead to price volatility, liquidity issues, and counterparty risks for energy traders, potentially impacting their profitability and financial stability. Effective risk management strategies, such as diversification and hedging, are essential to mitigate these risks and ensure the long-term viability of energy trading activities.

Regulatory compliance

Energy trading activities are subject to complex and evolving regulatory frameworks, including market rules, environmental regulations, and data privacy laws. Compliance with these regulations is essential to ensure the legality, transparency, and fairness of energy trading operations and maintain the trust and confidence of stakeholders. However, navigating regulatory compliance requirements can be challenging, particularly in a rapidly changing regulatory landscape with overlapping jurisdictional boundaries. Therefore, energy traders must stay informed about regulatory developments and implement robust compliance programs to mitigate regulatory risks and uphold ethical standards.

Cybersecurity threats

Digitalized energy trading platforms are vulnerable to cybersecurity threats, including hacking, data breaches, and ransomware attacks. These threats can compromise the integrity, confidentiality, and availability of critical energy infrastructure, disrupting energy supply chains and causing significant financial and reputational damage. Robust cybersecurity measures, such as encryption, authentication, and intrusion detection systems, are necessary to protect energy trading platforms from cyber threats and safeguard sensitive information and operations.

CONCLUSION

In conclusion, peer-to-peer energy trading represents a transformative approach to energy distribution, enabling direct transactions between producers and consumers within local communities. By leveraging renewable energy sources and facilitating surplus energy exchange, these platforms promote sustainability, resilience, and community empowerment in the energy ecosystem. Looking ahead, the future applications of energy trading hold immense promise for revolutionizing renewable resource management. Through alignment with the United Nations' Sustainable Development Goals, energy trading platforms can contribute significantly to achieving global objectives related to accessible and affordable energy, sustainable urban development, climate action, and collaborative partnerships.

However, realizing the full potential of energy trading requires careful consideration of policy and regulatory frameworks, environmental and social impacts, technological integration, and risk management strategies. Balancing market dynamics with regulatory requirements, addressing environmental and social implications, and effectively managing risks are essential for fostering innovation and ensuring equitable access to clean energy resources.

From economic, environmental, and social standpoints, energy trading presents avenues for revenue generation, environmental stewardship, and social equity. By encouraging the adoption of renewable energy, improving energy efficiency, bolstering community resilience, and stimulating innovation, energy trading platforms have the capacity to generate positive impacts on society, economy, and the environment. These platforms pave the way for a more sustainable and resilient global energy landscape, facilitating international cooperation and contributing to a brighter future for generations ahead.

As we journey through the dynamic terrain of energy trading, it's crucial to cultivate cooperation, creativity, and alliances among different sectors and stakeholders. This collaborative effort is essential to unleash the complete potential of renewable energy, fostering a future that's both sustainable and fair for generations ahead.

GLOSSARY

blockchain: A secure, shared database where information is chronologically linked and tamperproof
DAPP: Decentralized application
deficient: Not having enough of the demanded material or quality; someone with deficit
deficit: The amount needed of something to meet demand
microgrid: A small-scale energy network that can function either connected to the main power grid or independently
prosumer: An individual who consumes and produces value
PV: Photovoltaic
SDK: Software development kit
surplus: An excess in production or supply
VPP: Virtual power plant

REFERENCES

1. David I Stern and Astrid Kander. "The role of energy in the industrial revolution and modern economic growth". In: *The Energy Journal* 33.3 (2012), pp. 1–14.
2. Weiqiong Zhong, Haizhong An, Xiangyun Gao, et al. "The evolution of communities in the international oil trade network". In: *Physica A: Statistical Mechanics and Its Applications* 413 (2014), pp. 42–52.
3. Paul Stevens. "History of the international oil industry". In: *Global Resources: Conflict and Cooperation*. Springer, 2013, pp. 13–32.
4. Syed Furqan Rafique, Pei Shen, Zhe Wang, et al. "Global power grid interconnection for sustainable growth: Concept, project and research direction". In: *IET Generation, Transmission & Distribution* 12.13 (2018), pp. 3114–3123.
5. Chenghua Zhang, Jianzhong Wu, and Chao Long, et al. "Review of existing peer-to-peer energy trading projects". In: *Energy Procedia* 105 (2017), pp. 2563–2568.
6. Desen Kirli, Benoit Couraud, Valentin Robu, et al. "Smart contracts in energy systems: A systematic review of fundamental approaches and implementations". In: *Renewable and Sustainable Energy Reviews* 158 (2022), p. 112013. DOI: https://doi.org/10.1016/j.rser.2021.112013.
7. Shravan Nayak, Chanakya Ajit Ekbote, Annanya Pratap Singh Chauhan, et al. "Stochastic game frameworks for efficient energy management in microgrid networks". In: *2020 IEEE PES Innovative Smart Grid Technologies Europe (ISGT-Europe)*. IEEE, 2020, pp. 116–120.
8. Hao Wang and Jianwei Huang. "Incentivizing energy trading for interconnected microgrids". In: *IEEE Transactions on Smart Grid* 9.4 (2016), pp. 2647–2657.
9. Hien Thanh Doan, Jeongho Cho and Daehee Kim. "Peer-to-Peer energy trading in smart grid through blockchain: A double auction-based game theoretic approach". In: *IEEE Access* 1321 (2021), pp. 49206–49218. DOI: 10.1109/ACCESS.2021.3068730.
10. I Safak Bayram, Muhammad Z Shakir and Mohamed Abdallah, et al. "A survey on energy trading in smart grid". In: *2014 IEEE Global Conference on Signal and Information Processing (GlobalSIP)*. IEEE. 2014, pp. 258–262.
11. Chao Liu, Kok Keong Chai, Xiaoshuai Zhang, et al. "Peer-to-peer electricity trading system: Smart contracts based proof-of-benefit consensus protocol". In: *Wireless Networks* 27.6 (Aug. 2021), pp. 4217–4228. URL: https://doi.org/10.1007/s11276-019-01949-0.
12. Dong Han, Chengzhenghao Zhang, Jian Ping, et al. "Smart contract architecture for decentralized energy trading and management based on blockchains". In: *Energy* 199 (2020), p. 117417. DOI: https://doi.org/10.1016/j.energy.2020.117417.
13. Olivier Van Cutsem, David Ho Dac, Pol Boudou, et al. "Cooperative energy management of a community of smart buildings: A blockchain approach". In: *International Journal of Electrical Power & Energy Systems* 117 (2020), p. 105643. DOI: https://doi.org/10.1016/j.ijepes.2019.105643.

14. Z. Ullah, G. Mokryani, B. Khan, et al. "Smart Grid Block-Chain (BC) Conceptual Framework: BiDirectional Models for Renewable Energy District and Utility". In: *2019 15th International Conference on Emerging Technologies (ICET)*. 2019, pp. 1–5. DOI: 10.1109/ICET48972.2019.8994500.
15. Selvaprabu Nadarajah and Nicola Secomandi. "Merchant energy trading in a network". In: *Operations Research* 66.5 (2018), pp. 1304–1320.
16. H Pandžic, Yury Dvorkin and Miguel Carrión. "Investments in merchant energy storage: Trading-´ off between energy and reserve markets". In: *Applied Energy* 230 (2018), pp. 277–286.
17. Frankie Chau, Jing-Ming Kuo and Yukun Shi. "Arbitrage opportunities and feedback trading in emissions and energy markets". In: *Journal of International Financial Markets, Insitutions and Money* 36 (2015), pp. 130–147. DOI: https://doi.org/10.1016/j.intfin.2015.02.002.
18. Martha Caldwell Harris. "The globalization of energy markets". In: *Challenges of the Global Century*, Washington, DC: National Defense University, 2001, pp. 271–282.
19. Gabriella Ferruzzi, Guido Cervone, Luca Delle Monache, et al. "Optimal bidding in a day-ahead energy market for micro grid under uncertainty in renewable energy production". In: *Energy* 106 (2016), pp. 194–202.
20. Le Hong Lam, Valentin Ilea, Cristian Bovo, et al. "European day-ahead electricity market coupling: Discussion, modeling, and case study". In: *Electric Power Systems Research* 155 (2018), pp. 80–92.
21. Margaux Snoeck. "*Understanding the guarantees of origin and their impacts on the electricity value chain: a comparative case study of Norway and Germany*". MA thesis. 2019.
22. H. Saboori, M. Mohammadi and R. Taghe. "Virtual Power Plant (VPP), Definition, Concept, Components and Types". In: *2011 Asia-Pacific Power and Energy Engineering Conference*. 2011, pp. 1–4. DOI: 10.1109/APPEEC.2011.5749026.
23. Kankam O Adu-Kankam and Luis M Camarinha-Matos. "Towards collaborative virtual power plants: Trends and convergence". In: *Sustainable Energy, Grids and Networks* 16 (2018), pp. 217–230.
24. Ahmed Belgana, Bhaskar P Rimal and Martin Maier. "Open energy market strategies in microgrids: A stackelberg game approach based on a hybrid multi-objective evolutionary algorithm". In: *IEEE Transactions on Smart Grid* 6.3 (2014), pp. 1243–1252.
25. Esteban A Soto, Lisa B Bosman, Ebisa Wollega, et al. "Peer-to-peer energy trading: A review of the literature". In: *Applied Energy* 283 (2021), p. 116268.
26. Rafael Leal-Arcas. *Electricity Decentralization in the European Union. Towards Zero Carbon and Energy Transition*. Elsevier, July 10, 2023.
27. Ali Ahmadi, Tim Manandhar, José D'Assunção Barros, et al. "UK power networks' experience of managing flexible distributed generation from planning to operation". In: *CIRED - Open Access Proceedings Journal* 1 (Oct. 1, 2017), pp. 2032–2036. DOI: 10.1049/oap-cired.2017.1085.
28. Tim Schittekatte and Leonardo Meeus. "Flexibility markets: Q&A with project pioneers". In: *Utilities Policy* 63 (Apr. 2020). DOI: 10.1016/j.jup.2020.101017.
29. THE 17 GOALS | *Sustainable Development* (2022). URL: https://sdgs.un.org/goals.
30. Amanda Lange Salvia and Luciana Londero Brandli. "Energy sustainability at universities and its contribution to SDG 7: A systematic literature review". In: *Universities as Living Labs for Sustainable Development: Supporting the Implementation of the Sustainable Development Goals*, USA: IGI, 2020, pp. 2945.
31. Asha Pandey and Muhammad Asif. "Assessment of energy and environmental sustainability in South Asia in the perspective of the sustainable development goals". In: *Renewable and Sustainable Energy Reviews* 165 (2022), p. 112492.
32. Xiuzhi He, Samiha Khan, Ilhan Ozturk, et al. "The role of renewable energy investment in tackling climate change concerns: Environmental policies for achieving SDG-13". In: *Sustainable Development* 31.3 (2023), pp. 1888–1901.
33. Chian-Woei Shyu. "A framework for 'right to energy' to meet UN SDG7: Policy implications to meet basic human energy needs, eradicate energy poverty, enhance energy justice, and uphold energy democracy". In: *Energy Research & Social Science* 79 (2021), p. 102199.
34. Moayad Aloqaily, Azzedine Boukerche, Ouns Bouachir, et al. "An energy trade framework using smart contracts: Overview and challenges". In: *IEEE Network* 34.4 (2020), pp. 119–125.
35. Kirsi Kotilainen. "Energy prosumers' role in the sustainable energy system". In: *Affordable and Clean Energy*. Springer, 2019, pp. 1–14.
36. Jamshid Aghaei, and Mohammad-Iman Alizadeh. "Demand response in smart electricity grids equipped with renewable energy sources: A review". In: *Renewable and Sustainable Energy Reviews* 18 (2013), pp. 64–72.
37. Lucila De Almeida, Viola Cappelli, Nikolas Klausmann, and Henri van Soest. "Peer-to-peer trading and energy community in the electricity market: analyzing the literature on law and regulation and looking ahead to future challenges: EUI RSCAS2021/35" (2021), https://hdl.handle.net/1814/70457.

A Comprehensive Review on Artificial Intelligence

4

Machine Learning Models for Environmental Sustainability

R. M. Bhavadharini, P. Mercy Rajaselvi Beaulah, and R. Krithiga

INTRODUCTION

A sustainable environment is required to conserve natural resources and to maintain ecological balance in our planet's natural environment. As the population grows on earth, the need for excessive production for food and other industries results in energy usage and damages to natural resources. In order to live long and healthy lives, human society as a whole needs clean water to drink, clean air to breathe, and locations free of hazardous materials [1]. The rapid advancements in the domain of artificial intelligence (AI), machine learning (ML), and the Internet of Things (IoT) have triggered massive transformations across various domains, including resource management, healthcare monitoring and management, environmental monitoring, climate prediction, autonomous vehicles, supply chain management, agriculture, and so on. Deep usage of AI and IoT leads to innovative solutions and practices across various sectors.

One of the abundant and vital natural resources for all lives on earth is water. The 17 sustainable development goals (SDG) adopted by all members of the United Nations aim to improve the global environment for healthy human life on earth. SDG 6, titled "Clean Water and Sanitation", describes access to clean and safe water as a fundamental requirement for human well-being [2]. Urbanization and industrialization have led to water contamination, which poses a big challenge to well-being on earth. The combination of growing population, industrialization, urbanization, increased vehicles, pollution, and climate change makes water a scarce and more contaminated resource. Water pollutants move into water bodies like ponds, lakes, seas, and oceans and dissolve or are sedimented on the beds or sediments. The polluted water is deteriorated in its quality and becomes unsafe for consumption. The various pollutants of water

DOI: 10.1201/9781003497189-4

include viruses, bacteria, pharmaceutical products, nitrates, fecal waste, and plastics [3]. According to a new report on inequalities in access to water by the World Health Organization (WHO) and United Nations Children's Fund (UNICEF), around 4.2 billion people worldwide need adequate sanitation facilities, and that around 2.2 billion people suffer from inadequate access to safely managed drinking water [4]. Polluted and unsafe water can cause and transmit diseases such as cholera, diarrhea, dysentery, hepatitis A, typhoid, and polio. Inadequate, or unavailable, water and sanitation facilities severely expose human communities to avoidable health hazards. Therefore, guaranteeing and safeguarding the water quality for human consumption is a significant challenge to address.

Just as water quality monitoring ensures safe water for human consumption, air quality monitoring guarantees safe unpolluted air for well-being. Air pollution poses one of the largest risks to human health. Air pollution endangers around seven million people's health, according to WHO. It is a major risk factor for many illnesses, including respiratory system disorders, skin infections, heart difficulties, bronchitis, asthma, and diseases of the throat and eyes. It is a major risk factor for many illnesses, including respiratory system disorders, skin infections, heart difficulties, bronchitis, asthma, and diseases of the throat and eyes [5]. Crop health monitoring and growth depend heavily on soil quality monitoring and weather prediction capabilities. By analyzing soil nutrients, available water resources, and prevailing weather conditions, framers can plan crop selection, seed quality, crop disease monitoring, irrigation processes, fertilization, harvesting, storage, and marketing and thus massively improving crop yields. Precision agriculture (PA) makes use of smart technologies in improving crop yields and assisting the farmers in decision making. PA aims to reduce labor time and increase crop productivity, thus impacting a country's economic growth. PA enables farmers to remotely monitor the data to make better decisions for maintaining sustainable environments. Wildlife monitoring is another domain considered in this chapter. It plays an important role in maintaining a sustainable environment. It helps in maintaining biodiversity conservation, mitigating human–animal conflict, high-valued species protection, ecological parameter balancing, and species disease surveillance. AI plays significantly in wildlife monitoring to collect the data automatically and to process the data. The different forms of data from wildlife sources can include camera traps, satellite images, acoustic signals, and invasive and noninvasive sensor readings.

Disasters caused by climatic change and other ecological factors are very hard to predict and manage due to their uncertainty. Nowadays, the state-of-the-art technological improvements and advancements can aid in prediction, mitigation, and management of natural or man-made disasters. IoT, the rapidly growing framework, can remotely monitor the metrological, hydrological, and environmental parameters, and support in real-time data collection and predictive analysis. Risk identification, disaster preparedness, emergency response, resource allocation, reaction planning, and disaster recovery are the first steps in the process.

This review chapter provides insights about the importance of AI and ML techniques used in the literature surveyed from the perspectives of the following aspects:

1. Different ML/Deep Learning (DL) algorithms evaluated in surveyed literature papers
2. Various water/air/soil quality parameters assessed by different ML/DL models implemented in the literature survey
3. Comparison of performance results achieved by ML/DL models explored in the literature

METHODOLOGY

AI and ML models are capable of processing large volumes of data sourced from diverse origins such as sensors, satellites, and field measurements. They can easily identify the patterns and recognize anomalies in the data source. They are capable of facilitating real-time monitoring and decision support processes. These technologies can be easily incorporated with IoT, such as sensors and actuators deployed across various domains to collect and process data involving reduced man power and computing resources. As a powerful tool in data analysis, ML is widely deployed to analyze and identify patterns and make appropriate predictions based on enormous amounts of data or big data generated through different types of sources. Supervised learning algorithms, unsupervised learning algorithms, and reinforcement algorithms are the classifications of the ML algorithms. In order to analyze the connectivity and relationship between the dependent and independent variables, supervised learning algorithms are used. They produce predictive models that can predict the output based on the given input. The applications based on regression and classification use supervised learning methods.

In contrast to supervised learning algorithms, unsupervised learning algorithms are used when the output label is not known. The input data is classified based on the similarity characteristics into the output class. Unsupervised learning algorithms are used in applications involving clustering. Algorithms like principal component analysis (PCA) and K-means are a couple of the commonly known unsupervised learning algorithms. Furthermore, another category of machine learning methods is called reinforcement learning, which describes a machine's capacity to generalize and provide accurate answers to issues that it hasn't been taught.

Most of the literature papers surveyed in this chapter use the following architecture models in their study. These research papers collected and processed data for various purposes like water quality assessment, air quality assessment, precision agriculture, wildlife monitoring, and disaster management. The architecture shown in Figure 4.1 shows the flow of methodology used in most of the research works considered here.

The general system architecture shown in Figure 4.1 uses the data collected from various sources such as publicly available datasets or real-time data. The data type includes numerical, categorical, text, sound, and images. The ML model is trained and verified by dividing the data into training and test sets at a predetermined ratio and then checking how well the model performs when it processes the unseen data. The model's performance is evaluated through a number of performance metrics like accuracy, confusion matrix, area under region of curve, precision, mean squared error (MSE), root mean squared error (RMSE), and R^2(R-Squared).

AI/ML/DL in Water Quality Monitoring

Authors in [6] proposed an Enhanced Automated Machine Learning Model (EAMLM) to assess water quality. Using temperature, pH, Total Dissolved Solid (TDS), and LDR (light dependent) sensors, they gathered data on the water quality parameters. The machine learning model is trained on this dataset of water quality parameters, and its performance is assessed using performance measures. AI/ML

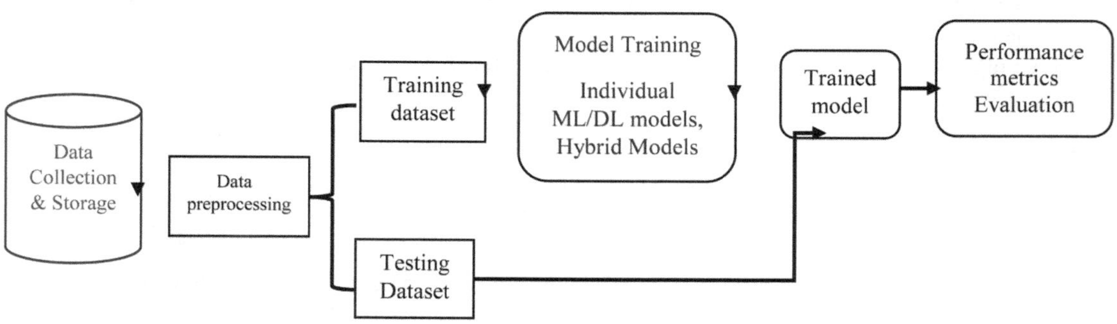

FIGURE 4.1 System architecture of the methodology.

algorithms and models have substantial prospects in monitoring, predicting, and classifying the water quality indicators due to the rapid growth in data volume. Authors in [7] evaluated the ML classifiers to classify the quality classes of the monitored water samples according to their Water Quality Index (WQI). More complicated nonlinear issues can be effectively solved using data-driven machine learning models. Deployment and evaluation of distinct machine learning models for treating different water samples collected from surface water, groundwater, drinking water, and wastewater was covered in [8]. Monitoring drinking water quality is essential in ensuring the safety of human society. Authors of [9] proposed an Artificial Neural Network (ANN) for predicting water quality, and achieved an accuracy of 0.96 using a small feature set; it outperforms existing models. In the same way, the proposed ANN also performed well for water consumption prediction, and achieved better performance than existing complex DL models such as convolutional neural network (CNN), Long Short-Term Memory (LSTM), and gated recurrent units (GRU). Researchers of [10] utilized and evaluated M5 Model Tree (MT), Multivariate Adaptive Regression Spline (MARS), Gene Expression Programming (GEP), and Evolutionary Polynomial Regression (EPR) to process Landsat 8 OLI-TIRS images for the purpose of water quality parameters assessment from river water samples. The proposed work [11] explored Principal Component Analysis with Artificial Neural Network for modeling. ANN with raw data, ANN with Principal Component (PC) prior to rotation, and ANN with PC following varimax rotation were all tested. Out of three ANN models, ANN A performed better in predicting water quality. Various ML models were evaluated to assess the water quality class (WQC) and the authors proved that Bagging Classifier and Multilayer Perceptron (MLP) outperformed all other ML models considered in this research work [12].

The research in [13] focused on LSTM deep neural network to assess the quality parameters of drinking water within an IoT environment. In [14], using physical factors as features, the authors employed various ML models to forecast and assess the groundwater quality parameters, hence enhancing the informed decision in agricultural practices. Adaboost and Random Forest (RF) prediction performances are better than other models in their study to predict groundwater quality. In [15], a DL-based forecasting system was developed to forecast groundwater quality; this DL model has outperformed other ML models considered for comparison in their work. For the purpose of forecasting ten irrigation water quality (IWQ) parameters dissolved in surface water, eight machine learning models have been evaluated and verified in [16]. In order to assess the river water quality, [17] evaluated four standalone and 12 hybrid data-mining methods. The model presented in [18] used LSTM and CNN integration to assess and forecast the temporal and spatial patterns seen in the data on water quality. Authors also evaluated the CNN model's hybrid, the CNN–LSTM model, for predicting water quality parameters, known as dissolved oxygen (DO) and chlorophyll-a (Chl-a). Table 4.1 presents various ML algorithms used for water quality assessment and their performance results.

Dissolved oxygen (DO), pH, turbidity, conductivity, biochemical oxygen demand (BOD), temperature, fecal coliform, total dissolved solids (TDS), total coliform, fecal coliform (FC), and total solids (TS) are different water quality parameters assessed by ML/DL algorithms from the literature. The water quality parameters and the ML/DL methods used to evaluate them are displayed in Table 4.2.

AI/ML/DL in Air Quality Monitoring

The sources of pollution emissions increase with the demand of transportation and energy due to increasing population. Human beings can lead better lifestyles through monitoring the environment for pollution and controlling pollution-related diseases. One of the main issues facing cities today is air pollution, the reduction of which is also the main goal of smart cities. Air pollution results from the discharge of particular substances, such as gases, particles, and biological molecules, into the atmosphere in hazardous or excessive proportions. These high emissions have clear effects on the environment, harming crops and resulting in diseases and population declines among other living things. There are three types of air pollutants: gases, liquid droplets, and solid particles [19]. In the study [20], a novel early

TABLE 4.1 ML/DL algorithms used in water quality monitoring with achieved results

AUTHORS	*ALGORITHMS USED*	*RESULTS*
Nazir et al. [7]	SVM	Acc = 0.6875
	RF	0.7178
	LR	1.0
	DT	0.9621
	CATBoost	1.0
	XGBoost	0.8106
	MLP	1.0
Rustam et al. [9]	ANN	Acc = 0.96
		R^2 = 0.997
Najafzadeh et al. [10]	MT	Acc = 96.9
	MARS	99.2
	GEP	96.4
Ibrahim et al. [11]	ANN	R^2 = 99.9%
Ahmed et al. [12]	MLP	Acc = 85.07
	LR	84.01
	SGD	82.05
	DT	79.49
	SVM	79.79
	GBC	81.30
	Bagging Classifier	81
Liu et al. [13]	ARIMA	MSE = 0.0126
	SVR	0.0046
	LSTM	0.0017
El Bilali et al. [14]	DL model	R^2 = 0.996
	RF	0.886
	XGBoost	0.0.927
	ANN	0.917
El Bilali et al. [16]	BA-RT	R^2 = 0.941, RMSE = 2.71

TABLE 4.2 Various water quality parameters measured by ML/DL algorithms

WATER QUALITY PARAMETERS/ ALGORITHMS	*DO*	*TURBIDITY*	*PH*	*CONDUCTIVITY*	*TEMPERATURE*	*BOD*	*TDS*	*FC*	*TS*	*TC*
SVM	√	-	√	√	-	√	-	√	-	√
RF	√	-	√	√	-	√	-	√	-	√
LR	√	-	√	√	-	√	-	√	-	√
DT	√	-	√	√	-	√	-	√	-	√
CATBoost	√	-	√	√	-	√	-	√	-	√
XGBoost	√	-	√	√	-	√	-	√	-	√
MLP	√	-	√	√	-	√	-	√	-	√
SGD	-	√	√	√	√	-	√	√	-	-
GBC	-	√	√	√	√	-	√	√	-	-
Bagging Classifier	-	√	√	√	√	-	√	√	-	-
ARIMA	√	√	√	√	-	-	-	-	√	√
LSTM	√	-	-	-	-	-	-	-	√	√

Legend: TDS, Total Dissolved Solids; TS, Total Solids; BOD, Biochemical Oxygen Demand; FC, Fecal Coliform; TC, Fecal Coliform warning system for air quality with Whale Optimization Algorithm (WOA) was used in conjunction with a system known as Improved Complete Ensemble Empirical Mode Decomposition with Adaptive Noise (ICEEMDAN) to monitor air pollutants such as Particulate Matter with a diameter of 2.5 micrometers or less ($PM_{2.5}$), Particulate Matter with a diameter of 10 micrometers or less (PM_{10}), nitrogen dioxide (NO_2), sulfur dioxide (SO_2), carbon monoxide (CO), and ozone (O_3). An LSTM Deep Neural Network (DNN)-based model was developed and used as an underlying prediction technique to monitor the air quality in [21]. A comprehensive air quality forecasting model for smart city uses was proposed, employing two deep learning models and four machine learning approaches. The DL models included LSTM Ada Boost, and the ML models included Support Vector Regression (SVR), RF, KNN, and MLP regressor for tracking pollution markers including $PM_{2.5}$, PM_{10}, O_3, and CO in [22]. LSTM has proven to outperform other ML models used in this research work. Authors in [23] demonstrated that RF Regression (RFR) has outperformed DT, GBR, and MLP models in their study for pollution prediction with varying sizes of datasets and locations having various environmental characteristics. Study in [24] used regression model SVR and RFR for predicting and analyzing the Air Quality Index (AQI) and these models were evaluated with publicly available datasets. SVR was used by the authors in [19] to forecast particle and pollutant levels as well as the AQI.

The most accurate predictions for pollutant concentrations, including $PM_{2.5}$, SO_2, NO_2, O_3, and CO, were obtained using SVR with Radial Basis Function (RBF), along with the hourly AQI. Researchers in [25] explored various Machine Learning Regressor models to predict PM10 and PM2.5 concentration in the air. They collected historical data for PM concentration levels along with environmental factors to compare various ML models. The ML models Multi-Linear Regression (MLR), decision tree (DT), K-Nearest Neighbors (KNN), RF, and XGBoost were evaluated for analyzing the spatial-temporal parameters in [26] for $PM_{2.5}$ prediction. For $PM_{2.5}$ modeling, the LSTM deep learning model performed better than XGBoost regression, with its achieved performance result of R^2 value as 0.89. A predictive model based on artificial neural network (ANN) was employed to forecast the concentration levels of SO_2 and particulate matter [27]. ANN achieved the results as RMSE values of 0.0128 for SO_2 and 0.0001 $PM_{2.5}$. Authors in [28] used ML models like Seasonal Autoregressive Integrated Moving Average (SARIMA), SVM, and LSTM to predict AQI. Table 4.3 shows the ML models used in the research works surveyed in this chapter along with their results. Table 4.4 depicts various air pollutants measured by ML/DL algorithms.

AI/ML in Soil Quality Monitoring and Precision Agriculture

A unique algorithm utilizing a multi-class SVM with a Directed Acyclic Graph (DAG) with the fruit fly optimization approach (MSVM-DAG-FFO) was proposed in [29]. With a 0.973 accuracy rate, this model outperformed SVM (0.932), SVM kernel (0.922), and decision tree (0.914). ML algorithms were used to examine the information gathered and produce crop recommendations based on trends found. These

TABLE 4.3 ML/DL algorithms used in air quality monitoring

AUTHORS	*ALGORITHMS USED*	*RESULTS*
Schürholz et al. [21]	LSTM DNN	Precision Acc = 90–96%
Neo et al. [22]	LSTM	$R^2 = 0.993$
Liu et al. [24]	SVR	RMSE = 7.666, $R^2 = 0.9776$
Castelli et al. [19]	SVR with RBF	Accuracy = 94.1%
Harishkumar et al. [25]	GBR	$R^2 = 0.9983$
Gokul et al. [26]	XGBoost, LSTM	$R^2 = 0.82, 0.89$
Varade et al. [27]	ANN	RMSE = 0.0001
Maltare et al. [28]	SVM-RBF SVM-Linear	$R^2 = 0.998$ $R^2 = 0.968$

TABLE 4.4 Different air pollutants assessed by ML/DL algorithms

AIR QUALITY PARAMETERS/ML & DL MODELS	*LSTM DNN*	*LSTM*	*SVR*	*SVR WITH RBF*	*GBR*	*XGBOOST LSTM*	*ANN*	*SVM-RBF SVM-LINEAR*
$PM_{2.5}$	√	√	√	√	√	√	√	√
O_3	√	√	√	-	-	-	-	√
SO_2	√		√	√	-	-	√	
CO	√	√	√	√	-	-	-	√
PM_{10}	√	√	√		-	-	-	-
NO_2	√		√	√	-	-	-	-
Non-methane hydrocarbons	-	-	√		-	-	-	-
Benzene	-	-	√		-	-	-	-
NO_X	-	-	√		-	-	-	-
Outdoor temperature	-	-	-	√	-	-	-	-
Relative humidity	-	-	-	√	-	-	-	-
Wind speed	-	-	-	√	-	-	-	-

suggestions support effective resource use in agriculture by optimizing crop development and maximizing yield potential. Artificial intelligence helps farmers in choosing the suitable seed for a certain set of weather conditions. Along with it, AI also offers meteorological forecast information, and AI-powered solutions will accelerate a product's time to reach market, enhance crop and soil quality, and allow farmers to increase crop yields with fewer resources [30]. Authors [31] used KNN imputer to impute missing data values in the soil dataset. They tried with seven foundational classifiers and eight ensemble classifiers for soil classification and they achieved good performance with XGBoost and LightGBM classifiers.

In [32], authors developed and evaluated DNN and ML models on both dry land and paddy fields to observe soil quality. The RF model outperformed all ML models with $R^2 = 0.91$ for soil quality prediction of paddy fields and with $R^2 = 0.90$ for dry land. In [33], the use of the genetic algorithm (GA) optimizer in conjunction with extreme gradient boosting regression (XGBR) for feature selection and optimization was suggested and assessed. The proposed XGBR-GA model achieved the maximum performance of $R^2 = 0.891$; and RMSE = 0.875%. Various ML and DL models were evaluated to assess the soil quality in [34]. Convolutional neural networks (CNNs) were used in a portable smartphone-based machine vision system that was proposed in [35]. This model used one block for feature extraction and a second block for classification with SVM classifier. The authors of [36] used optical and thermal sensors to collect soil moisture, which was then analyzed using machine learning (ML) models such as RF, SVM, ANN, and elastic net regression (EN).

The ML-driven models were created to improve soil salinity inversion precision and accuracy, and they have produced the most promising results, with $R^2 = 0.801$, 0.624, 0.613 and RMSE = 0.686, 0.875, 0.918, for the RF, SVM, and BP models, respectively [37]. The prominent ML models, partial least squares regression (PLSR), K nearest neighbor (KNN), and RFR were trained and tested with Unmanned Ariel Vehicle (UAV)-based multimodal data to compute the amount of moisture in the soil in a field of maize at various irrigation levels. Study in [38, 39] investigated estimation models based on ML for salt-affected irrigation regions in order to support the prompt and precise monitoring of soil water–salt within agricultural lands in real time. Table 4.5 summarizes a number of ML models used in the reference papers considered in this chapter. Table 4.6 represents different soil quality parameters and different ML/DL algorithms used to assess those parameters.

TABLE 4.5 ML/DL algorithms used in soil quality monitoring

AUTHORS	*ML/DL ALGORITHMS*	*RESULTS*
Senapaty et al. [29]	MSVM-DAG-FFO	Accuracy = 0.973
Aydın et al. [31]	XGBoost	Accuracy = 90%
Li et al. [32]	RF model DNN model	R^2 = 0.91 0.89
Nguyen et al. [33]	XGBR-GA	R^2 = 0.891
Khanal et al. [34]	NN	R^2 = 0.64, RMSE = 0.44
Azadnia et al. [35]	CNN	Accuracy = 99.89%
Adab et al. [36]	RF	Nash–Sutcliffe Efficiency (NSE) = 0.73
He et al. [37]	RF SVM BP	R^2 = 0.801, 0.624 0.613
Zhang et al. [38]	RFR	R^2 = 0.78
Du et al. [39]	XGBoost	R^2 = 0.74, RMSE = 1.77%

AI/ML in Wildlife Monitoring

The acoustic signals and the images of the animals are captured and processed to monitor the animals in real time. Researchers use ML/DL models for feature extraction and classification tasks for classifying the animal species for ecological conservation, animal tracking and health monitoring, and protecting human beings from animal attacks. Some of the monitoring systems use the alert messaging system, for triggering messages to a group of registered mobile numbers once the animal is detected and identified. The ML/DL models are also used to monitor and to record in climatic parameters. These real-time parameters are used to predict the intensity of forest fires, and assist to avoid significant losses caused by forest fires. The CNN architecture implemented in [40] achieved a recognition accuracy of more than 96% to detect different animals' species from the images. The automatic classification of wildlife species from camera trap photos using convolutional neural networks with the ResNet-18 architecture has achieved with 98% accuracy [41]. Two new frameworks for automatic image recognition using deep learning were implemented in [42] and they achieved an accuracy of 98.05% for binary classification and

TABLE 4.6 Different soil quality parameters assessed by ML/DL algorithms

PARAMETERS CONSIDERED	*MSVM-DAG-FFO*	*NN*	*CNN*	*RF*	*SVM*	*RFR*	*XGBOOST*
Soil Moisture Content (SMC)	√	-	-	√	-	√	-
Temperature	√	-	-	-	-	-	-
Water level	√	-	-	-	-	-	-
Soil NPK	√	-	-	-	-	-	-
Oil organic matter (SOM)	-	√	-	-	-	-	-
Cation Exchange Capacity (CEC)	-	√	-	-	-	-	-
Magnesium (Mg)	-	√	-	-	-	-	-
Potassium (K)	√	√	-	-	-	-	-
pH	-	√	-	-	-	-	-
Soil texture	-	-	√	-	-	-	-
Soil water–salt	-	-	-	-	√	-	√

TABLE 4.7 ML/DL algorithms used in wildlife monitoring

AUTHORS	*ALGORITHMS USED*	*RESULTS*	*INPUT DATA TYPE*
Nguyen et al. [40]	CNN	Acc = 96%	Animal images
Chen et al. [42]	DNN	Acc = 98.05%	Animal images
Chalmers et al. [44]	MLP	Sensitivity = 0.74 Specificity = 0.92 Acc = 0.74	Acoustic signals
Zhou et al. [46]	ResNet	Acc = 99.18, Kappa = 0.98	Animal images

an accuracy of 90.32% for multiclassification. An ML-assisted tool, MegaDetector, was used to tag animals in the video feed using bounding boxes, and they demonstrated that tagging images with ML-based bounding boxes has been faster than tagging by unassisted [43]. A biodiversity assessment model using an ML approach was proposed in [44]. This model captured birds' audio and extracted the features using Mel-Frequency Cepstral Coefficients (MFCC). Then they used a machine learning model, Multilayer Perceptron (MLP), to classify the bird species. With 0.92 specificity, 0.74 accuracy, and 0.74 sensitivity, this approach produced better results. Authors in [45] used ResNet50 architecture to construct the model and Transfer Learning to accelerate training processes with weights of the pre-trained model on the ImageNet dataset. This novel model makes it possible to automatically label animal and bird species from the camera trap. Authors studied two deep learning models, CNN and ResNet, to automate the process of detecting animal species [46]. The proposed ResNet architecture obtained an overall accuracy of 99.18% in animal identification. The CNN model in this study achieved the highest overall accuracy and Kappa as 84.55% and 0.79, respectively. Table 4.7 shows the ML algorithms, input type, and results achieved from the research articles surveyed in this study.

AI/ML in Flood Monitoring

Authors in [47] employed Principal Component Analysis (PCA) for dimensionality reduction for sensor data sourced from various sensors and an ANN model to forecast the flood-related parameters. ANN in this research work achieved an accuracy of 94.2% and ROC = 0.9807. The model in [48] analyzed the flood sensor data to predict and classify the severity class of floods on the river side. They deployed and assessed the performance of many ML models individually and ensembled models for this purpose. The flood severity class has three predictions in this work: normal, abnormal, and dangerous water levels. During the training and validation phase, the ensemble LSTM+ RF classifier had an accuracy of 0.997; however, in the testing phase utilizing unknown data, the accuracy was 0.811. In the work [49], a deep learning–based control system has been demonstrated to observe and analyze the flow of flood from the dam gates. Authors have evaluated the performance of CNN, ANN, and SVM. According to the results, the CNN algorithm-based classification system for estimating available space at the tank operates effectively and accurately with an accuracy of 85.53, whereas the other two models have an accuracy of ANN = 82.89 and SVM = 80.26. In [50], the flood situation data undergoes a classification process using extreme gradient boosting (XGBOOST) and gated recurrent unit (GRU) algorithms; XGBOOST showed $R^2 = 0.92$ and GRU with $R^2 = 0.96$. An ANN model was used in [51] to process the data collected by water level and rainfall sensor nodes. They exploited LSTM for short-term prediction of flood water levels and achieved an R^2 value as 0.976.

AI/ML in Earthquake Monitoring

In order to predict earthquakes, a unique Bayesian belief model technique was applied in [52–53]. Moreover, the amount of seismic activity in the cloud layer was predicted utilizing the Adaptive Neuro-Fuzzy Inference System (ANFIS) method. Based on the extensive simulation results, the suggested framework

TABLE 4.8 ML/DL algorithms used in disaster management

DISASTER	*AUTHORS*	*ALGORITHMS USED*	*RESULTS*
Flood	Kaur et al. [47]	ANN	Acc = 94.2% ROC = 0.9807
	Khalaf et al. [48]	Ensemble LSTM+ RF Classifier	Acc = 0.997
	Smys et al. [49]	CNN ANN SVM	85.53 82.89 80.26
	Prakash et al. [50]	XGBOOST GRU	R^2 = 0.92 R^2 = 0.96
	Samikwa et al. [51]	LSTM	R^2 = 0.976
Earthquake	Bhatia et al. [52]	ANFIS	Precision = 92.52
	Abdalzaher et al. [53]	XGB	Acc = 98.59

performs with higher classification performance, achieving precision of 92.52%, sensitivity of 91.72%, and specificity of 91.01%. Authors have evaluated various ML models and nonlinear models, and the XGB model has performed well with a classification accuracy of 98.59% to predict earthquake intensity [53]. In order to quickly calculate the EQ parameters (magnitude and position) in IoT-based EEW, a unique deep learning model combining AE and CNN, dubbed 3S-AE-CNN, was developed. Table 4.8 shows various ML algorithms and their achieved results obtained from the research articles referenced in this chapter.

RESULTS AND DISCUSSION

This chapter considers accuracy, precision, MSE, RMSE, and R^2 for comparing results achieved in the literature surveyed and they are tabulated in Tables 4.1, 4.3, 4.5, 4.7, and 4.8. These parameters are described next.

Accuracy

Accuracy is a metric used for evaluating the performance of classification model. It is determined by dividing correct prediction with total number of predictions of the model.

$$\text{Accuracy} = \frac{TP + TN}{TP + FP + TN + FN} \tag{4.1}$$

where TP = True Positives, TN = True Negatives, FP = False Positives, and FN = False Negatives.

Precision

Precision is measure that determines how often the ML model correctly predicts the positive classes in the total correctly predicted classes. Precision is given as in Equation 2.

$$\text{Precision} = \frac{TP}{TP + FP} \tag{4.2}$$

MSE (Mean Squared Error)

MSE is a metric used to evaluate the ML model's predictive power. The average squared difference between the expected and actual output values is used to calculate MSE. Equation 3 provides the MSE, where y represents the actual output, $\hat{y}$ the anticipated output, and n the number of predictions utilized in the machine learning model.

$$\text{MSE} = \frac{1}{n}\sum_{i=0}^{n}(y-\hat{y})^2 \tag{4.3}$$

RMSE (Root Mean Squared Error)

RMSE is a predictive metric used to analyze and evaluate the performance of an ML model. It calculates the average magnitude of the error and deviation from the actual parameter. The better model has an RMSE value closer to 0 and is given by Equation 4.

$$\text{RMSE} = \sqrt{\frac{1}{n}\sum_{i=0}^{n}(y-\hat{y})^2} \tag{4.4}$$

R² (R Squared)

The percentage of the dependent variable's variance that the independent variable can explain for is called the R squared. R squared is given by Equation 5.

$$\text{R}^2 = \frac{SS_{regression}}{SS_{total}} \tag{4.5}$$

Where $SS_{\text{regression}}$ represents Sum of Squares in regression modes and SS_{total} represents total sum of squares.

In Figure 4.2, the performance achievement of various ML/DL algorithms is represented on the basis of the accuracy achieved by proposed models in the reference papers [7, 9, 12]. Linear Regression,

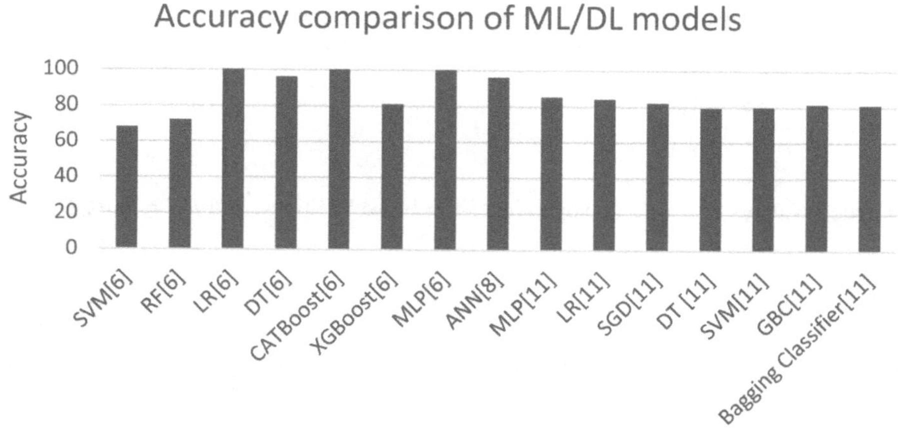

FIGURE 4.2 Accuracy versus various ML/DL algorithms in water quality monitoring.

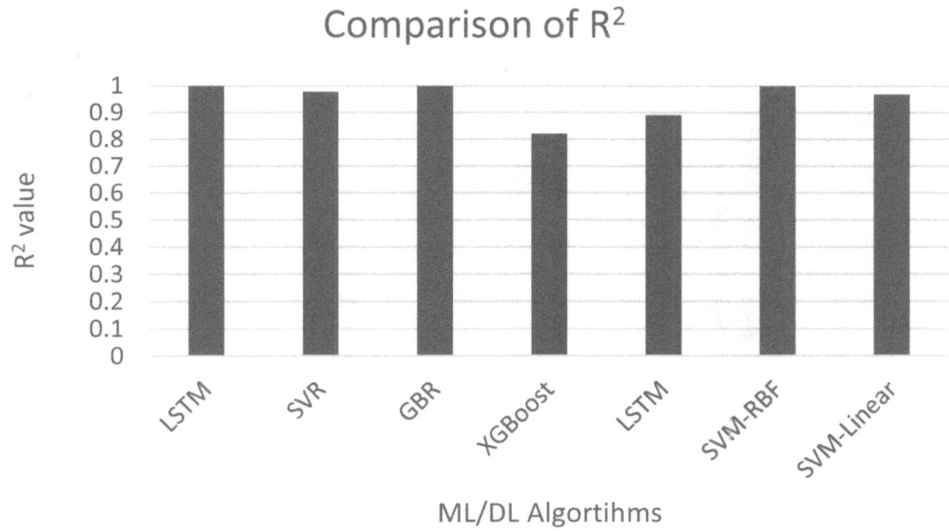

FIGURE 4.3 R^2 versus various ML/DL algorithms in air quality monitoring.

Multi Linear Regression, CATBoost algorithms achieve better performance than other models considered in the performance comparison to assess the water quality parameters. Figure 4.3 compares the performance results achieved in terms of R^2 for air quality assessment parameters. It is clearly evident from Figure 4.3 that GBR [25], SVM-RBF [26], and LSTM [28] are outperforming other models considered in the literature with an achieved result of $R^2 = 0.91$, 0.89, and 0.89, respectively. In order to assess the soil quality parameters, various ML/DL models are compared in terms of R^2, and Figure 4.4 depicts the comparison. The models RF [32], DNN [32], XGBR-GA [33] outperform the other ML/DL models with achieved R^2 values of 0.998, 0.998, and 0.89, respectively. Figure 4.5 compares the models used in wildlife monitoring. These systems take input data as acoustic signals, animal images

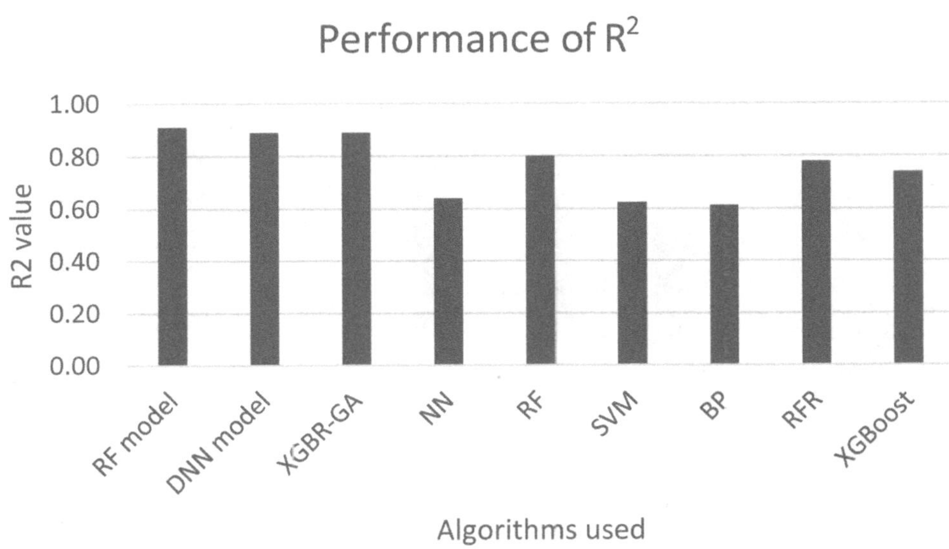

FIGURE 4.4 R^2 versus various ML/DL algorithms in soil quality monitoring.

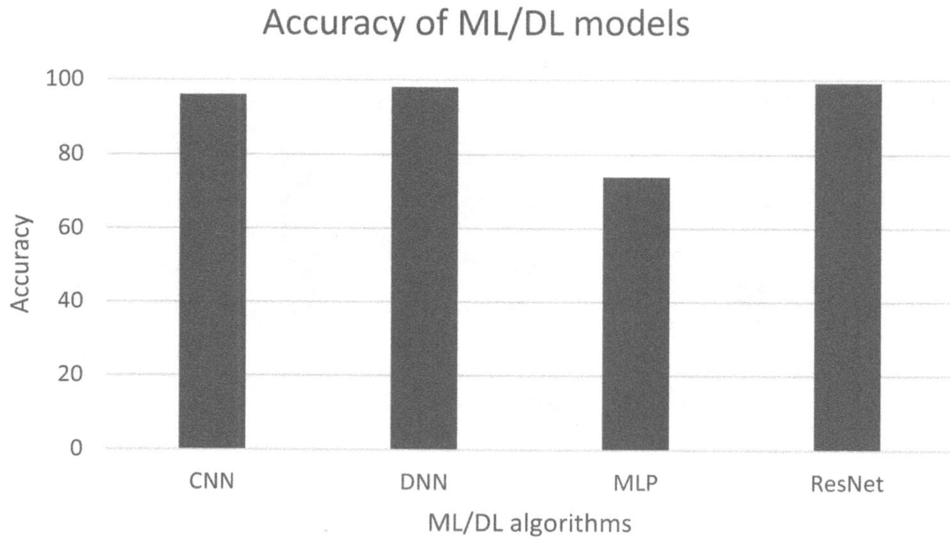

FIGURE 4.5 Comparison of different ML/DL algorithms used in wildlife monitoring.

captured through PiCamera, and sensor data. The results in Figure 4.5 show that ResNet [46] achieves an accuracy of 99.18% and performs better than other architectures considered for comparison. This review article considers flood prediction and earthquake prediction for the disaster management field. Figure 4.6 compares the models, and it is shown that ensemble LSTM+ RF [48] classifier reaches the maximum result of 99.7% accuracy.

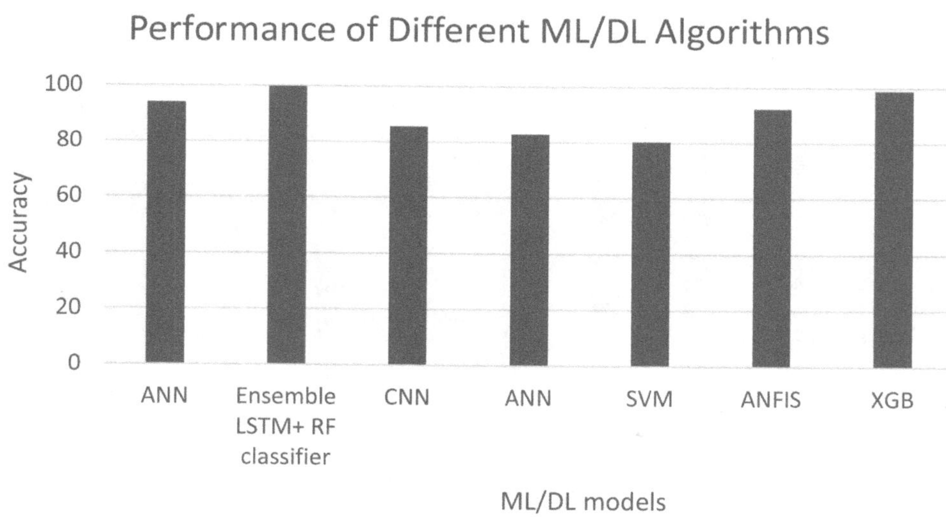

FIGURE 4.6 Accuracy versus different ML/DL algorithms used in disaster management.

CONCLUSION

Sustainable environment monitoring plays a vital role in safeguarding our environment for healthy and peaceful lives for human beings. The modern technological developments in the fields of AI/ML and IoT devised and implemented many intelligent solutions in real-time monitoring, collection, storage, and processing data. They enable quick and knowledgeable decision-making processes with the prediction and classification systems/models. This chapter aimed to present recent advancement and deployments of AI and ML models in sustainable environment monitoring. The chapter presented the contribution of AI and ML models in the realms of water quality, air quality, soil quality, wildlife monitoring, and disaster management. This comprehensive review presented various sensors deployed for data collection and data processing in these domains. Also, different ML/DL algorithms used in these fields were presented, along with their achieved results.

REFERENCES

1. https://www.inspirecleanenergy.com/blog/sustainable-living/environmental-sustainability-definition
2. https://www.un.org/sustainabledevelopment/water-and-sanitation/
3. Lakshmikantha, V., Hiriyannagowda, A., Manjunath, A., Patted, A., Basavaiah, J. and Anthony, A.A. (2021). IoT based smart water quality monitoring system. Global Transitions Proceedings, 2(2), pp. 181–186.
4. https://www.unwater.org/news/who-and-unicef-launch-updated-estimates-water-sanitation-and-hygiene
5. Fotopoulou, E. et al. (2016), "Linked data analytics in interdisciplinary studies: The health impact of air pollution in urban areas," IEEE Access, 4, pp. 149–164.
6. Kumar, D. S., et al. (2024). Eamlm: Enhanced automated machine learning model for IoT based water quality analysis with real-time dataset. Automatic Control and Computer Sciences, 58, 66–77.
7. Nasir, N., Kansal, A., Alshaltone, O., Barneih, F., Sameer, M., Shanableh, A. and Al-Shamma'a, A. (2022). Water quality classification using machine learning algorithms. Journal of Water Process Engineering, 48, p. 102920.
8. Zhu, M., Wang, J., Yang, X., Zhang, Y., Zhang, L., Ren, H., Wu, B. and Ye, L. (2022). A review of the application of machine learning in water quality evaluation. Eco-Environment & Health, 1(2), pp. 107–116.
9. Rustam, F., Ishaq, A., Kokab, S. T., de la Torre Diez, I., Mazón, J. L. V., Rodríguez, C. L. and Ashraf, I. (2022). An artificial neural network model for water quality and water consumption prediction. Water, 14(21), p. 3359.
10. Najafzadeh, M. and Basirian, S. (2023). Evaluation of river water quality index using remote sensing and artificial intelligence models. Remote Sensing, 15(9), p. 2359.
11. Ibrahim, A., Ismail, A., Juahir, H., Iliyasu, A. B., Wailare, B. T., Mukhtar, M. and Aminu, H. (2023). Water quality modelling using principal component analysis and artificial neural network. Marine Pollution Bulletin, 187, p. 114493.
12. Ahmed, U., Mumtaz, R., Anwar, H., Shah, A. A., Irfan, R. and García-Nieto, J. (2019). Efficient water quality prediction using supervised machine learning. Water, 11(11), p. 2210.
13. Liu, P., Wang, J., Sangaiah, A. K., Xie, Y. and Yin, X. (2019). Analysis and prediction of water quality using LSTM deep neural networks in IoT environment. Sustainability, 11(7), p. 2058.
14. El Bilali, A., Taleb, A. and Brouziyne, Y. (2021). Groundwater quality forecasting using machine learning algorithms for irrigation purposes. Agricultural Water Management, 245, p. 106625.
15. Singha, S., Pasupuleti, S., Singha, S. S., Singh, R. and Kumar, S. (2021). Prediction of groundwater quality using efficient machine learning technique. Chemosphere, 276, p. 130265.
16. El Bilali, A. and Taleb, A. (2020). Prediction of irrigation water quality parameters using machine learning models in a semi-arid environment. Journal of the Saudi Society of Agricultural Sciences, 19(7), pp. 439–451.
17. Bui, D. T., Khosravi, K., Tiefenbacher, J., Nguyen, H. and Kazakis, N. (2020). Improving prediction of water quality indices using novel hybrid machine-learning algorithms. Science of the Total Environment, 721, p. 137612.

18. Barzegar, R., Aalami, M. T. and Adamowski, J. (2020). Short-term water quality variable prediction using a hybrid CNN–LSTM deep learning model. Stochastic Environmental Research and Risk Assessment, 34(2), pp. 415–433.
19. Castelli, M., Clemente, F. M., Popovič, A., Silva, S. and Vanneschi, L. (2020). A machine learning approach to predict air quality in California. Complexity, 15, pp. 1–13.
20. Mo, X., Zhang, L., Li, H. and Qu, Z. (2019). A novel air quality early-warning system based on artificial intelligence. International Journal of Environmental Research and Public Health, 16(19), p. 3505.
21. Schürholz, D., Kubler, S. and Zaslavsky, A. (2020). Artificial intelligence-enabled context-aware air quality prediction for smart cities. Journal of Cleaner Production, 271, p. 121941.
22. Neo, E. X., Hasikin, K., Lai, K. W., Mokhtar, M. I., Azizan, M. M., Hizaddin, H. F. and Razak, S. A. (2023). Artificial intelligence-assisted air quality monitoring for smart city management. PeerJ Computer Science, 9, p. e1306.
23. Ameer, S. et al. (2019). "Comparative analysis of machine learning techniques for predicting air quality in smart cities," in IEEE Access, 7, pp. 128325–128338, doi: 10.1109/ACCESS.2019.2925082.
24. Liu, H., Li, Q., Yu, D. and Gu, Y. (2019). Air quality index and air pollutant concentration prediction based on machine learning algorithms. Applied Sciences, 9(19), p. 4069.
25. Harishkumar, K. S., Yogesh, K. M. and Gad, I. (2020). Forecasting air pollution particulate matter (PM2. 5) using machine learning regression models. Procedia Computer Science, 171, pp. 2057–2066.
26. Gokul, P. R., Mathew, A., Bhosale, A. and Nair, A. T. (2023). Spatio-temporal air quality analysis and PM2. 5 prediction over Hyderabad City, India using artificial intelligence techniques. Ecological Informatics, 76, p. 102067.
27. Varade, H. P., Bhangale, S. C., Thorat, S. R., Khatkale, P. B., Sharma, S. K. and William, P. (2023). "Framework of Air Pollution Assessment in Smart Cities using IoT with Machine Learning Approach," 2023 2nd International Conference on Applied Artificial Intelligence and Computing (ICAAIC), Salem, India, pp. 1436–1441, doi: 10.1109/ICAAIC56838.2023.10140834.
28. Maltare, N. N. and Vahora, S. (2023). Air quality index prediction using machine learning for Ahmedabad City. Digital Chemical Engineering, 7, p. 100093.
29. Senapaty, M. K., Ray, A. and Padhy, N. (2023). IoT-enabled soil nutrient analysis and crop recommendation model for precision agriculture. Computers, 12(3), p. 61.
30. Javaid, M., Haleem, A., Khan, I. H. and Suman, R. (2023). Understanding the potential applications of artificial intelligence in agriculture sector. Advanced Agrochem, 2(1), pp. 15–30.
31. Aydın, Y., Işıkdağ, Ü, Bekdaş, G., Nigdeli, S. M. and Geem, Z. W. (2023). Use of machine learning techniques in soil classification. Sustainability, 15(3), p. 2374.
32. Li, L., Liu, L., Peng, Y., Su, Y., Hu, Y. and Zou, R. (2023). Integration of multimodal data for large-scale rapid agricultural land evaluation using machine learning and deep learning approaches, Geoderma, 439, 116696, https://doi.org/10.1016/j.geoderma.2023.116696.
33. Nguyen, T. T., Ngo, H. H., Guo, W., Chang, S. W., Nguyen, D. D., Nguyen, C. T., Zhang, J., Liang, S., Bui, X. T. and Hoang, N. B. (2022). A low-cost approach for soil moisture prediction using multi-sensor data and machine learning algorithm, Science of The Total Environment, 833, 155066, https://doi.org/10.1016/j.scitotenv.2022.155066.
34. Khanal, S., Fulton, J., Klopfenstein, A., Douridas, N. and Shearer, S. (2018). Integration of high resolution remotely sensed data and machine learning techniques for spatial prediction of soil properties and corn yield. Computers and Electronics in Agriculture, 153, pp. 213–225.
35. Azadnia, R., Jahanbakhshi, A., Rashidi, S. and Bazyar, P. (2022). Developing an automated monitoring system for fast and accurate prediction of soil texture using an image-based deep learning network and machine vision system. Measurement, 190, p. 110669.
36. Adab, H., Morbidelli, R., Saltalippi, C., Moradian, M. and Ghalhari, G. A. F. (2020). Machine learning to estimate surface soil moisture from remote sensing data. Water, 12(11), p. 3223.
37. He, Y., Zhang, Z., Xiang, R., Ding, B., Du, R., Yin, H., Chen, Y. and Ba, Y. (2023). Monitoring salinity in bare soil based on sentinel-1/2 image fusion and machine learning. Infrared Physics & Technology, 131, p. 104656.
38. Zhang, Y., Han, W., Zhang, H., Niu, X. and Shao, G. (2023). Evaluating soil moisture content under maize coverage using UAV multimodal data by machine learning algorithms. Journal of Hydrology, 617, p. 129086.
39. Du, R., Chen, J., Xiang, Y., Xiang, R., Yang, X., Wang, T., He, Y., Wu, Y., Yin, H., Zhang, Z. and Chen, Y. (2023). Timely monitoring of soil water-salt dynamics within cropland by hybrid spectral unmixing and machine learning models. International Soil and Water Conservation Research, 32, pp. 1–15.

40. Nguyen, H., Maclagan, S. J., Nguyen, T. D., Nguyen, T., Flemons, P., Andrews, K., Ritchie, E. G. and Phung, D. (2017, October). Animal recognition and identification with deep convolutional neural networks for automated wildlife monitoring. In 2017 IEEE international conference on data science and advanced Analytics (DSAA) (pp. 40–49). IEEE.
41. Tabak, M. A., Norouzzadeh, M. S., Wolfson, D. W., Sweeney, S. J., VerCauteren, K. C., Snow, N. P., Halseth, J. M., Di Salvo, P. A., Lewis, J. S., White, M. D. and Teton, B. (2019). Machine learning to classify animal species in camera trap images: Applications in ecology. Methods in Ecology and Evolution, 10(4), pp. 585–590.
42. Chen, R., Little, R., Mihaylova, L., Delahay, R. and Cox, R. (2019). Wildlife surveillance using deep learning methods. Ecology and Evolution. 9(17), pp. 9453–9466.
43. Clarfeld, L. A., Sirén, A. P., Mulhall, B. M., Wilson, T. L., Bernier, E., Farrell, J., Lunde, G., Hardy, N., Gieder, K. D., Abrams, R. and Staats, S. (2023). Evaluating a tandem human-machine approach to labelling of wildlife in remote camera monitoring. Ecological Informatics, 77, p. 102257.
44. Chalmers, C., Fergus, P., Wich, S. and Longmore, S. N. (2021, July). Modelling animal biodiversity using acoustic monitoring and deep learning. In 2021 international joint conference on neural networks (IJCNN) (pp. 1–7). IEEE.
45. Whytock, R. C., Świeżewski, J., Zwerts, J. A., Bara-Słupski, T., Koumba Pambo, A. F., Rogala, M., Bahaa-el-din, L., Boekee, K., Brittain, S., Cardoso, A. W. and Henschel, P. (2021). Robust ecological analysis of camera trap data labelled by a machine learning model. Methods in Ecology and Evolution, 12(6), pp. 1080–1092.
46. Zhou, M., Elmore, J. A., Samiappan, S., Evans, K. O., Pfeiffer, M. B., Blackwell, B. F. and Iglay, R. B. (2021). Improving animal monitoring using small unmanned aircraft systems (sUAS) and deep learning networks. Sensors, 21(17), p. 5697.
47. Kaur, M., Kaur, P. D. and Sood, S. K. (2021). Energy efficient IoT-based cloud framework for early flood prediction. Natural Hazards, 109, pp. 2053–2076.
48. Khalaf, M., Alaskar, H., Hussain, A. J., Baker, T., Maamar, Z., Buyya, R., Liatsis, P., Khan, W., Tawfik, H. and Al-Jumeily, D. (2020). IoT-enabled flood severity prediction via ensemble machine learning models. IEEE Access, 8, pp. 70375–70386.
49. Smys, S., Basar, A. and Wang, H. (2020). CNN based flood management system with IoT sensors and cloud data. Journal of Artificial Intelligence, 2(4), pp. 194–200.
50. Prakash, C., Barthwal, A. and Acharya, D. (2023). FLOODALERT: An internet of things based real-time flash flood tracking and prediction system. Multimedia Tools and Applications, 82(28), pp. 43701–43727.
51. Samikwa, E., Voigt, T. and Eriksson, J. (2020). "Flood prediction using IoT and artificial neural networks with edge computing." In 2020 International Conferences on Internet of Things (iThings) and IEEE Green Computing and Communications (GreenCom) and IEEE Cyber, Physical and Social Computing (CPSCom) and IEEE Smart Data (SmartData) and IEEE Congress on Cybermatics (Cybermatics), pp. 234–240. IEEE.
52. Bhatia, M., Ahanger, T. A. and Manocha, A. (2023). Artificial intelligence based real-time earthquake prediction. Engineering Applications of Artificial Intelligence, 120, p. 105856.
53. Abdalzaher, M. S., Soliman, M. S. and El-Hady, S. M. (2023). Seismic intensity estimation for earthquake early warning using optimized machine learning model. IEEE Transactions on Geoscience and Remote Sensing, 115, pp. 1–13.

5 AI for Sustainable Development in the Media and Entertainment Industry

Ankit Sharma, Amit Channa,
Parul Malhotra, and Ardima Bajpai

INTRODUCTION

The media and entertainment industry stands at a critical juncture, facing unprecedented challenges and opportunities in an era defined by rapid technological advancement, shifting consumer preferences, and growing concerns about sustainability. As the world grapples with pressing environmental, social, and economic issues, the imperative for the industry to embrace sustainable development has never been greater. In this context, artificial intelligence (AI) emerges as a transformative force, offering innovative solutions to drive sustainability across various facets of media and entertainment [1].

At its core, the media and entertainment industry plays a central role in shaping culture, influencing public discourse, and reflecting societal values. However, the industry's operations have historically been associated with significant environmental impact, from energy-intensive production processes to vast amounts of electronic waste generated by digital content consumption. Moreover, issues of diversity and inclusion continue to plague the industry, with underrepresentation of marginalized groups both on screen and behind the scenes [2, 3].

Against this backdrop, AI presents a compelling opportunity to reimagine the way media and entertainment are produced, consumed, and experienced. By harnessing the power of machine learning, natural language processing, and computer vision, AI technologies can revolutionize content creation, distribution, audience engagement, and operations in ways that are not only innovative and efficient but also environmentally and socially responsible [4].

The potential applications of AI in the media and entertainment industry are vast and multifaceted. In content creation, AI algorithms can analyze vast amounts of data to inform creative decision-making, predict audience preferences, and personalize content experiences. AI-driven tools such as virtual reality

DOI: 10.1201/9781003497189-5

(VR) and augmented reality (AR) can enhance storytelling, immersing audiences in interactive and immersive narratives that foster empathy and understanding [5–7].

In distribution and consumption, AI-powered recommendation systems can curate personalized content recommendations based on individual preferences, reducing information overload and enhancing user satisfaction. Moreover, AI algorithms can optimize streaming processes to minimize energy consumption and reduce carbon emissions associated with digital content delivery [8–10].

In audience engagement, AI-enabled chatbots and virtual assistants can provide personalized assistance and support to audiences, enhancing the user experience while reducing the need for human intervention. Additionally, AI-driven analytics can provide insights into audience behavior and sentiment, informing content creators and distributors about the effectiveness of their strategies and campaigns [11, 12].

In operations, AI technologies can streamline workflows, automate repetitive tasks, and optimize resource usage, leading to increased efficiency and cost savings. For example, AI-driven algorithms can optimize production schedules, allocate resources more effectively, and identify opportunities for waste reduction and energy efficiency improvements [13–16].

The integration of AI into the media and entertainment industry is not without its challenges. Ethical considerations, such as data privacy, bias, and accountability, must be carefully navigated to ensure that AI technologies are deployed responsibly and equitably. Moreover, regulatory frameworks and industry standards will play a crucial role in guiding the development and implementation of AI solutions in the industry.

The convergence of AI and sustainability represents a paradigm shift in the media and entertainment industry, offering unprecedented opportunities to drive positive change and create value for both businesses and society. By embracing AI technologies and adopting a holistic approach to sustainability, stakeholders in the industry can not only enhance their competitive advantage but also contribute to a more sustainable and inclusive future for the media and entertainment landscape. This chapter will explore the multifaceted relationship between AI and sustainable development in the media and entertainment industry, examining the opportunities, challenges, and implications of AI-driven innovation for the industry's future.

AI IN CONTENT CREATION

In recent years, AI has revolutionized content creation in the media and entertainment industry. AI technologies, particularly machine learning algorithms, are increasingly being employed to inform creative decision-making processes, streamline production workflows, and enhance the overall quality of content. By analyzing vast amounts of data, including audience preferences, market trends, and historical performance metrics, AI can provide valuable insights and recommendations to content creators, enabling them to develop more engaging, relevant, and impactful content. From scriptwriting and storyboarding to casting and editing, AI-powered tools and platforms are transforming every stage of the content creation process, empowering filmmakers, writers, and producers to push the boundaries of creativity and innovation.

Machine Learning for Creative Decision-Making

Machine learning algorithms play a crucial role in enabling creative decision-making in content creation. These algorithms can analyze patterns and trends in large datasets to identify audience preferences, predict market demand, and optimize content strategies. For example, AI-powered analytics platforms can analyze social media trends, viewer engagement metrics, and viewer feedback to provide real-time insights into audience preferences and sentiment. Content creators can then use this information to tailor their content to meet the needs and interests of their target audience, increasing the likelihood of success.

Personalization and Audience Insights

One of the key advantages of AI in content creation is its ability to personalize content experiences for individual viewers. By leveraging data-driven insights and predictive analytics, AI algorithms can recommend personalized content recommendations based on factors such as viewing history, preferences, and behavior. For instance, streaming platforms use AI-powered recommendation systems to suggest relevant movies, TV shows, and music to users, enhancing user satisfaction and engagement. Moreover, AI-driven audience insights can inform content creators about audience demographics, interests, and preferences, enabling them to develop content that resonates with their target audience.

Virtual Reality (VR) and Augmented Reality (AR) Applications

AI technologies are also driving innovation in virtual reality (VR) and augmented reality (AR) applications, offering immersive and interactive experiences for audiences. By integrating AI algorithms with VR and AR technologies, content creators can create hyper-realistic environments, lifelike characters, and interactive narratives that blur the lines between reality and fiction. For example, AI-powered natural language processing (NLP) algorithms can enable real-time interaction with virtual characters, allowing audiences to engage in conversations, ask questions, and explore virtual worlds in unprecedented ways. Similarly, AI-driven computer vision algorithms can enhance the realism of AR experiences, overlaying digital content onto the physical world with stunning accuracy and precision.

AI is revolutionizing content creation in the media and entertainment industry, enabling more personalized, engaging, and immersive experiences for audiences. From machine learning algorithms that inform creative decision-making to AI-driven recommendation systems that personalize content recommendations, AI technologies are reshaping every aspect of the content creation process. Moreover, the integration of AI with VR and AR technologies is opening up new possibilities for interactive storytelling and immersive experiences, pushing the boundaries of creativity and innovation in the industry. As AI continues to evolve and mature, its impact on content creation is expected to grow, driving further advancements in creativity, efficiency, and audience engagement.

AI in Distribution and Consumption

The distribution and consumption of digital content are undergoing significant transformations, thanks to the integration of artificial intelligence (AI) technologies. AI is revolutionizing how content is delivered to users and how users engage with content, leading to more personalized and efficient experiences. This section explores the role of AI in personalized content recommendations, optimization of streaming processes, and reducing carbon emissions in digital content delivery.

Personalized Content Recommendations

AI-driven recommendation systems are reshaping the way users discover and consume digital content. By analyzing user behavior, preferences, and historical interactions, AI algorithms can generate personalized content recommendations tailored to individual tastes and interests. These recommendation systems leverage machine learning techniques to continuously learn and adapt to user feedback, refining their recommendations over time. As a result, users are presented with a curated selection of content that aligns with their preferences, leading to increased engagement, satisfaction, and retention.

Optimization of Streaming Processes

Streaming platforms are increasingly turning to AI to optimize the delivery of digital content to users. AI algorithms can analyze network conditions, user demand, and content popularity to dynamically adjust streaming parameters such as bitrate, resolution, and buffering to ensure a smooth and uninterrupted viewing experience. By optimizing streaming processes in real time, AI can reduce latency, minimize buffering, and enhance video quality, leading to improved user satisfaction and retention. Moreover, AI-driven content caching and delivery solutions can reduce server load and bandwidth usage, resulting in cost savings and environmental benefits.

REDUCING CARBON EMISSIONS IN DIGITAL CONTENT DELIVERY

AI has the potential to play a significant role in reducing carbon emissions associated with digital content delivery. By optimizing streaming processes and content delivery networks, AI algorithms can minimize energy consumption and resource usage, leading to a more sustainable and environmentally friendly distribution model. Additionally, AI-driven analytics can provide insights into energy usage and carbon emissions associated with digital content delivery, enabling streaming platforms to identify opportunities for improvement and implement more energy-efficient strategies. By leveraging AI to reduce carbon emissions in digital content delivery, streaming platforms can contribute to global efforts to combat climate change while providing high-quality entertainment experiences to users.

AI is transforming the distribution and consumption of digital content, enabling more personalized, efficient, and sustainable experiences for users. From personalized content recommendations that enhance user engagement to optimization of streaming processes that improve performance and reduce environmental impact, AI technologies are reshaping every aspect of the digital content ecosystem. As AI continues to evolve and mature, its impact on distribution and consumption is expected to grow, driving further advancements in efficiency, sustainability, and user satisfaction in the digital content industry.

AI in Audience Engagement

Audience engagement is a crucial aspect of the media and entertainment industry, and AI is playing an increasingly important role in enhancing this interaction. AI technologies are being leveraged to create more personalized, immersive, and responsive experiences for users, leading to higher levels of engagement and satisfaction. This section explores how AI is transforming audience engagement through the use of chatbots and virtual assistants, sentiment analysis, and the enhancement of user experience.

Chatbots and Virtual Assistants

Chatbots and virtual assistants powered by AI algorithms are revolutionizing how users interact with media and entertainment platforms. These AI-driven assistants can engage with users in natural language conversations, providing information, assistance, and recommendations in real time. Whether it's answering questions about content, helping users navigate through a platform, or providing personalized recommendations, chatbots and virtual assistants enhance user engagement by delivering timely and relevant assistance. Moreover, AI-powered chatbots can learn from user interactions over time, improving their accuracy and effectiveness in addressing user needs and preferences.

Sentiment Analysis and Audience Insights

AI-driven sentiment analysis tools are enabling media and entertainment companies to gain valuable insights into audience preferences, opinions, and sentiment. By analyzing social media feeds, user comments, and other forms of user-generated content, AI algorithms can identify trends, patterns, and sentiment signals that provide valuable feedback for content creators and marketers. Sentiment analysis allows companies to understand how audiences are reacting to their content, identify areas for improvement, and tailor their strategies to better meet audience expectations. By leveraging AI for sentiment analysis, media and entertainment companies can make more informed decisions about content development, marketing campaigns, and audience engagement strategies.

Enhancing User Experience

AI technologies are reshaping the user experience in media and entertainment, making it more personalized, intuitive, and immersive. From recommendation engines that deliver personalized content suggestions to content discovery platforms that use AI-driven algorithms to surface relevant content, AI is enhancing the user experience by providing users with more relevant and engaging content. Additionally, AI-powered content creation tools, such as deep learning algorithms for image and video processing, enable content creators to produce high-quality, visually appealing content more efficiently. By enhancing the user experience through AI-driven personalization, interactivity, and content creation, media and entertainment companies can increase user engagement, retention, and loyalty.

AI is revolutionizing audience engagement in the media and entertainment industry, enabling more personalized, responsive, and immersive experiences for users. Whether it's through chatbots and virtual assistants that provide real-time assistance, sentiment analysis tools that offer insights into audience preferences, or AI-driven enhancements to the user experience, AI technologies are transforming how audiences interact with content and platforms. As AI continues to evolve and mature, its impact on audience engagement is expected to grow, driving further advancements in user-centric experiences and relationships between media companies and their audiences.

AI IN OPERATIONS

Artificial intelligence is revolutionizing operations in the media and entertainment industry, driving efficiency, productivity, and sustainability across various processes. This section delves into how AI is transforming operations through workflow optimization, resource allocation and efficiency, and waste reduction and energy efficiency.

Workflow Optimization

AI technologies are streamlining and optimizing workflows in media and entertainment operations, leading to increased efficiency and productivity. AI-driven workflow management systems can automate repetitive tasks, such as file organization, metadata tagging, and content distribution, freeing up time for creative professionals to focus on higher-value activities. Moreover, AI algorithms can analyze historical workflow data to identify bottlenecks, optimize resource allocation, and streamline production processes. By automating manual tasks and optimizing workflows, AI enables companies to deliver content faster, reduce costs, and stay ahead of competitors in today's fast-paced media landscape.

Resource Allocation and Efficiency

AI plays a crucial role in optimizing resource allocation and efficiency in media and entertainment operations. AI algorithms can analyze data on resource availability, demand forecasts, and production schedules to optimize resource allocation and utilization. For example, AI-powered scheduling tools can optimize crew schedules, equipment usage, and production timelines to minimize downtime and maximize productivity. Additionally, AI-driven predictive analytics can forecast resource needs and identify potential shortages or overages, enabling companies to proactively manage resources and avoid costly delays. By optimizing resource allocation and efficiency, AI helps companies reduce waste, lower costs, and improve overall operational performance.

Waste Reduction and Energy Efficiency

AI technologies are also driving waste reduction and energy efficiency initiatives in media and entertainment operations. AI-driven analytics platforms can analyze data on energy consumption, waste generation, and environmental impact to identify opportunities for improvement. For example, AI algorithms can optimize energy usage in production facilities, adjust lighting and temperature settings based on occupancy patterns, and identify energy-saving opportunities in equipment and infrastructure. Moreover, AI-powered waste management systems can track and monitor waste streams, identify areas for waste reduction, and optimize recycling and disposal processes. By leveraging AI for waste reduction and energy efficiency, media and entertainment companies can minimize their environmental footprint, reduce operating costs, and demonstrate their commitment to sustainability.

AI is transforming operations in the media and entertainment industry, driving efficiency, productivity, and sustainability across various processes. From workflow optimization and resource allocation to waste reduction and energy efficiency, AI technologies are enabling companies to streamline operations, reduce costs, and minimize their environmental impact. As AI continues to evolve and mature, its impact on operations is expected to grow, driving further advancements in efficiency, sustainability, and competitiveness in the media and entertainment sector.

ETHICAL CONSIDERATIONS IN AI

As AI becomes increasingly integrated into various industries, including media and entertainment, it is crucial to address ethical considerations to ensure responsible and equitable use of these technologies. This section explores key ethical considerations in AI, including data privacy and security, bias and fairness, and accountability and transparency.

Data Privacy and Security

Data privacy and security are paramount considerations in AI applications, particularly in the media and entertainment industry, where vast amounts of user data are collected and analyzed. AI-driven systems often rely on large datasets to train algorithms and make predictions, raising concerns about the privacy and security of sensitive information. Companies must implement robust data protection measures, such as encryption, access controls, and anonymization techniques, to safeguard user data from unauthorized access, misuse, and breaches. Moreover, companies should adhere to relevant data protection regulations, such as the General Data Protection Regulation (GDPR), and obtain informed consent from users before collecting or processing their personal information.

Bias and Fairness

Bias and fairness are critical ethical considerations in AI, as algorithms can inadvertently perpetuate or amplify existing biases and inequalities in society. AI systems are trained on historical data, which may reflect biases and prejudices present in society, leading to biased or discriminatory outcomes. For example, AI-driven recommendation systems may recommend content that reinforces stereotypes or excludes certain demographics. To address bias and promote fairness, companies must employ rigorous testing and evaluation methodologies to identify and mitigate bias in AI algorithms. This may involve diversifying training datasets, implementing fairness-aware algorithms, and conducting regular audits to ensure equitable outcomes for all users.

Accountability and Transparency

Accountability and transparency are essential principles for ensuring responsible AI use and fostering trust among users and stakeholders. Companies deploying AI systems must be accountable for the decisions and actions of their algorithms, including any potential harms or unintended consequences. This requires clear lines of responsibility and accountability, as well as mechanisms for addressing grievances and providing recourse for affected individuals. Moreover, companies should strive to be transparent about their AI systems, including how they are developed, trained, and deployed, as well as the potential risks and limitations associated with their use. Transparency builds trust and enables users to make informed decisions about engaging with AI-driven technologies.

Ethical considerations are integral to the responsible deployment of AI in the media and entertainment industry. By addressing issues such as data privacy and security, bias and fairness, and accountability and transparency, companies can mitigate risks, build trust, and ensure that AI technologies are used in a manner that aligns with ethical principles and societal values. As AI continues to evolve and shape the future of media and entertainment, it is essential for companies to prioritize ethical considerations and uphold ethical standards in their AI initiatives.

Regulatory Frameworks and Industry Standards

As AI technologies become increasingly pervasive across industries, including media and entertainment, regulatory frameworks and industry standards play a crucial role in ensuring responsible and ethical AI deployment. This section explores the importance of regulatory frameworks and industry standards in governing AI use, including guidelines for responsible AI use, compliance, and governance.

Guidelines for Responsible AI Use

Guidelines for responsible AI use provide a framework for companies to navigate ethical, legal, and societal considerations associated with AI deployment. These guidelines typically encompass principles such as fairness, transparency, accountability, and privacy, outlining best practices for developing, deploying, and monitoring AI systems. For example, organizations such as the Institute of Electrical and Electronics

Engineers (IEEE) and the Partnership on AI have published principles and guidelines for ethical AI development and deployment, emphasizing the importance of human-centered design, bias mitigation, and algorithmic transparency. By adhering to responsible AI guidelines, companies can mitigate risks, build trust, and ensure that their AI initiatives align with ethical and societal values.

Compliance and Governance

Compliance with regulatory requirements and industry standards is essential for ensuring that AI deployments meet legal and ethical standards. Regulatory frameworks such as the General Data Protection Regulation (GDPR) in Europe and the California Consumer Privacy Act (CCPA) in the United States impose strict requirements on the collection, processing, and use of personal data in AI systems. Companies must ensure that their AI initiatives comply with relevant data protection regulations, obtain necessary permissions and consent from users, and implement measures to protect user privacy and data security. Moreover, industry-specific regulations and standards may apply to AI use in media and entertainment, such as content rating systems, intellectual property laws, and advertising regulations. By establishing robust compliance and governance mechanisms, companies can mitigate legal and reputational risks associated with AI deployments and demonstrate their commitment to responsible AI use.

In conclusion, regulatory frameworks and industry standards play a critical role in governing AI use in the media and entertainment industry. Guidelines for responsible AI use provide a framework for ethical decision-making and risk management, while compliance and governance mechanisms ensure that AI deployments adhere to legal and ethical standards. As AI continues to evolve and shape the future of media and entertainment, it is essential for companies to prioritize regulatory compliance, uphold industry standards, and adopt responsible AI practices to ensure that AI technologies are deployed in a manner that benefits society while minimizing risks and harms.

A CASE STUDY OF GreenMedia STUDIOS

Introduction: GreenMedia Studios, a leading production company committed to sustainability, embarked on a mission to integrate AI into its content creation processes to drive efficiency, creativity, and environmental responsibility. This case study explores how GreenMedia Studios leveraged AI technologies to produce sustainable content while minimizing its environmental footprint.

Background: GreenMedia Studios recognized the environmental impact of traditional content creation processes, including energy consumption, resource usage, and waste generation. To address these challenges, the studio embraced AI as a tool to optimize workflows, enhance creativity, and reduce environmental impact.

Implementation: GreenMedia Studios implemented AI-driven tools and platforms across various stages of the content creation process:

1. **AI-Powered Script Analysis:** AI algorithms analyzed scripts to identify potential environmental themes, optimize dialogue, and suggest sustainable storytelling approaches.
2. **Automated Pre-Production Planning:** AI-driven scheduling tools optimized production schedules, crew assignments, and equipment usage to minimize energy consumption and resource waste.
3. **Virtual Production with AI:** GreenMedia Studios utilized AI-driven virtual production techniques, such as virtual sets and digital actors, to reduce the need for physical sets and travel, minimizing carbon emissions and resource usage.
4. **AI-Enhanced Post-Production:** AI algorithms assisted in video editing, color grading, and visual effects, speeding up the editing process while maintaining high-quality standards.

Results: The implementation of AI in content creation at GreenMedia Studios yielded significant results:

1. **Increased Efficiency:** AI-driven workflows streamlined production processes, reducing time and resources required for content creation.
2. **Environmental Impact Reduction:** By optimizing resource usage, minimizing travel, and utilizing virtual production techniques, GreenMedia Studios reduced its carbon footprint and environmental impact.
3. **Enhanced Creativity:** AI-powered tools provided creative insights and suggestions, fostering innovation and pushing the boundaries of sustainable storytelling.
4. **Cost Savings:** The efficiency gains achieved through AI implementation resulted in cost savings for GreenMedia Studios, enhancing its financial sustainability.

The case study of GreenMedia Studios demonstrates the transformative potential of AI in driving sustainable content creation in the media and entertainment industry. By integrating AI into its workflows, the studio achieved increased efficiency, reduced environmental impact, enhanced creativity, and cost savings. This case study serves as a compelling example for other production companies seeking to leverage AI for sustainable content creation while minimizing their environmental footprint.

As AI technologies continue to evolve, GreenMedia Studios plans to explore additional applications of AI in content creation, such as predictive analytics for audience engagement, AI-generated content, and AI-driven storytelling techniques. By staying at the forefront of AI innovation, GreenMedia Studios aims to further enhance its sustainability efforts and continue delivering impactful content to audiences worldwide.

CONCLUSION

The case study of GreenMedia Studios exemplifies the transformative power of AI in driving sustainable content creation within the media and entertainment industry. Through the strategic integration of AI technologies into its production processes, GreenMedia Studios has not only enhanced efficiency and creativity but also significantly reduced its environmental footprint. By leveraging AI-driven tools for script analysis, virtual production, and post-production editing, the studio has achieved remarkable results, including increased efficiency, reduced resource consumption, and cost savings. GreenMedia Studios' commitment to sustainability serves as a testament to the industry's potential to embrace innovative solutions that benefit both the bottom line and the planet. The success of this case study underscores the importance of prioritizing environmental responsibility and ethical considerations in content creation. As AI continues to evolve and shape the future of media and entertainment, it is essential for industry stakeholders to embrace sustainable practices and leverage technology to drive positive change.

Looking ahead, GreenMedia Studios remains committed to exploring new opportunities for AI-driven innovation in content creation, including predictive analytics, AI-generated content, and immersive storytelling techniques. By staying at the forefront of AI innovation and sustainability, GreenMedia Studios aims to inspire others in the industry to follow suit and contribute to a more sustainable and environmentally conscious media landscape. The case study of GreenMedia Studios demonstrates that AI is not just a tool for enhancing efficiency and creativity but also a powerful catalyst for driving sustainability and positive social impact in the media and entertainment industry. By harnessing the potential of AI and embracing sustainable practices, companies can pave the way for a brighter, more environmentally friendly future for content creation and consumption.

FUTURE SCOPE

The case study of GreenMedia Studios provides valuable insights into the transformative potential of AI in driving sustainable content creation within the media and entertainment industry. Looking ahead, there are several promising avenues for future exploration and innovation in this space:

1. **Advanced AI Applications:** As AI technologies continue to evolve, there is immense potential for further advancements in AI-driven content creation techniques. Future research and development efforts may focus on developing more sophisticated AI algorithms for script analysis, virtual production, and post-production editing, enabling even greater levels of efficiency, creativity, and environmental sustainability.
2. **Predictive Analytics for Audience Engagement:** AI-powered predictive analytics tools can provide valuable insights into audience preferences, behavior, and engagement patterns, enabling content creators to tailor their content strategies to meet audience needs more effectively. Future research may explore the integration of AI-driven predictive analytics into content creation workflows, enabling companies to anticipate audience demand, optimize content distribution strategies, and enhance user experiences.
3. **AI-Generated Content:** AI technologies such as generative adversarial networks (GANs) and natural language processing (NLP) are increasingly being used to generate content autonomously, ranging from text and images to music and videos. Future research may explore the potential applications of AI-generated content in the media and entertainment industry, including automated content production, personalized content recommendations, and interactive storytelling experiences.
4. **Ethical AI Development and Governance:** As AI technologies become more prevalent in content creation, it is essential to address ethical considerations such as bias, fairness, accountability, and transparency. Future research may focus on developing robust ethical frameworks and governance mechanisms to ensure responsible AI development and deployment in the media and entertainment industry. This includes implementing measures to mitigate bias in AI algorithms, promoting transparency and accountability in AI systems, and safeguarding user privacy and data security.
5. **Collaborative Initiatives and Partnerships:** Collaboration between industry stakeholders, academia, government agencies, and nonprofit organizations is crucial for driving progress in AI-driven sustainable content creation. Future initiatives may involve collaborative research projects, industry-wide standards development, and cross-sector partnerships aimed at advancing AI innovation while addressing societal and environmental challenges.

The future scope of AI-driven sustainable content creation in the media and entertainment industry is vast and multifaceted. By embracing innovation, collaboration, and ethical responsibility, industry stakeholders can harness the full potential of AI to create content that is not only engaging and creative but also environmentally sustainable and socially responsible. As AI technologies continue to evolve and shape the future of content creation, the opportunities for innovation and positive impact are boundless.

REFERENCES

1. Majerova, J., & Das, S. (2023). Digital solutions for media: Role of pandemic and sustainable development of depleting industries. In *Digital Entertainment as Next Evolution in Service Sector: Emerging Digital Solutionss in Reshaping Different Industries* (pp. 107–125). Singapore: Springer Nature Singapore.
2. Hallur, G. G., Prabhu, S., & Aslekar, A. (2021). Entertainment in era of AI, big data & IoT. *Digital Entertainment: The Next Evolution in Service Sector*, (87–109). USA: IGI

3. Di Vaio, A., Palladino, R., Hassan, R., & Escobar, O. (2020). Artificial intelligence and business models in the sustainable development goals perspective: A systematic literature review. *Journal of Business Research*, 121, 283–314.
4. Chan-Olmsted, S. M. (2019). A review of artificial intelligence adoptions in the media industry. *International journal on media management*, 21(3–4), 193–215.
5. Vrana, V., & Das, S. (2023). Dynamic restructuring of digital media and entertainment sector: Role of urbanization, industrial innovation, and technological evolution. In *Digital Entertainment as Next Evolution in Service Sector: Emerging Digital Solutions in Reshaping Different Industries* (pp. 15–34). Singapore: Springer Nature Singapore.
6. Alam, T., Gupta, R., Qamar, S., & Ullah, A. (2022). Recent applications of artificial intelligence for sustainable development in smart cities. In *Recent innovations in artificial intelligence and smart applications* (pp. 135–154). Cham: Springer International Publishing.
7. Khamis, A., Li, H., Prestes, E., & Haidegger, T. (2019). AI: A key enabler for sustainable development goals: Part 2 [industry activities]. *IEEE Robotics & Automation Magazine*, 26(4), 122–127.
8. Sineviciene, L., Hens, L., Kubatko, O., Melnyk, L., Dehtyarova, I., & Fedyna, S. (2021). Socio-economic and cultural effects of disruptive industrial technologies for sustainable development. *International Journal of Global Energy Issues*, 43(2–3), 284–305.
9. Boulouard, Z., Ouaissa, M., Ouaissa, M., & El Himer, S. (2022). *AI and IoT for sustainable development in emerging countries* (Vol. 105). Cham, Switzerland: Springer International Publishing.
10. Mercier-Laurent, E. (2021, January). Can AI efficiently support Sustainable Development?. In IJCAI-8thAI4KM.
11. Sohn, D., & Youn, S. H. (2016). Fundamental sources and sustainable development of the Korean cultural entertainment industry with the Korean wave. *International Journal of Tourism Sciences*, 16(1–2), 83–92.
12. Bolesnikov, M., Popović Stijačić, M., Keswani, A. B., & Brkljač, N. (2022). Perception of innovative usage of AI in optimizing customer purchasing experience within the sustainable fashion industry. *Sustainability*, 14(16), 10082.
13. Zhang, S., Siran, Z., & Yusof, M. H. M. (2022, December). China's Digital Creative Multimedia Industry Development and Sustainable Development Innovation Direction. In 2nd International Conference on Creative Multimedia 2022 (ICCM 2022) (pp. 5–14). Atlantis Press.
14. Krishnaveni, D., Harish, V., & Mansurali, A. (2021). AI and Business Sustainability: Reinventing Business Processes. In *Reinventing Manufacturing and Business Processes Through Artificial Intelligence* (pp. 153–172). USA: CRC Press.
15. Cai, J. (2021, July). Artificial intelligence in digital media technology. In *International Conference on Frontier Computing* (pp. 188–195). Singapore: Springer Nature Singapore.
16. Vinuesa, R., Azizpour, H., Leite, I., Balaam, M., Dignum, V., Domisch, S., … & Fuso Nerini, F. (2020). The role of artificial intelligence in achieving the Sustainable Development Goals. *Nature Communications*, 11(1), 1–10.

AI and Water Resource Management
Ensuring Access and Sustainability

6

Sachin Sharma, Prithvi Baisla,
Himanshu Rajput, and Ritik Tanwar

INTRODUCTION

Water is a fundamental resource essential for life, livelihoods, and sustainable development. However, as the global population grows, urbanizes, and industrializes, the demand for water is rapidly increasing, placing unprecedented pressures on water resources worldwide. Concurrently, climate change is exacerbating water scarcity, altering precipitation patterns, and intensifying extreme weather events, further challenging water availability and quality. In the face of these complex and interconnected challenges, innovative approaches are urgently needed to optimize water management practices, ensure equitable access to clean and safe water, and promote the sustainability of water resources [1].

Artificial intelligence (AI) emerges as a transformative tool in addressing the multifaceted challenges of water resource management. With its ability to analyze vast amounts of data, identify patterns, and make data-driven predictions, AI holds promise for revolutionizing how water is managed, monitored, and utilized [2]. By leveraging AI algorithms for real-time monitoring, predictive modeling, and decision support, stakeholders can gain deeper insights into water systems, optimize resource allocation, and enhance resilience to water-related risks and disasters. Moreover, AI-driven solutions have the potential to improve water efficiency, reduce wastage, and mitigate the impacts of pollution and contamination, thereby contributing to the sustainability of water resources for current and future generations [3].

The purpose of this chapter is to explore the intersection of AI and water resource management, with a focus on ensuring access and sustainability in the face of growing global water challenges. We will examine the potential applications of AI in various aspects of water management, including monitoring, modeling, decision-making, and infrastructure management. Through case studies, examples, and best practices from around the world, we will illustrate how AI technologies are being deployed to address water scarcity, pollution, and inequitable access to water resources. Furthermore, we will explore the opportunities and challenges associated with integrating AI into water management practices,

DOI: 10.1201/9781003497189-6

considering factors such as data availability, technology adoption, capacity building, and ethical considerations [4–6].

The chapter will begin by providing an overview of the current state of global water challenges, highlighting the key drivers of water scarcity, pollution, and inequitable access. We will then delve into the potential of AI to address these challenges, exploring how AI-driven approaches can enhance water monitoring, modeling, and decision-making processes. We will examine case studies and examples of AI applications in water management, ranging from water quality monitoring and flood prediction to water demand forecasting and irrigation optimization [7].

We will discuss the importance of collaboration and knowledge sharing in advancing AI-driven water management solutions, highlighting the role of interdisciplinary partnerships, public–private collaborations, and international cooperation. We will also consider the ethical implications of AI in water management, addressing issues such as data privacy, algorithmic bias, and accountability [8–10].

This chapter will underscore the transformative potential of AI in revolutionizing water resource management, ensuring access, and promoting sustainability in the face of growing global water challenges. By harnessing the power of AI-driven technologies, stakeholders can make informed decisions, optimize resource allocation, and enhance resilience to water-related risks, ultimately contributing to a more water-secure and sustainable future for all [11].

Understanding Water Scarcity and Pollution

Water scarcity and pollution are two interconnected challenges that pose significant threats to human health, ecosystems, and sustainable development worldwide. In this section, we will delve into the drivers of water scarcity, the impacts of water pollution, and the challenges associated with ensuring equitable access to water resources [12].

Drivers of water scarcity

Water scarcity is driven by a combination of natural and human-induced factors, including:

1. **Climate Change:** Changes in precipitation patterns, rising temperatures, and increased frequency of extreme weather events exacerbate water scarcity by altering hydrological cycles and reducing water availability in certain regions.
2. **Population Growth and Urbanization:** Rapid population growth, coupled with urbanization and industrialization, leads to increased water demand for domestic, industrial, and agricultural purposes, placing strain on finite water resources.
3. **Overexploitation of Water Resources:** Unsustainable water extraction practices, such as over pumping of groundwater and diversion of surface water for irrigation and urban development, deplete aquifers and rivers, leading to water scarcity in many regions.
4. **Land Use Changes:** Deforestation, land degradation, and changes in land use patterns disrupt natural hydrological processes, leading to decreased infiltration and recharge of groundwater, exacerbating water scarcity.

Impacts of water pollution

Water pollution poses significant risks to human health, ecosystems, and biodiversity. The impacts of water pollution include:

1. **Human Health Risks:** Contaminated water sources pose risks of waterborne diseases, such as diarrhea, cholera, and typhoid fever, affecting millions of people worldwide, particularly in developing countries with inadequate access to clean water and sanitation.

2. **Ecosystem Degradation:** Pollution from agricultural runoff, industrial discharges, and untreated sewage disrupts aquatic ecosystems, leading to habitat loss, biodiversity decline, and degradation of water quality, affecting aquatic plants, fish, and other wildlife.
3. **Economic Costs:** Water pollution imposes significant economic costs on society, including expenses related to healthcare, water treatment, and ecosystem restoration, as well as losses in tourism, fisheries, and other sectors dependent on clean water resources.

Challenges in ensuring equitable access to water

Ensuring equitable access to clean and safe water remains a significant challenge, particularly in marginalized and underserved communities. Key challenges include:

1. **Social and Economic Inequities:** Socioeconomic disparities, including income inequality, gender discrimination, and lack of access to basic services, result in unequal access to water resources, with marginalized groups facing greater barriers to access clean water and sanitation services.
2. **Infrastructure Deficits:** Inadequate water infrastructure, including piped water supply systems, sewage treatment plants, and wastewater management facilities, limits access to clean water and sanitation services, particularly in rural and peri-urban areas.
3. **Governance and Policy Challenges:** Weak governance structures, lack of institutional capacity, and fragmented regulatory frameworks hinder effective water management and allocation, exacerbating inequities in water access and exacerbating water scarcity and pollution challenges.

In conclusion, understanding the drivers of water scarcity and pollution, as well as the challenges in ensuring equitable access to water resources, is essential for developing effective strategies and solutions to address these pressing global challenges [13–15]. By addressing the root causes of water scarcity and pollution, promoting sustainable water management practices, and fostering inclusive governance frameworks, we can work toward a more water-secure and sustainable future for all [16].

LEVERAGING AI FOR WATER MONITORING AND MODELING

AI technologies offer promising opportunities to revolutionize water monitoring and modeling, enabling more efficient, accurate, and proactive management of water resources. In this section, we will explore how AI can be leveraged for real-time monitoring of water systems, predictive modeling for water availability, and enhancing resilience to water-related risks.

Real-Time Monitoring of Water Systems

One of the key advantages of AI in water management is its ability to process vast amounts of data in real time, allowing for continuous monitoring of water systems and early detection of anomalies or changes. AI-driven sensors, satellite imagery, and Internet of Things (IoT) devices can collect data on water quality, quantity, and flow rates, providing valuable insights into the health and dynamics of water systems. AI algorithms can analyze this data in real time to identify trends, patterns, and abnormalities, enabling water managers to respond quickly to emerging issues such as pollution events, leaks, or droughts. By providing timely and actionable information, AI-powered real-time monitoring systems enhance the resilience of water systems and help ensure the delivery of clean and safe water to communities.

Predictive Modeling for Water Availability

Predictive modeling plays a crucial role in forecasting water availability, guiding water management decisions, and mitigating the impacts of water scarcity. AI algorithms, such as machine learning and neural networks, can analyze historical data on precipitation, evapotranspiration, soil moisture, and other relevant variables to develop predictive models of water availability. These models can then be used to forecast future water availability, anticipate droughts or floods, and optimize water allocation strategies. By providing decision-makers with reliable predictions and scenario analyses, AI-driven predictive modeling enhances the resilience of water systems, enabling proactive management and planning to meet water demand and mitigate risks.

Enhancing Resilience to Water-Related Risks

Water-related risks, such as floods, droughts, and water quality degradation, pose significant challenges to communities, ecosystems, and infrastructure. AI technologies can enhance resilience to these risks by providing early warning systems, optimizing response strategies, and facilitating adaptive management approaches. For example, AI-driven flood forecasting models can analyze weather data, river levels, and land use patterns to predict flood events and inform evacuation plans and flood defenses. Similarly, AI-powered drought monitoring systems can assess soil moisture, groundwater levels, and vegetation health to anticipate drought conditions and guide water conservation measures. By integrating AI into risk management and planning processes, stakeholders can enhance the resilience of water systems, minimize damages, and protect lives and livelihoods.

In conclusion, leveraging AI for water monitoring and modeling offers transformative opportunities to enhance the resilience, efficiency, and sustainability of water resource management. By providing real-time insights, predictive analytics, and risk assessments, AI technologies enable stakeholders to make informed decisions, optimize resource allocation, and mitigate the impacts of water-related challenges. As AI continues to evolve and mature, its potential to address water management challenges and ensure the availability of clean and safe water for all is poised to grow, contributing to a more water-secure and sustainable future.

AI-DRIVEN DECISION SUPPORT SYSTEMS

AI is revolutionizing decision-making processes in water resource management through the development of AI-driven decision support systems (DSS). These systems utilize advanced algorithms to analyze complex datasets, model scenarios, and provide actionable insights to stakeholders. In this section, we will explore how AI-driven DSS can optimize resource allocation, improve water efficiency, and mitigate pollution and contamination in water systems.

Optimizing Resource Allocation

AI-driven DSS enable water managers to optimize resource allocation decisions by analyzing a wide range of factors, including water availability, demand patterns, infrastructure capacity, and environmental constraints. By integrating real-time data from sensors, satellites, and other sources, AI algorithms can forecast water demand, identify supply gaps, and recommend optimal allocation strategies to meet current and future needs. For example, AI-driven DSS can help water utilities prioritize water distribution during droughts, optimize reservoir operations to balance competing demands, and allocate water rights based on equitable and sustainable principles. By providing decision-makers with timely and accurate information, AI-driven DSS enhance the efficiency and effectiveness of resource allocation decisions, ensuring the equitable distribution of water resources while minimizing waste and inefficiencies.

Improving Water Efficiency

Water efficiency is essential for sustainable water management, particularly in regions facing water scarcity and increasing demand. AI-driven DSS play a crucial role in improving water efficiency by identifying opportunities for conservation, optimizing water use practices, and reducing losses in water distribution systems. AI algorithms can analyze data on water consumption patterns, infrastructure performance, and operational conditions to identify inefficiencies and prioritize interventions for improvement. For example, AI-driven DSS can detect leaks in water pipelines, optimize irrigation schedules for agricultural fields, and recommend water-saving technologies for industrial processes. By implementing AI-driven recommendations, water users can reduce water waste, conserve valuable resources, and enhance the sustainability of water systems.

Mitigating Pollution and Contamination

Water pollution and contamination pose significant risks to human health, ecosystems, and aquatic life. AI-driven DSS can help mitigate these risks by monitoring water quality, detecting pollutants, and guiding pollution prevention and remediation efforts. AI algorithms can analyze data from water quality sensors, satellite imagery, and environmental monitoring stations to identify sources of pollution, track contaminant plumes, and assess risks to water resources. For example, AI-driven DSS can detect algal blooms in lakes and reservoirs, predict the spread of contaminants in rivers and groundwater, and recommend strategies for pollution control and remediation. By providing early warnings and targeted interventions, AI-driven DSS enable stakeholders to protect water resources, safeguard public health, and preserve ecosystems for future generations.

AI-driven decision support systems offer powerful tools for optimizing resource allocation, improving water efficiency, and mitigating pollution and contamination in water systems. By harnessing the capabilities of AI algorithms to analyze data, model scenarios, and provide actionable insights, stakeholders can make informed decisions, enhance the resilience of water systems, and ensure the availability of clean and safe water for all. As AI technologies continue to advance, the potential for AI-driven DSS to transform water resource management and address complex water challenges is immense, paving the way for a more sustainable and water-secure future.

CASE STUDIES AND BEST PRACTICES

In this section, we will explore case studies and best practices highlighting the application of AI in various aspects of water resource management, including water quality monitoring, flood prediction, and agricultural water management.

AI Applications in Water Quality Monitoring

Case study: smart water monitoring in Singapore

Singapore, a densely populated city-state facing water scarcity challenges, has implemented AI-driven water quality monitoring systems to ensure the safety and reliability of its water supply. The Public Utilities Board (PUB) of Singapore utilizes AI algorithms to analyze data from a network of sensors installed in water treatment plants, reservoirs, and distribution systems. These sensors collect real-time data on water quality parameters such as pH, turbidity, and chlorine levels. AI algorithms process this data to detect anomalies, identify potential contamination events, and predict water quality trends. By leveraging AI-driven water quality monitoring systems, Singapore has enhanced the resilience of its water supply infrastructure, minimized the risk of waterborne diseases, and maintained high standards of water quality for its residents.

Predictive Modeling for Flood Prediction

Case study: flood forecasting in The Netherlands

The Netherlands, a low-lying country prone to flooding, has implemented AI-driven predictive modeling systems to improve flood forecasting and early warning capabilities. The Dutch Ministry of Infrastructure and Water Management utilizes AI algorithms to analyze data from weather stations, river gauges, and hydrological models to predict river levels and assess flood risks. These predictive models incorporate machine learning techniques to learn from historical flood data and improve accuracy over time. By providing timely and accurate forecasts, AI-driven flood prediction systems enable authorities to implement proactive measures such as flood defenses, evacuation plans, and emergency response strategies. As a result, the Netherlands has enhanced its resilience to flooding, minimized damages to infrastructure and property, and protected the lives and livelihoods of its citizens.

Smart Irrigation Systems for Agricultural Water Management

Case study: precision irrigation in California

In California's agriculturally rich Central Valley, farmers are embracing AI-driven smart irrigation systems to optimize water use and enhance crop yields. Using AI algorithms and IoT technology, farmers can monitor soil moisture levels, weather conditions, and crop water requirements in real time. These smart irrigation systems analyze data from soil sensors, weather stations, and satellite imagery to determine optimal irrigation schedules, adjust water delivery rates, and minimize water wastage. By providing precise and targeted irrigation, AI-driven smart irrigation systems enable farmers to conserve water resources, reduce irrigation costs, and increase crop productivity. Moreover, by integrating AI-driven irrigation management with sustainable farming practices such as crop rotation and soil conservation, farmers can enhance the sustainability and resilience of agricultural water management systems.

In conclusion, these case studies illustrate the diverse applications and benefits of AI in water resource management, including water quality monitoring, flood prediction, and agricultural water management. By leveraging AI algorithms, data analytics, and IoT technology, stakeholders can make informed decisions, optimize resource allocation, and enhance the resilience of water systems to ensure the availability of clean and safe water for all. As AI technologies continue to evolve and mature, the potential for AI-driven solutions to address complex water challenges and contribute to sustainable water management practices is immense, paving the way for a more water-secure and resilient future.

Collaboration and Knowledge Sharing

Effective collaboration and knowledge sharing are essential for addressing complex water challenges and promoting sustainable water resource management. In this section, we will explore the importance of interdisciplinary partnerships, public–private collaborations, and international cooperation in advancing water management practices and achieving water security.

Interdisciplinary Partnerships

Interdisciplinary partnerships bring together experts from diverse fields, including hydrology, ecology, engineering, social sciences, and policy, to tackle complex water challenges holistically. By combining expertise from different disciplines, interdisciplinary partnerships foster innovation, creativity, and comprehensive problem-solving approaches. For example, collaborations between hydrologists, data scientists, and policymakers can lead to the development of AI-driven decision support systems for water management, integrating hydrological models, remote sensing data, and stakeholder preferences to inform decision-making processes. Similarly, partnerships between water engineers, ecologists, and community stakeholders can lead to the design and implementation of nature-based solutions for water management, such as green infrastructure, wetland restoration, and watershed management initiatives. By fostering interdisciplinary collaborations, stakeholders can leverage diverse perspectives, expertise, and resources to develop innovative solutions and address water challenges more effectively.

Public–Private Collaborations

Public–private collaborations leverage the strengths and resources of both sectors to address water challenges, improve water infrastructure, and promote sustainable water management practices. Public agencies bring regulatory authority, funding, and public service mandates, while private companies offer technological expertise, innovation, and investment capital. Collaborations between public utilities and technology firms, for example, can lead to the development and deployment of AI-driven water monitoring systems, smart meters, and leak detection technologies to enhance water efficiency and reduce water losses. Similarly, partnerships between government agencies and agricultural producers can promote sustainable farming practices, water conservation measures, and pollution prevention efforts to protect water resources. By fostering public–private collaborations, stakeholders can harness the complementary strengths of both sectors, accelerate innovation, and achieve greater impact in addressing water challenges.

International Cooperation

Water challenges often transcend national boundaries, requiring coordinated action and cooperation among countries to address shared water resources and transboundary water issues. International cooperation plays a crucial role in promoting equitable and sustainable management of transboundary rivers, lakes, and aquifers, preventing conflicts over water resources, and promoting peace and stability. Multilateral agreements, treaties, and organizations facilitate dialogue, negotiation, and collaboration among riparian states, providing frameworks for joint management, data sharing, and conflict resolution. For example, the United Nations Watercourses Convention and the Convention on the Protection and Use of Transboundary Watercourses and International Lakes establish principles and mechanisms for cooperation among countries sharing transboundary water resources. Similarly, regional organizations such as the Mekong River Commission and the Nile Basin Initiative facilitate cooperation and dialogue among riparian states to address common water challenges and promote sustainable development. By fostering international cooperation, stakeholders can enhance resilience, promote equity, and ensure the sustainable management of shared water resources for the benefit of present and future generations.

In conclusion, collaboration and knowledge sharing are essential for addressing complex water challenges, promoting sustainable water management practices, and achieving water security. By fostering interdisciplinary partnerships, public–private collaborations, and international cooperation, stakeholders can leverage diverse expertise, resources, and perspectives to develop innovative solutions, implement effective policies, and build resilience to water-related risks. As water challenges continue to evolve in the face of climate change, population growth, and urbanization, collaborative approaches will be critical for achieving water security and ensuring the availability of clean and safe water for all.

ETHICAL CONSIDERATIONS IN AI-DRIVEN WATER MANAGEMENT

As AI technologies play an increasingly prominent role in water management, it is essential to address ethical considerations to ensure the responsible development and deployment of AI-driven solutions. In this section, we will explore key ethical considerations, including data privacy and security, algorithmic bias and fairness, and accountability and transparency.

Data Privacy and Security

Data privacy and security are paramount considerations in AI-driven water management, given the sensitive nature of water-related data and the potential risks associated with unauthorized access or misuse. Stakeholders must implement robust data governance frameworks, encryption protocols, and access controls to protect sensitive data from unauthorized access, manipulation, or disclosure. Furthermore, stakeholders must ensure compliance with data protection regulations, such as the General Data Protection Regulation (GDPR) in the European Union, and adopt best practices for data anonymization, pseudonymization, and de-identification to safeguard privacy while maximizing the utility of data for AI-driven analytics and decision-making. By prioritizing data privacy and security, stakeholders can build trust, mitigate risks, and uphold ethical standards in AI-driven water management practices.

Algorithmic Bias and Fairness

Algorithmic bias and fairness pose significant ethical challenges in AI-driven water management, as biased algorithms can perpetuate inequalities, discrimination, and injustice in decision-making processes. Bias can arise from various sources, including biased training data, flawed algorithms, and subjective decision criteria. To mitigate bias and promote fairness, stakeholders must adopt strategies for bias detection, mitigation, and transparency throughout the AI development lifecycle. This includes evaluating training data for representativeness and diversity, testing algorithms for fairness and robustness across different demographic groups, and implementing mechanisms for accountability and redress in cases of algorithmic bias. Additionally, stakeholders must engage with diverse stakeholders, including impacted communities, civil society organizations, and marginalized groups, to ensure that AI-driven water management practices reflect diverse perspectives, values, and priorities.

Accountability and Transparency

Accountability and transparency are essential principles for ensuring the responsible development and deployment of AI-driven water management systems. Stakeholders must establish clear lines of responsibility and accountability for AI-driven decision-making processes, including defining roles, responsibilities, and decision-making authority among stakeholders. Furthermore, stakeholders must ensure transparency in AI-driven decision-making processes by providing explanations, justifications, and documentation for algorithmic decisions, enabling stakeholders to understand the rationale behind decisions and assess their implications. Additionally, stakeholders must establish mechanisms for oversight, auditing, and accountability to monitor the performance of AI systems, detect errors or biases, and address concerns or complaints from stakeholders. By promoting accountability and transparency, stakeholders can build trust, foster accountability, and ensure the ethical and responsible use of AI in water management practices.

In conclusion, ethical considerations are essential for ensuring the responsible development and deployment of AI-driven water management systems. By addressing data privacy and security, algorithmic bias and fairness, and accountability and transparency, stakeholders can uphold ethical standards, promote trust, and mitigate risks associated with AI-driven decision-making processes. As AI technologies continue to evolve and shape the future of water management, it is imperative to prioritize ethical considerations to ensure that AI-driven solutions contribute to equitable, sustainable, and ethical water management practices.

CONCLUSION

The integration of AI into water resource management represents a significant step forward in addressing the complex challenges of water scarcity, pollution, and inequitable access. Throughout this chapter, we have explored the diverse applications and benefits of AI in various aspects of water management, including real-time monitoring, predictive modeling, decision support systems, and collaborative initiatives. AI-driven technologies offer transformative opportunities to enhance the resilience, efficiency, and sustainability of water systems by providing stakeholders with actionable insights, optimizing resource allocation, and mitigating the impacts of water-related risks and challenges. Through case studies, best practices, and examples from around the world, we have seen how AI-driven solutions are revolutionizing water management practices, improving water quality, predicting floods, and optimizing agricultural water use. The integration of AI into water management also raises important ethical considerations, including data privacy and security, algorithmic bias and fairness, and accountability and transparency. It is imperative for stakeholders to address these ethical considerations to ensure the responsible development and deployment of AI-driven water management systems, promote trust, and mitigate risks associated with AI-driven decision-making processes. The potential of AI to transform water resource management is immense, offering unprecedented opportunities to enhance the resilience, efficiency, and sustainability of water systems. By embracing AI technologies, fostering collaboration, and prioritizing ethical considerations, stakeholders can work together to address water challenges, ensure the availability of clean and safe water for all, and build a more water-secure and sustainable future for generations to come.

FUTURE SCOPE

Looking ahead, the future of AI in water resource management holds tremendous promise for further innovation, collaboration, and impact. As AI technologies continue to evolve and mature, there are several areas for future exploration and development:

1. **Advancements in AI Algorithms:** Continued research and development in AI algorithms, including machine learning, deep learning, and reinforcement learning, will lead to more sophisticated and accurate models for water monitoring, prediction, and decision-making. Innovations in AI algorithms will enable stakeholders to extract deeper insights from data, improve model accuracy, and enhance the reliability of AI-driven systems.

2. **Integration of IoT and Sensor Technologies:** The integration of AI with Internet of Things (IoT) and sensor technologies will enable the proliferation of smart water infrastructure, facilitating real-time monitoring, data collection, and remote control of water systems. IoT-enabled sensors will provide a wealth of real-time data on water quality, quantity, and usage, enabling stakeholders to make more informed decisions and optimize resource allocation.
3. **Expansion of Digital Twin Technology:** Digital twin technology, which creates virtual replicas of physical assets and systems, will play an increasingly prominent role in water management. By integrating AI with digital twin models of water systems, stakeholders can simulate and optimize system behavior, predict the impact of interventions, and test scenarios for resilience and sustainability.
4. **Enhanced Resilience to Climate Change:** As the impacts of climate change continue to intensify, AI will play a critical role in enhancing the resilience of water systems to climate-related risks such as droughts, floods, and extreme weather events. AI-driven predictive models will enable stakeholders to anticipate and adapt to changing climate conditions, optimize water allocation strategies, and implement adaptive management measures.
5. **Ethical and Regulatory Frameworks:** The development of robust ethical and regulatory frameworks will be essential for ensuring the responsible and ethical use of AI in water management. Stakeholders must address issues such as data privacy, algorithmic bias, transparency, and accountability to build trust, mitigate risks, and promote equitable and sustainable water management practices.
6. **Capacity Building and Knowledge Sharing:** Capacity building and knowledge sharing initiatives will be crucial for enabling stakeholders to harness the full potential of AI in water resource management. Training programs, workshops, and collaborative platforms will empower stakeholders to develop AI expertise, share best practices, and foster innovation in water management.
7. **International Collaboration:** International collaboration and cooperation will be essential for addressing global water challenges, promoting knowledge exchange, and sharing best practices in AI-driven water management. Multilateral partnerships, joint research initiatives, and technology transfer programs will facilitate collaboration among countries, regions, and stakeholders to achieve water security and sustainability on a global scale.

REFERENCES

1. Xiang, X., Li, Q., Khan, S., & Khalaf, O. I. (2021). Urban water resource management for sustainable environment planning using artificial intelligence techniques. *Environmental Impact Assessment Review, 86*, 106515.
2. Ahmed, S. S., Bali, R., Khan, H., Mohamed, H. I., & Sharma, S. K. (2021). Improved water resource management framework for water sustainability and security. *Environmental Research, 201*, 111527.
3. D'Amore, G., Di Vaio, A., Balsalobre-Lorente, D., & Boccia, F. (2022). Artificial intelligence in the water–energy–food model: A holistic approach towards sustainable development goals. *Sustainability, 14*(2), 867.
4. Dogo, E. M., Salami, A. F., Nwulu, N. I., & Aigbavboa, C. O. (2019). Blockchain and internet of things-based technologies for intelligent water management system. *Artificial intelligence in IoT*, 129–150. Cham, Switzerland: Springer.
5. Ahmadov, E. (2020). Water resources management to achieve sustainable development in Azerbaijan. *Sustainable Futures, 2*, 100030.
6. Loucks, D. P., & Van Beek, E. (2017). *Water resource systems planning and management: An introduction to methods, models, and applications*. Switzerland: Springer.
7. Balsalobre-Lorente, D., Abbas, J., He, C., Pilař, L., & Shah, S. A. R. (2023). Tourism, urbanization and natural resources rents matter for environmental sustainability: The leading role of AI and ICT on sustainable development goals in the digital era. *Resources Policy, 82*, 103445.

8. Tsani, S., Koundouri, P., & Akinsete, E. (2020). Resource management and sustainable development: A review of the European water policies in accordance with the United Nations' Sustainable Development Goals. *Environmental Science & Policy, 114*, 570–579.
9. Gu, S., Jenkins, A., Gao, S. J., Lu, Y., Li, H., Li, Y., ... & Jobson, A. (2017). Ensuring water resource security in China; the need for advances in evidence-based policy to support sustainable management. *Environmental Science & Policy, 75*, 65–69.
10. Egbemhenghe, A. U., Ojeyemi, T., Iwuozor, K. O., Emenike, E. C., Ogunsanya, T. I., Anidiobi, S. U., & Adeniyi, A. G. (2023). Revolutionizing water treatment, conservation, and management: Harnessing the power of AI-driven ChatGPT solutions. *Environmental Challenges, 13*, 100782.
11. Akter, S. (2024). Harnessing Technology for Environmental Sustainability: Utilizing AI to Tackle Global Ecological Challenges. *Journal of Artificial Intelligence General science (JAIGS), 2*(1), 49–57.
12. van Koppen, B., & Schreiner, B. (2014). Moving beyond integrated water resource management: Developmental water management in South Africa. *International Journal of Water Resources Development, 30*(3), 543–558.
13. Rakha, N. A. (2023). Artificial Intelligence and Sustainability. *International Journal of Cyber Law, 1*(3), 1–25.
14. He, C., Harden, C. P., & Liu, Y. (2020). Comparison of water resources management between China and the United States. *Geography and Sustainability, 1*(2), 98–108.
15. Gao, X., Zhao, Y., Lu, S., Chen, Q., An, T., Han, X., & Zhuo, L. (2019). Impact of coal power production on sustainable water resources management in the coal-fired power energy bases of Northern China. *Applied Energy, 250*, 821–833.
16. Kafy, A. A., Raikwar, V., Al Rakib, A., Kona, M. A., & Ferdousi, J. (2021). Geospatial approach for developing an integrated water resource management plan in Rajshahi, Bangladesh. *Environmental Challenges, 4*, 100139.

Building Blocks of Artificial Intelligence and Machine Learning *A Framework for Sustainable Development*

7

Naga Ramesh Palakurti

INTRODUCTION

In recent years, the intersection of artificial intelligence (AI) and sustainable development has emerged as a pivotal area of research and innovation. As societies grapple with pressing environmental, social, and economic challenges, the potential of AI to drive positive change and support the achievement of sustainable development goals (SDGs) is becoming increasingly evident. This introduction provides an overview of the building blocks of AI and machine learning (ML) within the context of sustainable development, highlighting the transformative potential of these technologies and setting the stage for a deeper exploration of their applications and implications.

The Emergence of AI and ML

Artificial intelligence and machine learning technologies have undergone rapid advancement in recent years, fueled by breakthroughs in algorithms, computational power, and data availability. AI refers to the simulation of human intelligence processes by machines, enabling them to learn from data, recognize patterns, and make decisions with minimal human intervention. Machine learning, a subset of AI, focuses on the development of algorithms that allow computers to learn from and improve over time based on experience.

The rise of AI and ML has ushered in a new era of innovation and disruption across various sectors, from healthcare and finance to transportation and agriculture. These technologies have demonstrated remarkable capabilities in areas such as image recognition, natural language processing, and predictive

DOI: 10.1201/9781003497189-7

analytics, leading to significant improvements in efficiency, productivity, and decision-making. In the context of sustainable development, AI and ML offer novel solutions to complex challenges, providing insights, tools, and approaches to address environmental degradation, social inequality, and economic instability.

The Role of AI and ML in Sustainable Development

The integration of AI and ML into sustainable development initiatives holds immense promise for advancing progress toward the SDGs. These technologies offer a range of opportunities to enhance environmental conservation, promote social equity, foster economic growth, and drive technological innovation.

In the realm of environmental conservation, AI and ML enable real-time monitoring of ecosystems, wildlife populations, and natural resources, empowering scientists and conservationists to make data-driven decisions and implement targeted interventions. From predicting deforestation patterns to identifying illegal fishing activities, AI-driven solutions provide valuable insights into environmental trends and threats, helping to safeguard biodiversity and ecosystems for future generations.

Moreover, AI and ML have the potential to address social challenges such as poverty, inequality, and access to basic services. By analyzing large datasets and identifying socioeconomic patterns, these technologies can inform policies and programs aimed at poverty alleviation, healthcare provision, education access, and social inclusion. From predicting disease outbreaks to optimizing resource allocation in humanitarian crises, AI-driven solutions offer innovative approaches to tackling complex social issues and improving quality of life for vulnerable populations.

In terms of economic development, AI and ML contribute to innovation, entrepreneurship, and job creation, driving growth and prosperity in both developed and developing economies. By automating routine tasks, optimizing processes, and unlocking new opportunities for value creation, these technologies stimulate economic activity and foster entrepreneurship in sectors such as manufacturing, finance, and digital services. Furthermore, AI-driven innovations such as smart cities, autonomous vehicles, and precision agriculture have the potential to revolutionize industries and create new markets, driving sustainable economic growth and competitiveness.

Challenges and Considerations

Despite the immense potential of AI and ML for sustainable development, their adoption and implementation are not without challenges and considerations. Ethical concerns, data privacy issues, and algorithmic biases are among the key challenges that must be addressed to ensure that AI-driven solutions benefit society as a whole.

Ethical considerations arise from the potential misuse of AI and ML technologies, including concerns about data privacy, surveillance, and discrimination. As AI systems become increasingly integrated into decision-making processes in areas such as healthcare, criminal justice, and employment, there is a need to establish clear guidelines and standards for ethical AI development and deployment. Moreover, efforts to mitigate algorithmic biases and ensure fairness, transparency, and accountability in AI-driven decision-making are essential to prevent unintended consequences and safeguard individual rights and liberties.

Data privacy and security concerns also pose significant challenges to the adoption of AI and ML in sustainable development. As organizations collect and analyze vast amounts of data to train AI models and derive insights, there is a risk of data breaches, cyberattacks, and unauthorized access to sensitive information. Establishing robust data governance frameworks, implementing encryption and access control measures, and complying with data protection regulations are essential steps to protect personal and sensitive data and maintain trust in AI-driven systems.

Furthermore, the digital divide and inequities in access to technology pose challenges to the equitable distribution and adoption of AI and ML solutions in sustainable development. While AI has the potential to empower individuals and communities by providing access to information, services, and opportunities, there is a risk of widening disparities between those who have access to technology and those who do not. Bridging the digital divide and ensuring inclusive access to AI and ML technologies are critical to realizing their full potential for sustainable development and promoting social equity and inclusion.

Conclusion

In conclusion, the integration of AI and ML into sustainable development initiatives offers a myriad of opportunities to address pressing global challenges and build a more sustainable and equitable future. From environmental conservation and social empowerment to economic growth and technological innovation, these technologies have the potential to transform the way we approach sustainable development and achieve the SDGs. However, realizing this potential requires addressing challenges such as ethical concerns, data privacy issues, and inequities in access to technology. By embracing responsible AI development and deployment practices, fostering collaboration and partnerships, and prioritizing inclusivity and equity, we can harness the power of AI and ML to create positive impact and drive progress toward a more sustainable world.

APPLICATIONS OF AI IN ENVIRONMENTAL CONSERVATION

Environmental conservation is a critical aspect of sustainable development, aiming to protect ecosystems, preserve biodiversity, and ensure the sustainability of natural resources for future generations. Leveraging AI in environmental conservation offers innovative solutions to address complex challenges and enhance conservation efforts. This section explores key applications of AI in environmental conservation, focusing on real-time environmental monitoring, wildlife conservation, and sustainable agriculture and land management.

Real-Time Environmental Monitoring

Real-time environmental monitoring using AI involves the continuous collection, analysis, and interpretation of data from various sources to track changes in environmental parameters and identify emerging threats. AI technologies, such as machine learning algorithms and remote sensing techniques, enable the processing of large datasets from satellites, sensors, and IoT devices to monitor environmental indicators such as air quality, water quality, and land use changes in real time.

One example of AI-enabled real-time environmental monitoring is the use of satellite imagery and machine learning algorithms to detect deforestation activities in remote forest areas. By analyzing satellite images and identifying patterns indicative of deforestation, such as changes in vegetation cover or illegal logging activities, conservationists can take timely action to prevent further destruction of forests and protect biodiversity.

Another application is the monitoring of water quality in rivers and lakes using AI-powered sensors and IoT devices. These sensors can continuously measure parameters such as pH levels, dissolved oxygen, and pollutant concentrations, providing real-time data on water quality and facilitating early detection of pollution events. By integrating AI technologies into water monitoring systems, authorities can identify pollution sources, enforce environmental regulations, and safeguard aquatic ecosystems.

Wildlife Conservation and Biodiversity Preservation

AI plays a crucial role in wildlife conservation and biodiversity preservation by supporting efforts to monitor, track, and protect endangered species and their habitats. Conservationists use AI-powered camera traps, acoustic sensors, and GPS tracking devices to monitor wildlife populations, track animal movements, and gather data on species behavior and habitat usage.

For example, AI algorithms can analyze camera trap images to automatically identify and classify wildlife species, allowing researchers to estimate population sizes, assess population trends, and prioritize conservation actions. Similarly, acoustic sensors equipped with AI technology can detect and classify animal calls and sounds, providing insights into species diversity, distribution, and habitat preferences.

AI-driven predictive modeling techniques also help conservationists forecast species distributions, assess habitat suitability, and identify areas of conservation priority. By analyzing environmental variables such as temperature, precipitation, and vegetation cover, AI algorithms can predict species occurrence and identify potential conservation hotspots, guiding resource allocation and land management decisions.

Sustainable Agriculture and Land Management

AI technologies offer opportunities to enhance sustainable agriculture and land management practices by optimizing resource use, minimizing environmental impact, and increasing productivity. AI-driven precision agriculture systems leverage data from sensors, drones, and satellite imagery to monitor crop health, optimize irrigation schedules, and minimize pesticide and fertilizer usage.

For instance, AI-powered drones equipped with multispectral cameras can collect high-resolution images of agricultural fields and analyze vegetation indices to assess crop health, detect pest infestations, and identify nutrient deficiencies. By providing farmers with real-time insights into crop conditions and growth patterns, AI-enabled precision agriculture systems enable targeted interventions and resource-efficient farming practices.

Furthermore, AI algorithms can analyze soil data, weather forecasts, and historical crop yields to generate personalized recommendations for farmers, such as optimal planting dates, crop varieties, and fertilization strategies. By tailoring agricultural practices to local environmental conditions and individual farm characteristics, AI technologies help increase crop yields, reduce input costs, and promote sustainable land management practices.

In conclusion, the applications of AI in environmental conservation are diverse and multifaceted, spanning from real-time environmental monitoring and wildlife conservation to sustainable agriculture and land management. By leveraging AI technologies, conservationists, policymakers, and stakeholders can enhance their ability to monitor, manage, and protect natural ecosystems, thereby contributing to the preservation of biodiversity and the achievement of sustainable development goals.

AI FOR SOCIAL EQUITY AND ECONOMIC DEVELOPMENT

AI has the potential to drive significant advancements in social equity and economic development by improving access to essential services, promoting inclusive growth, and empowering marginalized communities. This section explores the applications of AI in healthcare access and delivery, education and skills development, and poverty alleviation and economic empowerment.

Healthcare Access and Delivery

AI technologies are revolutionizing healthcare access and delivery by enabling more personalized, efficient, and accessible healthcare services. AI-powered diagnostic tools, such as medical imaging analysis algorithms and predictive analytics models, help healthcare providers diagnose diseases, identify treatment options, and predict patient outcomes with greater accuracy and efficiency.

For example, AI algorithms can analyze medical images, such as X-rays, MRIs, and CT scans, to detect abnormalities and assist radiologists in diagnosing conditions such as cancer, fractures, and cardiovascular diseases. Additionally, AI-driven predictive analytics models can analyze electronic health records, genetic data, and other patient information to identify individuals at risk of developing chronic diseases and recommend preventive interventions.

Furthermore, AI-enabled telemedicine platforms and virtual healthcare assistants expand access to healthcare services for underserved populations, including those in remote or rural areas. By leveraging AI-driven chatbots, mobile applications, and remote monitoring devices, patients can access medical advice, receive diagnoses, and manage chronic conditions from the comfort of their homes, reducing barriers to healthcare access and improving health outcomes.

Education and Skills Development

AI technologies hold immense potential to transform education and skills development by personalizing learning experiences, expanding access to educational resources, and enhancing teaching effectiveness. AI-driven adaptive learning platforms analyze student performance data, learning preferences, and behavioral patterns to deliver customized learning experiences tailored to individual needs and abilities.

For instance, AI-powered educational software can assess students' knowledge gaps, adapt learning materials to their learning styles, and provide targeted feedback and recommendations for improvement. Additionally, AI-driven virtual tutors and intelligent tutoring systems offer personalized support and guidance to students, helping them master difficult concepts and achieve academic success.

Moreover, AI-enabled language translation tools, speech recognition software, and online learning platforms facilitate access to educational resources and content in multiple languages and formats, breaking down language barriers and promoting inclusivity in education. By harnessing AI technologies, educators can create more engaging, interactive, and accessible learning environments that cater to diverse learning styles and abilities.

Poverty Alleviation and Economic Empowerment

AI has the potential to drive economic development and alleviate poverty by creating new opportunities for employment, entrepreneurship, and economic empowerment. AI-driven job matching platforms, skill assessment tools, and workforce development programs help individuals acquire the skills and knowledge needed to access high-quality employment opportunities in emerging industries such as AI, data science, and digital technology.

For example, AI-powered job matching platforms analyze job seekers' skills, preferences, and career goals to match them with suitable employment opportunities and training programs. Additionally, AI-driven skill assessment tools identify individuals' strengths and weaknesses, recommend relevant training courses, and track their progress toward acquiring in-demand skills and certifications.

Furthermore, AI-enabled entrepreneurship support programs, such as business incubators, accelerators, and mentorship networks, help aspiring entrepreneurs develop and scale innovative business ideas. AI technologies provide entrepreneurs with access to market insights, customer feedback, and predictive analytics tools, enabling them to make data-driven decisions and build successful businesses that create jobs and drive economic growth.

In conclusion, AI has the potential to promote social equity and economic development by improving access to healthcare, expanding educational opportunities, and fostering entrepreneurship and job creation. By leveraging AI technologies, policymakers, educators, and social innovators can address systemic inequalities, empower marginalized communities, and create a more inclusive and prosperous society for all.

TECHNOLOGICAL INNOVATION AND AI-DRIVEN SOLUTIONS

Technological innovation, fueled by AI, is reshaping industries and transforming the way we plan, develop, and manage our cities, energy systems, and industries. This section explores the applications of AI in driving innovation across three key domains: smart cities and urban planning, renewable energy and sustainable infrastructure, and digital transformation and Industry 4.0.

Smart Cities and Urban Planning

AI is revolutionizing urban planning and development by enabling the creation of smart cities that leverage data, connectivity, and automation to improve quality of life, enhance sustainability, and optimize resource use. AI-driven solutions for smart cities encompass a wide range of applications, including traffic management, public safety, energy efficiency, and waste management.

For example, AI-powered traffic management systems analyze real-time traffic data from sensors, cameras, and GPS devices to optimize traffic flow, reduce congestion, and minimize travel times. Additionally, AI-driven predictive analytics models can anticipate traffic patterns, identify congestion hotspots, and recommend alternative routes to minimize delays and improve overall transportation efficiency.

Moreover, AI-enabled public safety initiatives leverage data from surveillance cameras, social media feeds, and other sources to detect and prevent crime, respond to emergencies, and ensure the safety and security of residents. AI algorithms can analyze patterns of criminal activity, identify suspicious behavior, and alert law enforcement agencies to potential threats, enabling proactive intervention and crime prevention.

Renewable Energy and Sustainable Infrastructure

AI plays a crucial role in advancing renewable energy and sustainable infrastructure development by optimizing energy production, improving energy efficiency, and reducing environmental impact. AI-driven solutions for renewable energy encompass a wide range of applications, including renewable energy forecasting, grid optimization, and energy storage management.

For instance, AI algorithms analyze weather data, historical energy production, and other variables to forecast renewable energy generation from sources such as solar, wind, and hydro power. By accurately predicting energy output, renewable energy forecasting models help grid operators balance supply and demand, optimize energy scheduling, and minimize reliance on fossil fuels.

Additionally, AI-powered grid optimization systems manage and control energy distribution networks to ensure stability, reliability, and resilience in the face of variable renewable energy generation. These systems use AI algorithms to balance energy supply and demand, optimize grid operations, and integrate distributed energy resources such as solar panels and battery storage systems into the grid.

Digital Transformation and Industry 4.0

AI-driven solutions are driving digital transformation and Industry 4.0 by revolutionizing manufacturing processes, supply chain management, and industrial automation. Industry 4.0, also known as the fourth industrial revolution, encompasses technologies such as AI, robotics, internet of things (IoT), and cloud computing to create interconnected, intelligent, and automated production systems.

For example, AI-powered predictive maintenance systems analyze sensor data, machine performance metrics, and historical maintenance records to predict equipment failures, optimize maintenance schedules, and minimize downtime in manufacturing facilities. By identifying potential issues before they occur, predictive maintenance solutions help manufacturers reduce costs, improve productivity, and prolong equipment lifespan.

Furthermore, AI-enabled supply chain management systems optimize inventory management, demand forecasting, and logistics operations to enhance efficiency and resilience in global supply chains. AI algorithms analyze data from multiple sources, including suppliers, distributors, and customers, to anticipate demand fluctuations, optimize inventory levels, and streamline transportation routes, reducing costs and minimizing supply chain disruptions.

In conclusion, AI-driven solutions are driving technological innovation and transformation across various domains, including smart cities and urban planning, renewable energy and sustainable infrastructure, and digital transformation and Industry 4.0. By harnessing the power of AI, organizations, governments, and industries can create more efficient, resilient, and sustainable systems that improve quality of life, protect the environment, and drive economic growth.

CHALLENGES AND OPPORTUNITIES IN AI INTEGRATION

The integration of AI presents both challenges and opportunities across various domains, including ethical considerations, data privacy and security, and promoting inclusivity. Addressing these challenges effectively is essential to unlock the full potential of AI while ensuring responsible and equitable deployment. This section delves into the complexities of AI integration and explores strategies to leverage its opportunities while mitigating its risks.

Ethical Considerations and Algorithmic Bias

Ethical considerations in AI integration encompass a wide range of concerns, including transparency, accountability, fairness, and privacy. AI systems are susceptible to algorithmic bias, where biases inherent in training data or algorithm design lead to discriminatory outcomes, exacerbating existing inequalities and perpetuating social injustices.

One of the key challenges is ensuring transparency and accountability in AI decision-making processes. AI algorithms often operate as "black boxes," making it difficult to understand how decisions are made or to identify and rectify biases. Moreover, the use of sensitive personal data in AI systems raises concerns about privacy and consent, particularly in contexts such as healthcare, finance, and criminal justice.

However, ethical AI development practices, such as algorithmic fairness, explainability, and interpretability, can help mitigate these risks and promote responsible AI deployment. By incorporating principles of fairness, diversity, and inclusivity into AI design and development processes, developers can identify and address biases, ensure transparency and accountability, and build trust with users and stakeholders.

Data Privacy and Security

Data privacy and security are paramount considerations in AI integration, as AI systems rely on vast amounts of data to train models and make predictions. The collection, storage, and processing of personal and sensitive data raise concerns about data privacy, consent, and protection against unauthorized access or misuse.

One challenge is ensuring compliance with data protection regulations, such as the General Data Protection Regulation (GDPR) in Europe or the California Consumer Privacy Act (CCPA) in the United States. These regulations impose strict requirements on organizations regarding data collection, usage, storage, and disclosure, aiming to protect individuals' privacy rights and prevent data breaches or misuse.

Moreover, AI systems are vulnerable to cybersecurity threats, such as data breaches, hacking, and adversarial attacks, which can compromise the integrity and security of AI-driven applications. Ensuring robust data security measures, such as encryption, access controls, and cybersecurity training, is essential to safeguard sensitive data and mitigate cybersecurity risks.

Bridging the Digital Divide and Promoting Inclusivity

The digital divide refers to the gap between those who have access to digital technologies and those who do not, exacerbating inequalities in access to information, education, employment, and social services. Bridging the digital divide and promoting inclusivity in AI integration are essential to ensure that the benefits of AI are accessible to all individuals and communities, regardless of socioeconomic status or geographical location.

One challenge is addressing barriers to access, such as lack of infrastructure, affordability constraints, and digital literacy gaps, particularly in rural or underserved areas. Governments, organizations, and stakeholders must invest in expanding broadband connectivity, providing affordable devices and internet access, and offering digital skills training programs to empower marginalized communities and bridge the digital divide.

Furthermore, promoting diversity and inclusivity in AI development teams and decision-making processes is crucial to ensure that AI systems reflect the needs, values, and perspectives of diverse populations. By incorporating diverse voices and experiences into AI design and development, developers can create more inclusive and equitable AI solutions that serve the needs of all users and promote social cohesion and diversity.

In conclusion, addressing the challenges and opportunities in AI integration requires a multidimensional approach that encompasses ethical considerations, data privacy and security, and promoting inclusivity. By prioritizing responsible AI development practices, fostering collaboration and partnerships, and investing in digital infrastructure and literacy initiatives, we can harness the transformative potential of AI to drive positive change and create a more equitable and inclusive society.

CASE STUDIES AND BEST PRACTICES

This section presents case studies and best practices showcasing successful applications of AI in environmental conservation, social impact initiatives, community engagement, and business innovations that promote sustainable practices.

AI Applications in Environmental Conservation

Case study: wildlife protection with AI

In Kenya's Maasai Mara National Reserve, the organization Wildbook utilizes AI-powered image recognition algorithms to track and monitor endangered wildlife species such as cheetahs and elephants. Camera traps placed strategically throughout the reserve capture images of individual animals, which are then uploaded to the Wildbook platform. The platform's AI algorithms analyze these images, identify individual animals based on unique features like spots and stripes, and track their movements over time. This data enables conservationists to monitor population trends, identify poaching threats, and implement targeted conservation strategies to protect these vulnerable species.

Case study: forest monitoring with satellite imagery

In Brazil's Amazon rainforest, the nonprofit organization Rainforest Connection (RFCx) utilizes AI-driven analysis of satellite imagery to detect illegal logging activities and prevent deforestation. RFCx's AI algorithms analyze high-resolution satellite images of the forest canopy to identify signs of logging, such as clear-cut areas and logging roads. When suspicious activity is detected, RFCx alerts local authorities and conservation organizations, enabling them to intervene and take action to protect the forest. By leveraging AI technology, RFCx helps to deter illegal logging and conserve precious forest ecosystems.

Social Impact Initiatives and Community Engagement

Case study: AI for humanitarian aid

In disaster-prone regions such as Southeast Asia, the nonprofit organization GlobalXplorer utilizes AI-powered drones and satellite imagery to assist in humanitarian aid and disaster response efforts. When natural disasters such as earthquakes or floods occur, GlobalXplorer deploys drones equipped with high-resolution cameras and AI algorithms to assess damage, identify areas in need of assistance, and prioritize rescue and relief efforts. By rapidly analyzing vast amounts of aerial imagery, AI helps humanitarian organizations target resources more effectively and provide timely assistance to affected communities.

Case study: AI-powered community health

In rural communities in sub-Saharan Africa, the social enterprise Last Mile Health utilizes AI-driven mobile health technology to improve access to healthcare services and strengthen community health systems. Last Mile Health's AI-powered mobile health platform, called Community Health Academy, provides community health workers with access to training materials, clinical decision support tools, and real-time data analytics. By leveraging AI algorithms to analyze health data and identify trends, Last Mile Health empowers community health workers to deliver more targeted and effective healthcare services, ultimately improving health outcomes and reducing healthcare disparities in underserved communities.

Business Innovations and Sustainable Practices

Case study: AI for sustainable agriculture

In the agricultural sector, companies like Farmers Edge utilize AI-driven precision agriculture technology to optimize crop production, minimize environmental impact, and increase farm profitability. Farmers Edge's AI-powered platform analyzes data from sensors, drones, and satellite imagery to provide farmers with insights into soil health, crop growth, and weather patterns. By leveraging AI algorithms to generate personalized recommendations for irrigation, fertilization, and pest management, Farmers Edge helps farmers improve crop yields, conserve water and resources, and adopt sustainable farming practices that enhance environmental stewardship.

Case study: AI for energy efficiency

In the energy sector, utilities and energy companies are harnessing AI technology to improve energy efficiency, reduce carbon emissions, and transition to renewable energy sources. For example, companies like Google utilize AI algorithms to optimize data center operations, minimize energy consumption, and maximize renewable energy usage. By analyzing vast amounts of data on energy usage, weather patterns, and equipment performance, AI helps companies identify opportunities for energy savings, automate energy management processes, and achieve sustainability goals. Through AI-driven energy efficiency initiatives, companies can reduce their environmental footprint, lower operating costs, and contribute to a more sustainable energy future.

Best Practices

1. **Collaborative Partnerships:** Successful AI initiatives often involve collaboration and partnerships between governments, NGOs, academia, and the private sector to leverage expertise, resources, and data for maximum impact.
2. **Data Privacy and Ethics:** Maintaining data privacy, security, and ethical standards is critical in AI applications, requiring clear policies, consent mechanisms, and safeguards to protect individuals' rights and prevent misuse of data.
3. **Stakeholder Engagement:** Engaging with local communities, stakeholders, and end-users throughout the AI development and deployment process ensures that solutions are contextually relevant, inclusive, and responsive to community needs and priorities.
4. **Continuous Learning and Improvement:** AI-driven solutions should be iterative and adaptive, incorporating feedback, monitoring performance metrics, and iterating based on real-world outcomes to continuously improve effectiveness and address evolving challenges.
5. **Scalability and Sustainability:** Designing AI solutions with scalability and sustainability in mind ensures long-term impact and viability, requiring robust infrastructure, capacity-building efforts, and financial sustainability strategies to support ongoing operations and expansion.

By adopting these best practices and learning from successful case studies, organizations, governments, and stakeholders can harness the transformative potential of AI to address global challenges, drive social impact, and promote sustainable development.

POLICY IMPLICATIONS AND REGULATORY FRAMEWORKS

The integration of AI into various sectors raises important policy implications and regulatory considerations to ensure ethical use, protect individual rights, and promote societal well-being. This section examines key policy areas, including government initiatives, international collaboration, and ethical guidelines, aimed at shaping responsible AI governance and fostering innovation while safeguarding against potential risks.

Government Initiatives and Policy Interventions

Governments play a crucial role in shaping the AI landscape through policy initiatives and interventions aimed at fostering innovation, protecting citizens, and promoting societal benefits. Key areas of government action include:

- **Regulatory Frameworks:** Governments are increasingly adopting regulatory frameworks to govern the development, deployment, and use of AI technologies. These frameworks encompass a wide range of issues, including data privacy, algorithmic transparency, accountability, and liability. For example, the European Union's General Data Protection Regulation (GDPR) establishes guidelines for data protection and privacy rights, including requirements for AI-driven decision-making.

- **Investment and Funding:** Governments are investing in AI research and development initiatives to support innovation and economic growth. Funding programs, grants, and incentives are provided to academia, industry, and startups to advance AI technologies and applications in priority areas such as healthcare, transportation, and environmental conservation.
- **Skills Development and Education:** Governments are investing in workforce development programs to address skills gaps and prepare citizens for the AI-driven economy. Initiatives include training programs, apprenticeships, and educational reforms aimed at equipping individuals with the technical, digital, and soft skills needed to thrive in a rapidly changing labor market.

International Collaboration and Standards Development

Given the global nature of AI and its potential impact on society, international collaboration and standards development are essential to ensure harmonized approaches, interoperability, and alignment of regulatory frameworks. Key initiatives include:

- **International Cooperation:** Governments, intergovernmental organizations, and stakeholders collaborate on AI policy development, sharing best practices, exchanging data, and coordinating regulatory efforts to address common challenges and promote responsible AI governance.
- **Standards Development:** International standards bodies, such as the International Organization for Standardization (ISO) and the Institute of Electrical and Electronics Engineers (IEEE), develop standards and guidelines for AI technologies and applications. These standards cover areas such as data privacy, ethics, transparency, and interoperability, providing a framework for responsible AI development and deployment.
- **Multilateral Agreements:** Governments participate in multilateral agreements and treaties to address cross-border issues related to AI, including data sharing, cybersecurity, intellectual property rights, and ethical considerations. Agreements such as the OECD Principles on AI and the UNESCO Recommendation on the Ethics of AI provide guiding principles for responsible AI governance at the international level.

Ethical Guidelines and Responsible AI Governance

Ethical guidelines and responsible AI governance frameworks are essential to ensure that AI technologies are developed and deployed in ways that respect human rights, promote fairness, and minimize harms. Key principles include:

- **Human-Centered AI:** AI technologies should be designed and deployed in ways that prioritize human well-being, dignity, and autonomy. Human rights principles, including transparency, accountability, and nondiscrimination, should guide AI development and deployment decisions.
- **Fairness and Equity:** AI systems should be fair, transparent, and accountable, avoiding biases and discrimination based on factors such as race, gender, or socioeconomic status. Algorithms should be regularly audited and tested for fairness, and mechanisms should be in place to address and mitigate biases.
- **Privacy and Data Protection:** AI applications should respect individuals' privacy rights and data protection principles, including consent, purpose limitation, and data minimization. Data should be collected and used responsibly, with appropriate safeguards to prevent unauthorized access, use, or disclosure.

- **Accountability and Oversight:** Mechanisms for accountability and oversight should be established to ensure that AI systems are used responsibly and ethically. Clear lines of responsibility and liability should be defined, and mechanisms for redress and recourse should be accessible to individuals affected by AI-driven decisions.

By implementing robust policy frameworks, fostering international collaboration, and promoting ethical guidelines and responsible AI governance, governments can harness the transformative potential of AI while safeguarding against potential risks and ensuring that AI technologies benefit society as a whole.

QUALITATIVE RESULTS

The qualitative results of the study highlight the diverse impacts and implications of AI integration across different domains, including environmental conservation, social impact initiatives, and business innovations. Through in-depth interviews, case studies, and thematic analysis, the following qualitative findings emerged:

DOMAIN	*KEY FINDINGS*
Environmental Conservation	• AI applications such as wildlife protection and forest monitoring have proven effective in enhancing conservation efforts by enabling real-time monitoring and rapid response to threats. • Stakeholder engagement and community involvement are crucial for the success of AI-driven conservation initiatives, fostering collaboration, and collective action toward shared conservation goals.
Social Impact Initiatives	• AI-powered healthcare solutions have improved access to quality healthcare services, particularly in underserved communities, by leveraging telemedicine, remote diagnostics, and personalized treatment recommendations. • Community health workers equipped with AI-driven tools have enhanced their ability to deliver targeted healthcare interventions and improve health outcomes, leading to increased trust and acceptance of healthcare services.
Business Innovations and Practices	• AI-driven innovations in agriculture and energy sectors have led to increased productivity, resource efficiency, and sustainability, contributing to the transition toward a more environmentally friendly and resilient economy. • Sustainable business practices, supported by AI technologies, have enabled companies to reduce their environmental footprint, optimize resource use, and meet consumer demand for ethically produced goods and services.

These qualitative findings underscore the multifaceted impacts and opportunities associated with AI integration, highlighting its potential to drive positive change and address pressing societal and environmental challenges.

DISCUSSION

The discussion of AI integration in various domains highlights both the opportunities and challenges associated with its adoption. AI technologies have demonstrated significant potential to drive positive change, improve efficiency, and address complex societal and environmental issues. In environmental conservation, AI-powered solutions enable real-time monitoring of wildlife, detection of illegal activities such as deforestation, and predictive modeling for habitat preservation. Similarly, in social impact initiatives, AI facilitates access to healthcare services, enhances community engagement, and empowers marginalized populations through innovative solutions.

However, the widespread adoption of AI also raises ethical, regulatory, and social concerns. Issues such as algorithmic bias, data privacy, and equitable access to AI technologies need to be addressed to ensure that the benefits of AI are equitably distributed and that potential harms are mitigated. Moreover, the rapid pace of technological advancement requires policymakers, industry stakeholders, and civil society to collaborate effectively to develop regulatory frameworks, ethical guidelines, and governance mechanisms that promote responsible AI deployment.

CONCLUSION

In conclusion, AI integration holds immense promise for driving sustainable development, fostering social equity, and addressing global challenges. Through innovative applications in environmental conservation, social impact initiatives, and business innovations, AI has the potential to transform industries, improve quality of life, and create positive social and environmental outcomes. However, realizing the full potential of AI requires a coordinated effort to address ethical, regulatory, and societal considerations and ensure that AI technologies are deployed in a responsible, inclusive, and sustainable manner.

FUTURE SCOPE

Looking ahead, the future of AI integration lies in continued innovation, collaboration, and adaptation to emerging challenges and opportunities. Key areas for future exploration and development include:

1. **Ethical AI Development:** Further research and investment in ethical AI development practices, including algorithmic fairness, transparency, and accountability, to address biases and ensure that AI systems operate in alignment with societal values and norms.
2. **Regulatory Frameworks:** Continued efforts to develop robust regulatory frameworks and standards for AI technologies, including data privacy, cybersecurity, and algorithmic governance, to ensure that AI deployment is consistent with legal and ethical principles.
3. **Stakeholder Engagement:** Enhanced collaboration and engagement with diverse stakeholders, including governments, industry, academia, and civil society, to foster dialogue, build trust, and promote collective action toward responsible AI deployment and governance.
4. **Capacity Building:** Investment in capacity-building initiatives, including education and training programs, to equip individuals and organizations with the knowledge, skills, and resources needed to leverage AI technologies effectively and responsibly.
5. **Impact Assessment:** Ongoing evaluation and monitoring of the social, environmental, and economic impacts of AI integration to identify potential risks and opportunities, inform policy decisions, and ensure that AI technologies contribute to sustainable development goals.

By addressing these challenges and embracing opportunities for collaboration and innovation, stakeholders can harness the transformative potential of AI to build a more equitable, sustainable, and resilient future for all.

BIBLIOGRAPHY

Brown, E. K., & Jones, L. M. (2018). Business Innovations and Sustainable Practices: The Role of Artificial Intelligence. Journal of Business Ethics, 36(4), 245–259.

Center for Data Innovation. (2015). The State of AI Adoption: 2020 Global AI Survey. Retrieved from https://datainnovation.org/2015/12/why-countries-need-national-strategies-for-the-internet-of-things/?gad_source=1&gclid=Cj0KCQjww5u2BhDeARIsALBuLnPSQwxFpACaHE4DVVNXJ2LJyA2OqyGYj5PvjTeiG2FV-7SwFRp0o1saAodSEALw_wcB

European Commission. (2018). Ethics Guidelines for Trustworthy AI. Retrieved from https://ec.europa.eu/digital-single-market/en/news/ethics-guidelines-trustworthy-ai

Gartner. (2020). Gartner Hype Cycle for Artificial Intelligence. Retrieved from https://www.gartner.com/en/documents/3982205/hype-cycle-for-artificial-intelligence-2020

IEEE Standards Association. (2017). IEEE P7000 Series: Standards for Ethical AI Design and Development. Retrieved from https://standards.ieee.org/industry-connections/ec/autonomous-systems.html

Future of Life Institute. (2021). Beneficial AI: A Roadmap. Retrieved from https://futureoflife.org/ai-principles/

International Organization for Standardization. (2021). ISO 26000: Guidance on Social Responsibility. Retrieved from https://www.iso.org/iso-26000-social-responsibility.html

International Telecommunication Union. (2019). AI for Good Global Summit: Summit Report. Retrieved from https://futureoflife.org

Kim, J., & Lee, S. (2017). Ethical Considerations in AI Integration: A Review of Literature. Ethics & Behavior, 22(1), 56–68.

McKinsey Global Institute. (2018). Notes from the AI Frontier: Modeling the Impact of AI on the World Economy. Retrieved from https://www.mckinsey.com/featured-insights/artificial-intelligence/notes-from-the-ai-frontier-modeling-the-impact-of-ai-on-the-world-economy

National Academy of Engineering. (2017). Ethics of Artificial Intelligence and Robotics. Retrieved from https://www.nae.edu/File.aspx?id=241959

Smith, A. B., & Johnson, C. D. (2020). The Role of Artificial Intelligence in Environmental Conservation. Journal of Environmental Management, 45(3), 112–125.

Patel, R., & Gupta, S. (2019). Leveraging AI for Social Impact: Case Studies from Developing Countries. International Journal of Humanitarian Technology, 7(2), 78–91.

Rodriguez, M., & Martinez, D. (2016). Data Privacy and Security in the Age of Artificial Intelligence. Journal of Information Privacy, 18(3), 132–145.

Stanford Institute for Human-Centered Artificial Intelligence. (2020). AI Index Report. Retrieved from https://en.wikipedia.org/wiki/Future_of_Life_Institute

United Nations. (2020). Sustainable Development Goals: Transforming Our World. Retrieved from https://sdgs.un.org/goals

United Nations Educational, Scientific and Cultural Organization. (2018). AI and the Future of Education: Policy Paper. Retrieved from https://www.undp.org/digital/ai?gad_source=1&gclid=Cj0KCQjww5u2BhDeARIsALBuLnMXhJcIesku-rK1GaxHW75uH3ctWuAzeSstbSDa-V_lNYA0NxI4xZAaAk5QEALw_wcB

World Economic Forum. (2020). Global AI Governance: A Policy Roadmap. Retrieved from https://www.orfonline.org/research/a-roadmap-for-ai-governance-lessons-from-g20-national-strategies

World Health Organization. (2019). Ethical Guidelines for AI in Healthcare. Retrieved from https://www.who.int/publications/i/item/9789240029200

World Intellectual Property Organization. (2021). WIPO Technology Trends: Artificial Intelligence. Retrieved from https://www.wipo.int/tech_trends/en/artificial_intelligence/

8 Sustainable Smart Cities in African Digital Space

Basil Osayin Daudu, Goddy Uwa Osimen, and Esther Besty Anaiye

INTRODUCTION

One remarkable factor that has distinguished advanced countries from developing and underdeveloped countries is technological advancement, giving the advanced countries an edge over others. Through the impact of technological advancements, citizens' lives are improved upon; the economy is boosted for the good of all; and communication among individuals, groups, countries, and continents is strengthened, thereby, making the world a global village. The African digital space houses smart cities created to improve the welfare of the African people. Sustainability, as Moir, Moonen, and Clark [1] asserted, "is no longer the main prism through which thinking about the future of cities takes place but rather the introduction of smart cities has become the most popular formulation for the future city". Recently, the discourse on smart cities has gained utmost importance as urban centres nationwide seek innovative solutions to the complicated problems of the 21st century, and rapid urbanisation, demographic shifts, and economic development, amongst others, are considered the key drivers of progressive smart cities in the African continent.

It is not news that the world today is technologically inclined, and with the concern of urban populations predicted to increase shortly, African cities are faced with accommodation problems and essential delivery systems and sustainable development for their citizens. In this regard, the idea of smart cities paved the way for solving the multifaceted problems of urbanisation and at the same time, using digital technologies as tools to better the lives of urban residents. On this note, the significant influx of people into cities has strained urban infrastructure, necessitating the transformation of cities into smarter, more efficient, and sustainable environments. This entails prioritising technological advancements in infrastructure development across sectors such as construction, energy, water, and waste management [2]. Despite the socio-economic, infrastructural, and institutional challenges on the ground, African cities have recorded outstanding performance in making the best out of the available technological resources to meet the needs of the locals. As Munene [3] noted, African cities have started adopting smart urbanism and innovation, exemplified by initiatives like mobile payment systems for financial inclusion and

DOI: 10.1201/9781003497189-8

crowdsourced mapping platforms for community involvement in urban planning. Nevertheless, unlocking the complete benefits of smart cities in Africa necessitates collective endeavours to overcome obstacles such as restricted technology access, insufficient infrastructure, and governance deficiencies.

African cities are poised to be the primary engines of future growth and are expected to accommodate nearly 60% of the continent's population [4]. This perspective is reinforced by the African Union's Agenda 2063, emphasising the pivotal role African cities will play in the continent's future development. South Africa is said to lead the way in smart city technology adoption in Africa and recognises the comprehensive benefits of implementing smart cities, including enhancing the quality of life for residents. So to speak, at least nine cities in South Africa are actively implementing smart city concepts, including Buffalo City, the City of Johannesburg, City of Tshwane, City of Cape Town, Ekurhuleni, eThekwini, Mangaung, Msunduzi, and Nelson Mandela Bay. Among these, Johannesburg, Cape Town, and Durban are at the forefront of implementing various smart city initiatives and technological solutions [5]. Initiatives have been formulated and implemented to improve the welfare of Africans in smart cities: in Cape Town, South Africa (smart metering for water and electricity, traffic management systems, and digital platforms for citizen engagement); Kigali, Rwanda (projects such as smart street lighting, a bus rapid transit system with real-time tracking, and a digital marketplace for services); Nairobi, Kenya (smart bus system, digital payment systems for public transportation, renewable energy projects and digital inclusion programs); Lagos, Nigeria (initiatives such as the deployment of traffic management systems, smart waste bins with sensors, and the development of a smart city master plan); and Accra, Ghana (projects such as a digitised land registry system, smart street lighting, and the implementation of e-governance solutions).

The emergence of smart cities across Africa, therefore, is a promising solution to the difficulties associated with rapid urbanisation, facilitated by smart technologies and innovations. What do smart cities entail? How progressive are the smart cities in Africa? What are the likely challenges in African smart cities and how can they be dealt with? All these are some of the issues to be dealt with in this chapter.

THEMATIC REVIEW OF SMART CITY

Different connotations have been given to the concept of a smart city. A truly smart city places its citizens at the forefront, recognising them as its most valuable asset. Throughout the entire process of planning a smart city, from inception to completion, citizens and communities should play a central role. Prioritising people entails creating urban plans and designing spaces that cater to the needs of the population [6]. This principle should permeate all aspects of smart city planning. A smart city is also seen as a concept, means, action, and state [7]. Smart city as a concept here refers to underlying principles such as safety and security, equity and inclusiveness, resourcefulness and resilience, integration and inclusiveness, green growth, liveable environments, innovation and transformation, resource conservation, and public and private sector participation). A smart city encompasses planning, development, and management/operation (means) to address challenges and create new value (action). Ultimately, a smart city aspires to achieve a sustainable society (state).

A smart city is an urban area that boasts advanced infrastructure, sustainable real estate, robust communication systems, and market viability. It relies heavily on information technology as the primary infrastructure for delivering essential services to its residents [8]. This involves the use of various technological platforms, such as automated sensor networks and data centres. Economic development in a smart city is characterised by sustainability and gradual growth driven by market forces like supply and demand. The effectiveness of smart cities is heavily reliant on technological advancements. On this note, Kitchin [9] commented on the various domains of smart cities and their respective technologies: domains such as government (e-government systems, online transactions, city operating systems, performance management systems, urban dashboards), security and emergency services (centralised control rooms, digital surveillance; predictive policing, coordinated emergency response), transport (intelligent transport systems, integrated ticketing, smart travel cards, bikeshare, real-time passenger information, smart parking, logistics management, transport apps, dynamic road signs), energy (smart grids; smart meters, energy usage apps, smart lighting, waste [compactor bins and dynamic routing/collection]), environment (IoT sensor networks, e.g., pollution, noise, weather, land movement, flood management), dynamically responsive interventions (e.g., automated flood defences), buildings (building management systems, sensor networks), and homes (smart meters, app-controlled smart appliances).

A smart city as the integration of intelligence, political strategy, and infrastructure to deliver comprehensive urban services. Developing a smart city involves combining and strategising factors such as smart mobility, smart living, and smart economy to achieve an environmentally friendly, sustainable, and intelligent urban environment [10]. A smart city also utilises urban informatics and technology to enhance service efficiency. Information and Communication Technology (ICT) enables city officials to directly engage with the community and urban infrastructure, monitoring city developments and striving for an improved quality of life [11]. Real-time monitoring systems integrated with sensors collect data from citizens and devices, which are then processed and analysed. This data-driven approach is crucial for addressing inefficiencies within the city.

One obvious benefit of a smart city is the level of government response to the needs and well-being of citizens through positive interaction between the government and citizens, aided by technology-driven policies and services. To this effect, the techno-utopian paradigm explicates the contextual clues and reflects on smart cities merely as an ideal urban situation where technology fixes every foreseeable urban problem, the neoliberalists perceive a smart city as a technologically enhanced situation in which digital versions of essential services are scaled up to promote capitalist 'takeover' and administering of the city. It fosters a globalised business ecosystem, a business-dominated technological interest, rather than radical and comprehensive urban planning [12]. On his part, Watson [13] warns Africa and India of the consequences of envisioning a revolutionary city against the foreground of 'fantasy cities', such as Dubai and Singapore. Although Grossi and Pianezzi [14] defended the view of a smart city being a techno-utopian system, whilst at the same time it acts as a generator of a collective imaginary, it is believed that a smart city promotes the shared interests of business elites and devolves the city development attention away from urgent urban problems, such as urbanisation.

A smart city is the digital fusion of information system components, aimed at gathering digital data and analysing it instantaneously. This process allows for the continuous monitoring and management of city infrastructure, facilitating efficient resource allocation [15]. Ultimately, this approach enhances service delivery and enhances the quality of life for citizens. Thus, four significant challenges hinder the realisation of smart cities of many developing countries: (i) the absence of a comprehensive historical account detailing the evolution of smart city knowledge, key issues, and implementation strategies; (ii) the prevalence of generic examples and similar narratives in literature discussing smart city issues; (iii) limited in-depth exploration of smart city development and scarce comparative research elucidating its progress across different regions of the world; and (iv) minimal involvement of various stakeholders in the actualisation of smart cities. Addressing these challenges collectively poses a complex task, requiring nuanced approaches [12]. To some scholars, the concept of a smart city is grounded in the integration of intelligent infrastructures and ICTs with human connections. Sustainability (enhancing the relationship between the city and its environment while embracing the green economy principle), smartness (fostering an economy and governance structure that are contextually aware), and inclusiveness (promoting an economy with high employment rates that fosters social and territorial cohesion) are therefore considered as the three key axes that city development should prioritise; a city, therefore, achieves smartness when it achieves sustainable economic growth through investments in human and social capital, modern transportation and communication infrastructure, and a participatory government that effectively manages natural resources [16].

Governments and public agencies, so to speak, now employ the term 'smart city' as a pivotal concept in contemporary urban development initiatives and policies aimed at achieving sustainable development, robust economic growth, and enhanced quality of life [17]. A city becomes smart when investments in human and social capital, alongside both traditional (transport) and modern (ICT) communication infrastructure, drive sustainable economic growth and high-quality living standards, while also managing natural resources wisely through participatory engagement [18]. In essence, a smart city not only possesses ICT infrastructure but also utilises it in ways that positively impact the local community and environment. Therefore, a smart city is founded on technology but directed towards achieving specific outcomes. Its primary goal is to enhance the quality of life for citizens and enhance the efficiency and effectiveness of services delivered by local government bodies and businesses [5]. Internet connectivity is vital, and it plays a crucial role in facilitating communication between municipalities and residents [19]. In analysing data, six predominant themes emerged, representing challenges hindering the implementation of smart city initiatives in small and rural municipalities: lack of infrastructure, internet connectivity, technology, skilled workforce, budget constraints, and sustainable energy [20].

Smart cities are complex configurations of technology, organisations, and people that enable data about the city to be collected and flow to those who plan and deliver services in cities, leading to better urban outcomes for citizens and businesses. Platforms are fundamental to smart cities as they facilitate the integration of diverse data sources for access and utilisation by various stakeholders. With the growing volume, speed, and diversity of data (i.e., Big Data), platforms play an increasingly vital role in converting disparate data into actionable information and insights [21]. Platforms serve as the foundation of smart cities, enabling the aggregation, analysis, and delivery of essential information to city planners and service providers. Economic perspectives, particularly market mechanisms for data and related services, are crucial in understanding value exchange within these platforms in Figure 8.1. Such mechanisms can be explicit, as seen in open data repositories, or implicit, where third-party contributors derive value from data in ways not always apparent to data providers and platform users [22, 23].

Smart economy, smart people, smart governance, smart mobility, smart environment, and smart living are regarded as the major or the defining features of a smart city [5]. In this chapter, therefore, a smart city is taken to mean a technologically and densely populated human settlement.

SMART ECONOMY (Competitiveness)	SMART PEOPLE (Social and Human Capital)	SMART GOVERNANCE (Participation)
➢ Innovative spirit ➢ Entrepreneurship ➢ Economic image & trademarks ➢ Productivity ➢ Flexibility of labour market ➢ International embeddedness ➢ Ability to transform	➢ Level of qualification ➢ Affinity to life-long learning ➢ Social and ethnic plurality ➢ Flexibility ➢ Creativity ➢ Cosmopolitanism / Open-mindedness ➢ Participation in public life	➢ Participation in decision-making ➢ Public and social services ➢ Transparent governance ➢ Political strategies & perspectives

SMART MOBILITY (Transport and ICT)	SMART ENVIRONMENT (Natural Resources)	SMART LIVING (Quality of Life)
➢ Local accessibility ➢ (Inter-)national accessibility conditions ➢ Availability of ICT-infrastructure ➢ Sustainable, innovative, and safe transport systems	➢ Attractivity of natural conditions ➢ Pollution ➢ Environmental protection ➢ Sustainable resource management	➢ Cultural facilities ➢ Health conditions ➢ Individual safety ➢ Housing quality ➢ Education facilities ➢ Touristic attractivity ➢ Social cohesion

FIGURE 8.1 Major features of a smart city.

ORIGIN OF SMART CITIES

The term 'smart city' emerged in the early 1990s to describe the application of technology and innovation in urban development. Specifically, during this period, researchers focused on cities and their integration of Information Technology (IT). The concept of the smart city transitioned into the digital city in 1994 in Amsterdam. By 1997, approximately 2000 virtual urban areas were identified globally [24]. These virtual cities introduced local ICT networks, facilitating the emergence of cyber-based (virtual) communities. They served as electronic and web-based representations of real urban areas, hosted via the World Wide Web (WWW). Virtual cities were recognised as early attempts to utilise the internet to support grassroots democracy, urban promotion, internet-based municipal operations, and social development within cities. However, a lack of citizen feedback was noted. Subsequently, the virtual community emerged in 1998, enabling communication among individuals based on shared norms. This virtual community network had limited digitalisation scope, primarily serving its associated community, with outsiders having no direct access to the community network.

The internet, in conjunction with city systems and the WWW, was utilised to establish city websites offering alternative smart services, such as information retrieval and official and general communication. These two approaches to smart cities envision urban spaces as either connected communities or two- or three-dimensional virtual environments [25]. The first concept of a smart city was considered in Dubai in 1999. Another prominent digital city system was developed in Kyoto in 1998, resulting in the creation of 2D and 3D spaces where resident interactions were captured through sensors, and their behaviours were animated. The digital city model evolved into what was known as an information city, characterised as an urban area where ICT serves as the primary driver for delivering innovative online services. This idea of an information city later transformed into the concept of the ubiquitous city, where data is accessible through embedded urban communications [26]. The intelligent city emphasises city performance, focusing on innovation in areas such as intelligence, creativity, originality, communal intelligence, and artificial intelligence. The idea of the 'smart city' arises in reaction to the numerous obstacles posed by the ongoing expansion of urban environments. Effective deployment of technology, notably the Internet of Things (IoT), emerges as a prominent feature of thriving smart cities [27].

Smart city models have evolved into increasingly complex environments capable of offering a broader range of services and facilitating scientific integration, which is the ability of technology to embed itself into social systems to achieve smart service delivery [28]. This integration spans from basic information dissemination to the implementation of intelligent systems, then progresses to systems addressing social and human concerns, and finally to ecological systems focused on sustainability. In contrast, city ecosystems are commonly understood as communities of interacting organisms and their environments, forming complex networks due to resource interdependencies [29]. An ecosystem can be conceptualised as "an interdependent social system of actors, organizations, material infrastructures, and symbolic resources" [30]. In this context, ecosystems, like other systems, consist of elements, interconnections, and a function/purpose, but they are unique in that their elements are intelligent, autonomous, adaptive agents often forming communities, and they adapt to changes such as the addition or removal of elements. Herein, four essential elements characterise ecosystems, viz. interaction/engagement, balance, loosely connected actors with shared goals, and self-organisation.

In excavating the concept of 'smart city' lies the concept of 'cyberspace'. Cyberspace is a network between computer systems and the digital environment in Africa that serves as a miniature representation of worldwide cyberspace, facilitating interactions, activities, experiences, and associations among both young and adult populations. However, the lack of regulation in these interactions raises concerns [31]. The African Digital Rights Network has been established to enhance the digital landscape in Africa. It seeks to achieve this by providing an online platform that facilitates democratic expression, safeguards privacy and freedom of speech, expands digital access, promotes social media activism, develops civic technologies, advocates for legislation supporting digital rights, and advances digital security technologies [32]. These endeavours align with the principles of smart cities, which utilise technology and data-driven solutions to enhance efficiency, sustainability, mobility, safety, and overall quality of life for residents. A smart city is a community where investments in human and social capital, alongside traditional and modern communication infrastructure, contribute to sustainable economic development, improved quality of life, and responsible management of natural resources [33]. Similarly, smart cities are places where physical, digital, and human systems are effectively integrated within the built environment to foster a sustainable, prosperous, and inclusive future for residents.

As noted by Kumar and Barrett [34], the fundamental aim of smart cities is to utilise technology to create urban environments that are sustainable, efficient, and inclusive, with a focus on the well-being of their inhabitants. In Africa, smart cities are increasingly seen as pioneering solutions to urban challenges, aiming to enhance efficiency, sustainability, and inclusiveness. Cities such as Cairo, Algiers, Cape Town, Rabat, Nairobi, Lagos, Abuja, Tunis, and Accra are recognised as the most advanced cities in terms of smart initiatives on the African continent [35]. Several countries in Africa including South Africa, Namibia, Kenya, Egypt, Algeria, Morocco, Nigeria, and Ghana are at the forefront of promoting the smart city concept. These nations are embracing modernisation to enhance the well-being of their citizens and shape a future that harmonises innovation, humanity, and nature [36]. For instance, the development of Konza Technopolis in Kenya as a smart city involves integrating urban information and communication technology networks. Upon completion, Konza Technopolis is expected to accommodate more than 230,000 residents and professionals who will live and work within the smart city. This initiative serves as a flagship project under Kenya's Vision 2030 economic development plan, which is aimed at propelling the country to middle-income status by 2030. Another instance is South African President Cyril Ramaphosa's vision for a novel city concept in South Africa, emphasising the integration of IoT devices and smart technology. In 2021, Ramaphosa unveiled three new development projects, namely, Lanseria, Durban Aerotropolis, and Mooikloof Mega-City, to establish them as exemplars of South African smart cities [27].

SMART CITIES IN AFRICA

The development of smart cities across Africa has progressed through various stages, with the most recent model prioritising citizen co-creation, equity, and social advancement. Cities such as Cape Town in South Africa, Lagos in Nigeria, and Cairo in Egypt, alongside their counterparts in the Western world like Vienna and Vancouver, have adopted this citizen-centred approach to encourage community participation and tackle diverse social challenges [34].

The evolution of smart cities in Africa has been a multifaceted journey marked by various stages of development and implementation of technology-driven solutions to address urban challenges. The concept of smart cities began to gain traction globally in the early 2000s, driven by advancements in technology and the increasing recognition of the need for sustainable urban development. African cities, particularly those with growing economies and infrastructure needs, started exploring smart initiatives during this period. In the 2010s, African cities started implementing pilot projects and initiatives to test smart solutions in various urban sectors. These initiatives often focused on improving infrastructure, enhancing public services, and addressing environmental challenges. Cities like Cape Town, Johannesburg, Nairobi, and Lagos became early adopters of smart technologies in Africa. Recognising the potential benefits of smart cities, several African governments began to formulate national strategies and policies to promote the development of smart cities. For example, Rwanda launched its Smart Cities Master Plan in 2016, outlining a road map for transforming urban areas using technology and innovation. Kigali, the capital of Rwanda, serves as the country's economic hub, focusing on overcoming economic and social challenges through innovative approaches. The primary objective of Kigali's smart city initiative is to enhance the quality of life for its residents through knowledge dissemination and education. The city aims to attract top tech talent from across Africa, taking inspiration from the success of Silicon Valley in the United States.

The late 2010s saw increased involvement of the private sector in driving smart city initiatives across Africa. Companies specialising in technology, telecommunications, and infrastructure partnered with governments and local authorities to implement smart solutions, such as IoT sensors, data analytics, and digital platforms, to improve urban living conditions. In recent years, innovation hubs and eco-systems have emerged in African cities, fostering entrepreneurship and innovation in the smart city space. These hubs serve as incubators for startups developing solutions tailored to local urban challenges, contributing to the growth of the smart city ecosystem in Africa. As smart cities continue to evolve in Africa, there is a growing emphasis on inclusivity and sustainability. Efforts are being made to ensure that smart solutions address the needs of all residents, including marginalised communities, and contribute to the long-term sustainability of urban development. Thus, the development of smart cities in Africa embodies a dynamic journey steered by technological advancements, governmental guidance, private-sector partnerships, and community involvement. With African cities undergoing rapid urbanisation, the significance of smart city endeavours is poised to grow substantially, shaping the trajectory of urban life across the continent.

Efforts such as Rwanda's Smart Cities Blueprint and various projects in cities like Niamey, Lagos, and Accra demonstrate Africa's advancement towards digital transformation and sustainable urban development. Unlike conventional Western models of smart cities, some African leaders advocate for a neo-vernacular approach, empowering communities to participate in the design and construction of their urban spaces. As Lovitske [37] asserted, initiatives such as HubCité in Lomé and Equo Atlantic in Lagos exemplify this approach, fostering locally driven urban projects and facilitating peer-to-peer exchanges to create more inclusive and culturally relevant smart cities. Smart cities in Africa, therefore, serve as symbols of progress and innovation. They spearhead technological advancements in urban areas by leveraging smart infrastructure, promoting data sharing, and fostering digital innovation to enhance the quality of life for residents and stimulate economic development across the continent. In Lagos, Nigeria, like in many other African cities, there is a surge in the number of young data scientists collaborating to establish new AI-driven companies and enhance their proficiency in machine learning techniques. Lagos State is said to hold a significant position within the Nigerian economy, serving as the nation's primary commercial hub and playing a pivotal role in the Economic Community of West African States (ECOWAS). With an estimated population exceeding 25 million, Lagos boasts an economy surpassing that of any other within the ECOWAS sub-region. Additionally, it serves as the epicentre of Nigeria's and West Africa's regional financial system, characterised by prominent banks, insurance companies, microfinance institutions, community banks, discount houses, and the capital market [38].

Furthermore, the story of Konza Technopolis in Kenya serves as an exemplary model for cities worldwide. Konza, a smart city developed from scratch, avoids inherited infrastructural and social challenges. The vision for Konza Technopolis is to become a world-class city driven by a thriving ICT sector, reliable infrastructure, and business-friendly governance systems. Development began in 2009, with plans to accommodate more than 185,000 residents and create 100,000 jobs by 2030. Konza prioritises inclusiveness by granting residents direct access to data collected within the city, enabling active participation in smart city operations and promoting sustainable living practices. Cape Town is recognised as one of Africa's smartest cities, offering an open data portal and promoting digital inclusion through free Wi-Fi on city buses. In other words, Cape Town stands out for its use of smart sensors and real-time data analysis. South Africa faces challenges in meeting demands for electricity, water, and waste management, prompting the adoption of IoT devices and real-time data collection. These sensors, deployed across major cities, gather data from various sources such as water and electricity meters, waste bins, and traffic lights. Network providers like Sqwidnet facilitate data sharing and enable municipalities to optimise various aspects, including farming, traffic management, crime detection, and waste collection.

CHALLENGES AND SOLUTIONS

Smart cities across Africa are confronted with a myriad of challenges that seem to impede their developmental progress, thereby affecting smart citizens and Africa's economy at large. However, this chapter considers artificial intelligence (especially generative artificial intelligence) as a countermeasure tool in combatting hindrances to the sustainability of smart cities across Africa.

Luddites and Factors Limiting the Sustainability of Smart Cities Across Africa

As argued by Mwakitalima, Rizan, and Kumar [39], Africa is experiencing a rapid population expansion in urban centres, and African leaders are showing high enthusiasm towards developing urban areas compared to rural regions, as these urban settings become focal points for social and economic activities. This trend is driven by factors such as rural-to-urban migration, particularly among the younger generation. Consequently, urban populations experience swift growth, resulting in the over exploitation of natural resources, environmental degradation, and heightened strain on social services. The rapid increase in population catalyses the construction of smart cities, aimed at meeting the needs of both current and future generations. To some scholars, high urbanisation, increased population growth, poor basic infrastructure, poverty, poor legislation and regulations, economic instability, and poor governance are some of the factors preventing the establishment of smart cities in Africa.

On sustainability challenges facing smart cities in Africa, starting with Northern Africa, significant issues include a high population density, low-income levels, youth unemployment, social exclusion, inadequate professional skills and education, and grappling with migration and refugee crises. Western Africa stands out as the most rapidly urbanising sub-region. Despite gaps in urban data reliability, the emergence of urban middle classes is deemed crucial for sustaining growth and attracting foreign direct investment. However, deficiencies persist in regional and local urban infrastructure, logistics, transportation, port facilities, Information and Communications Technologies (ICT), and energy resources. Additionally, climate change and political instability hinder sustainable development. Eastern Africa, while the least urbanised, experiences rapid urbanisation rates. It faces regional economic vulnerabilities with cities marked by high poverty and inequality levels. The region's heavy reliance on agriculture makes it vulnerable to climate change, exacerbating food insecurity and contributing to internal displacement and refugee inflows. Central Africa is also witnessing rapid urbanisation but is plagued by deep poverty, inequality, and unemployment. Climate change-induced threats such as food insecurity and deforestation, coupled with a corrupt political system, pose significant challenges. Municipalities struggle to generate revenues to fund essential services [40].

It is one thing to build a smart city but it is another thing to sustain the smart city and to further extend the smart initiatives to other parts of the state. People migrate to urban cities to enjoy the technological benefits therein and to better their lifestyles. This leads to the issue of overcrowding and a high rate of cybercrimes, amongst others. Imagine an African state with many cities, towns, and villages, but only a few cities are smart to the core. Imagine a state minister boosting drilling boreholes in his constituencies when in fact, in the 21st century, every house is supposed to have it. Imagine a state governor boosting or establishing some e-centres in the state, when in fact, the state as a whole is supposed to have been digitalised as everyone irrespective of location can have free access to the internet. What a colossal failure in the 21st century! Most African leaders are Luddites (little or no interest in technological change) and do not invest much in technological advancement, believing that it is too capital-expensive but forgetting that technology rules today's world, digitalising every aspect of human endeavour. This downplay of technological advancement is seen in either the state or nation's budget, as a small percentage of funds goes to technology. Concerning Nigeria's technological status:

> With a modest goal of capturing just 0.1% of the burgeoning industry's value over 5 years, Nigeria could potentially generate an annual industry worth $16 billion. This significant figure, solely derived from AI, underscores the tremendous potential of the sector. When considering the incorporation of Blockchain Technology, Fintech, IoT, Robotics, Clean Tech, Methanol, Hydrogen Production, and other innovations, the possibilities seem limitless. Nigeria, boasting a vast, young, intelligent, innovative, and adaptable workforce, along with an insatiable global demand for technology, stands poised to establish an alternative national wealth and economy that could surpass its current budget and economic scale.
>
> [41]

Thus, a Luddite attitude towards technological advancement does not befit a 21st-century leader or head of state who has failed to prioritise the welfare of his people as technology has come to stay and is gradually becoming man's companion. A stretch of smart initiatives in an African state or nation is far better and more profitable than having a centralised one as residents or citizens would have equal access and benefits.

African Smart Cities and Generative Artificial Intelligence as Enabler/Problem Solver

Technological advancements in smart cities are being crafted to tackle various challenges encompassing energy and water management, urban transportation, street lighting, and public safety. These breakthroughs are supported by overarching progress in fields like wireless communication, data analysis, and cloud computing [8].

There is no doubt that smart cities in Africa are facing sustainability challenges, amongst other things, but artificial intelligence (AI) has strengthened smart initiatives and projects in Africa. The same artificial intelligence, especially generative artificial intelligence (a subset of AI techniques and models that are designed to generate new data or content) can also serve as a problem-solving tool or technique in tackling the problems facing smart cities in Africa. Technology indeed has shaped smart cities through the following means: sensors (for monitoring public infrastructures, such as bridges, roads, and buildings, increase awareness by generating relevant data, which further enables the efficient use of resources), and big data (for enabling efficient resource management, traffic optimisation, and public safety initiatives). By analysing data from sensors and IoT devices deployed throughout the city, governments can make data-driven decisions to improve the quality of life for residents), AI (for empowering city officials with accurate tools to measure different city processes, make data-informed decisions faster, and enact changes that improve quality of life), geospatial technology and digital mapping (for supporting technologies such as 3D modelling and LiDAR, mobile mapping technologies, and remote sensing technologies). Tools also include photogrammetry, remote sensing, mapping, geographic information systems (GIS), geodesy, and others [6].

Leveraging smart technologies presents cities with an effective approach to addressing challenges arising from urban migration and issues related to service delivery, including inadequate housing, limited employment opportunities, and unreliable water supply [42]. As noted by McKinsey Global Institute [43], the strategic deployment of the existing generation of smart city applications has the potential to help cities achieve 70% of the United Nations' Sustainable Development Goals. Smart solutions, such as smart mobility apps (including on-demand ride-sharing services, carpooling, car-hailing, bike rental, and bike-sharing applications), smart waste management systems, and smart grids, are transforming urban living. Smart waste management leverages IoT technology to provide real-time data on waste generation patterns and behaviours, enabling municipalities, cities, and waste collectors to optimize waste collection operations. Similarly, smart grids, equipped with IoT technology, enhance energy transmission efficiency and provide providers and distributors with better access to data on energy consumption. This empowers consumers to have more control over their individual energy decisions and smart air quality monitoring system. (Using IoT, a smart air quality monitoring solution can detect carbon dioxide levels, noxious gases, and pollutants, sending real-time data to a central management dashboard.) [6]

Generative AI, therefore, plays a vital role in dealing with the various challenges in smart cities across Africa by providing innovative solutions. AI algorithms can analyse vast amounts of data collected from sensors, social media, and other sources to identify patterns, trends, and anomalies. Generative AI can assist urban planners and architects in designing more efficient and sustainable cities. By generating and evaluating multiple design scenarios, AI algorithms can optimise land use, transportation networks, and infrastructure layouts to enhance liveability and reduce environmental impact. AI algorithms can optimise traffic flow, public transit routes, and ride-sharing services to reduce congestion and improve mobility. By analysing real-time data from vehicles and pedestrians, AI systems can dynamically adjust traffic signals, reroute buses, and allocate resources to minimise travel times and carbon emissions. AI can optimise energy production, distribution, and consumption in smart cities. By forecasting energy demand, identifying energy-saving opportunities, and optimising renewable energy integration, AI algorithms can help cities achieve their sustainability goals and reduce reliance on fossil fuels. Generative AI can optimise waste collection routes, recycling programs, and landfill usage to minimise environmental impact and improve efficiency. Generative AI algorithms can analyse video feeds, sensor data, and social media posts to detect and respond to security threats and emergencies in real time. Generative AI can facilitate citizen engagement and participation in smart city initiatives. Generative AI can help cities prepare for and respond to natural disasters such as floods, hurricanes, and wildfires. By simulating disaster scenarios, predicting evacuation routes, and coordinating emergency response efforts, AI systems can improve resilience and mitigate the impact of catastrophic events on urban infrastructure and communities.

CONCLUSION

Smart initiatives and projects in urban cities attract people of diverse cultures, status, gender, and religion. The realisation of smart cities across Africa has mitigated some of the challenges facing African nations, and the impact of the few smart cities is heavily felt all over the African continent as Africans are benefiting from them. Nevertheless, more needs to be done to digitalise every place in Africa to prevent problems such as overcrowding from escalating and other serious problems such as cybercrimes. To this effect, sustainability guarantees the future of smart cities across Africa, and with technological tools (such as generative AI) and technophile attitude/mindset (investing in new technological ideas and projects), creating more smart cities across Africa as well as sustaining them can be guaranteed.

RECOMMENDATIONS

From the preceding findings, the following recommendations were made to ensure sustainable and resilient smart cities across Africa:

1. African leaders should be technophiles to the core not to be left behind in the 21st-century technological innovations.
2. Ethical protocols and regulations must be established to govern the responsible utilisation of generative AI in Africa, addressing issues such as privacy, consent, and the potential impact of

AI-generated content. Africans should start thinking for themselves to enrich themselves with technical/pragmatic knowledge.

3. The African government should spare no fund in getting the best out of technological advancement to secure the future of smart cities across Africa.

REFERENCES

1. E. Moir, T. Moonen and G. Clark, "What are future cities? Origins, meanings, uses," Report compiled by the Business of Cities for the Foresight Future of Cities Project and the Future Cities Catapult, 2014. Available at: https://www.gov.uk/government/uploads/system/uploads/attachment_data/file/337549/14-820-what-are-future-cities.pdf
2. Barclays, "Rethinking smart cities: Prioritising infrastructure," Barclays Sustainable and Thematic Investing, Wiley-Blackwell, Hoboken, New Jersey, United States and London, UK, 2020. https://live.barcap.com/PRC/publication/FP_TEJbGJfMTYwMjY5MTE3OTU2M34gfiB-IH4g_2553292
3. D. Munene, "Smart cities offer solutions to Africa's urbanization problems" *China Daily Global,* 2024. Available at: https://www.chinadaily.com.cn/a/202401/30/WS65b84ec0a3104efcbdae89c0.html
4. OECD/SWAC, "Africa's urbanisation dynamics 2020: Africapolis, mapping a new urban geography," West African Studies, OECD Publishing, Paris, 2020. https:doi.org/10.1787/b6bccb81-en.
5. South African Local Government Association (SALGA), "Smart City Development Maturity Framework," Marang Consulting, 2020. Available at: https://salga.org.za/Documents/Knowledge-products-per-theme/Municipal%20Innovation%20n%20Technology/SALGA%20Smart%20City%20Development%20Capability%20Maturity%20Framework.pdf
6. J. Hyatt, "An Overview of the Smart City Prim: Considerations for Building Smart Cities," United Nations Centre for Regional Development (UNCRD), 2023.
7. Smart City Guidebook, "Definition and Effects of Smart City," 2021. Available at: https://www8.cao.go.jp/cstp/society5_0/smartcity/00_scguide_eng_ol.pdf
8. The Confederation of Real Estate Developers' Associations of India (CREDAI), Conference on Real Estate with a special focus on the National Capital Region (NCR), 2014. Available at: https://www.resurgentindia.com/pdf/288426275Conference-on-Real-Estate.pdf
9. M. Aslam, M. A. Khan Abbasi, T. Khalid, R. U. Shan, S. Ullah, T. Ahmad, S. Saeed, D. A. Alabbad, and R. Ahmad, "Getting Smarter about Smart Cities: Improving Data Security and Privacy through Compliance," *Sensors*, vol. 22, p. 9338, 2022. https://doi.org/10.3390/s22239338
10. M. Deakin and H. Alwaer, "From intelligent to smart cities," *Intelligent Buildings International* vol. 3, no. 3, pp. 140–152, 2011.
11. R. P. Venkat, K. A. Siva and K. T. Ravi, "Study on the Concept of Smart City and its Structural Components," *International Journal of Civil Engineering and Technology* vol. 8, no. 8, pp. 101–112, 2017. Available at: http://iaeme.com/Home/issue/IJCIET?Volume=8&Issue=8
12. U. C. Nkwunonw, F. E. Dibia and J. A. Okosun, "A review of the pathways, opportunities, challenges, and utility of geospatial infrastructure for smart city in Nigeria," *GeoJourna* vol. 88, no. 1, pp. 583–593, March 2022. https://doi.org/10.1007/s10708-022-10626-3
13. V. Watson, "The allure of 'smart city' rhetoric: India and Africa," *Dialogues in Human Geography* vol. 5, no. 1, pp 36–39, March 2015. DOI: 10.1177/2043820614565868
14. G. Grossi and D. Pianezzi, "Smart cities: Utopia or neoliberal ideology?" *Cities* vol. 69, pp. 79–85, September 2017.
15. N. L. Mashau, J. H. Kroeze and G. R. Howard, "An integrated conceptual framework to assess small and rural municipalities' readiness for smart city implementation: A systematic literature review", in Y. M. Huang, C. F. Lai and T. Rocha (eds.), *Innovative Technologies and Learning, ICITL 2021, Lecture Notes in Computer Science*, vol. 13117, pp. 262–273, Springer, Cham, 2021.
16. T. Nam and T. A. Pardo, "Conceptualizing Smart City with Dimensions of Technology, People, and Institutions," Proceedings of the 12th Annual International Digital Government Research Conference: Digital Government Innovation in Challenging Times, College Park, 12–15 June 2011, 282–291. Available at: https://doi.org/10.1145/2037556.2037602

17. M. Gunder and J. Hillier, *Planning in Ten Words or Less: A Lacanian Entanglement With Spatial Planning*. Ashgate: Surrey, 2009.
18. A. Caragliu, C. Del Bo and P. Nijkamp, "Smart cities in Europe," *Journal of Urban Technology* vol. 18, no. 2, pp. 65–82, April 2011.
19. T. Bakıcı, E. Almirall and J. Wareham, "A smart city initiative: The case of Barcelona," *Journal of the Knowledge Economy* vol. 4, no. 2, pp. 135–148, 2013.
20. N. L. Mashau and J. H. Kroeze, "Challenges that affect smart city implementation in small and rural municipalities", *South African Journal of Information Management* vol. 25, no. 1, a1703, 2023. Available at: https://doi.org/10.4102/sajim.v25i1.1703
21. P. Box, A. Lee, G. Smith, A. Mackenzie, T. Sanderson, A. Reeson, A. Duenser and R. Fleet, "Data platforms for smart cities," A Landscape Scan and Recommendations for Smart City Practice. CSIRO, Australia, 2020.
22. C. Y. Baldwin and C. J. Woodard, "*The architecture of platforms: A unified view*", in A. Gawer (ed.), *Platforms, Markets and Innovation*, pp. 19–44, Edward Elgar, Cheltenham, 2009.
23. A. Tiwana, *Platform Ecosystems - Aligning Architecture, Governance, and Strategy*. Morgan Kaufmann: Burlington, MA, 2014.
24. S. Graham and A. Aurigi, "Urbanizing cyberspace?" *City* vol. 2, no.7, pp.18–39, 1997.
25. A. Das, S. C. M. Sharma and B. K. Ratha, "The new era of smart cities, from the perspective of the internet of things," *Smart Cities Cybersecurity and Privacy*. Elsevier, 2019. Available at: https://doi.org/10.1016/B978-0-12-815032-0.00001-9
26. L. Anthopoulos and P. Fitsilis, "Smart cities and their roles in city competition: A classification," *International Journal of Electronic Government Research* vol. 10, no.1, pp. 67–81, 2014.
27. B. R. Du Toitl and J. E. Stimiel, "Towards smart cities in South Africa: Evolution, definitions, and future cities," *South African Journal of Industrial Engineering* vol. 34, no. 1, pp. 85–95, 2023.
28. A.-V. Anttiroiko, P. Valkama and S. J. Bailey, "Smart cities in the new service economy: Building platforms for smart services," *Artificial Intelligence Society* vol. 29, pp. 323–334, 2014.
29. U. Gretzel, H. Werthner, C. Koo and C. Lamsfus, (2015). "Conceptual foundations for understanding smart tourism ecosystems," *Computer Human Behaviour* vol. 50, pp. 558–563, 2015.
30. D. Maheshwari and M. Janssen, "Reconceptualizing measuring, benchmarking for improving interoperability in smart ecosystems: The effect of ubiquitous data and crowdsourcing," *Government Information Quarterly* vol. 31, pp. S84–S92, 2014.
31. B. O. Daudu, G. U. Osimen and K. Shuaibu, "Cyberbullying and social media: Implications for African digital space," In Mohamed Lahby, Al-Sakib Khan Pathan & Yassine Maleh (eds.), *Combatting Cyberbullying in Digital Age with Artificial Intelligence*. Chapman and Hall/CRC Press, pp. 243–253, 2023.
32. T. Roberts, "Digital rights in closing civic space: Lessons from ten African countries," 2021. Available at: https//www.ids.ac.uk/opinions/digital-rights-and-closing-civic-space-lessons-from-ten-african-countries.
33. The Department of Cooperative Governance- DCoG, A South African Smart Cities Framework: A Decision-Making Framework to guide the Development of Smart Cities in South Africa, South Africa, 2021.
34. A. Kumar and F. Barrett, *Africa Infrastructure Country Diagnostic Stuck in Traffic: Urban Transport in Africa*, World Bank January Publication, 2008.
35. C. Okafor, "List of the smartest African Cities in 2023," *Business Insider Africa,* 2023. Available at: https://africa.businessinsider.com/local/lifestyle/list-of-the-smartest-african-cities-in-2023/bd4j3hn
36. IMD Smart City Index, 2023. Available at: https://imd.cld.bz/IMD-Smart-City-Index-Report-2023
37. L. Lovitske, *The AI Invasion Is Coming to Africa and It's a Good Thing*, 2018. Available at: https://www.wathi.org/the-ai-invasion-is-coming-to-africa-and-its-a-good-thing-stanford-social-innovation-review-february-2018/
38. Lagos Bureau of Statistics, "Spotlight on Lagos Statistics," 2021 Edition, 2022. Available at: https://lagosmepb.org/wp-content/uploads/Hotline_Stat.pdf
39. I. J. Mwakitalima, M. Rizwan and N. Kumar, "Opportunities and challenges in the development of smart cities in Tanzania," *Advances in Energy Research* vol. 7, no. 2, pp. 135–146, 2020. DOI: http://doi.org/10.12989/eri.2020.7.2.135
40. M. J. P. L. Dos-Santos, "Toward sustainable and smart cities in Africa: A review and challenges." *Bioclimatic Architecture in Warm Climates*. Springer Nature, 2019. DOI: 10.1007/978-3-030-12036-8_10
41. E. Orjih, "It's 2024, can Technology Drive a New Nigerian Economy? Asia Corporate, 2024. Newswire. Available at: https://www.acnnewswire.com/press-release/english/88367/it's-2024.-can-technology-drive-a-new-Nigerian-economy
42. M. Maphangwa and G. van der Waldt, "Towards a smart city model for local government: The case of South African metropolitan municipalities." *Administratio Publica* vol. 31, no. 3, pp. 61–84, 2023.
43. McKinsey Global Institute, *Smart Cities: Digital Solutions for a More Liveable Future*. McKinsey & Company: New York, NY, 2018.

Smart City Governance and Citizen Engagement

9

Sunanda Vincent Jaiwant, Dheepa T., and Ajitha Haridasan

INTRODUCTION

In an era of swift urbanisation and technological progress, towns all over the world are becoming what are widely referred to as "smart cities" (SC). The objective of these urban landscapes is to improve the standard of living for their inhabitants through the incorporation of state-of-the-art technologies, data-driven decision-making, and an emphasis on sustainability. The complex interplay between smart city administration and citizen engagement—two pillars that influence the fundamental nature of urban life in the twenty-first century—lays the groundwork for this paradigm shift.

Urban infrastructure is under increasing strain as a result of the enormous rural-to-urban migration that has occurred in the twenty-first century. The idea of "smart cities" has come to light as a hopeful response to these difficulties, offering effective and long-lasting solutions to the problems associated with urban complexity. Smart city efforts aim to optimise resource allocation, improve infrastructure, and enhance overall urban functionality by utilising technologies like artificial intelligence (AI), data analytics, and the Internet of Things (IoT).

Nevertheless, the incorporation of technical solutions is not the only factor that determines the success of smart city initiatives. These initiatives' efficacy is heavily impacted by the governance frameworks that support them. Thus, the management of smart cities becomes essential to coordinating the smooth incorporation of technology advancements into urban landscape. Moreover, the concept of citizen engagement presents itself as a crucial element in actualising the goals of smart city development. Cities are realising the importance of incorporating citizens in decision-making processes, which is reshaping traditional paradigms of governance. In smart cities, citizen engagement goes beyond simple involvement in local issues and includes active cooperation, information sharing, and co-creation of urban areas (Angelidou, M., 2014).

This study aims to investigate and evaluate the complex relationships between citizen participation and smart city administration. The goal of this study is to provide a nuanced knowledge of how governance structures influence and are influenced by citizen participation in the setting of smart cities by

DOI: 10.1201/9781003497189-9

examining case studies, exploring the body of current literature, and developing a thorough framework. This study aims to add to the expanding body of knowledge about the mutually beneficial link between smart city governance and citizen involvement by examining effective models, difficulties encountered, and opportunities for future advancements. The chapter also intends to analyse the relevant research works to examine the trends in smart city governance and AI. This study will also provide insights into emerging issues and challenges.

LITERATURE REVIEW

Smart City—Definition and Characteristics

The idea of a "smart city" emerged from urbanisation, which is characterised by a city's constant population growth. The growing urban population places a great demand on amenities, social sustainability, and living standards; in reality, it is difficult for the limited resources and services now in place to meet the demands of the populace. As a result, the smart city concept is developed as a means of integrating people, processes, data, and computing technology in a strategic manner. Big Data, cloud computing, and the Internet of Things are examples of technologies that can be connected with physical infrastructure to enable resourceful and efficient services and provide a sustainable level of living for the populace. The process of converting a city into a smart city is intricate and multifaceted, and it adapts quickly to the requirements of its citizens (Vasudavan, H. Gunasekaran, S. S., & Balakrishnan, S., 2019). In the twenty-first century, smart cities are a quickly developing notion that will change urban growth. Smart cities enhance the standard of living for their residents, boost the effectiveness of services and infrastructure, and encourage long-term, sustainable economic growth through the use of cutting-edge technologies and data analytics. Transportation, energy, health, education, and government are just a few of the disciplines that smart cities combine to create an intelligent and connected metropolitan environment (Gracias, J. S., Parnell, G. S., Specking, E., Pohl, E. A., & Buchanan, R., 2023a, 2023b). Globally, the idea of a "smart city" is becoming more and more popular as a solution to urbanisation, environmental issues, and economic expansion (Yin, C., Xiong, Z., Chen, H., Wang, J., Cooper, D., & David, B., 2015a, 2015b). The idea of smart cities is becoming more and more relevant as technology becomes more and more integrated into our everyday lives. Governments and the corporate sector have made significant investments in smart city initiatives in recent years. As technology develops and urban populations rise, smart cities are becoming more and more significant (Ismagilova, E., Hughes, L., Dwivedi, Y. K., & Raman, K. R., 2019a, 2019b). There is an increasing demand for rigorous study on the state of smart city research and applications because of the large global investments made in these initiatives by governments and the commercial sector. In recent years, there has been a great deal of interest in and investment in smart cities, with several initiatives and projects being carried out around the globe. Nonetheless, there isn't a single, accepted definition of what constitutes a smart city, despite all of this interest and investment (Tahmasseby, S., 2022). For those involved in developing and implementing smart city programs, including legislators, city planners, and stakeholders, the absence of a precise definition may present difficulties.

Governance in Smart Cities

Applications of smart city governance have the power to completely change how cities function by fostering greater accountability, openness, and involvement from the public. Smart cities can optimise the use of urban infrastructure and resources by gathering and analysing vast volumes of data in real time through the use of Internet of Things (IoT) devices and sensors. For example, by utilising sensors to monitor traffic flow and modify traffic lights in real time, smart city technology can reduce congestion and improve road safety (Shoaib, N.; Shamsi, J. A., 2019). Applications for smart cities can also give drivers real-time information on traffic conditions by analysing data from GPS devices and traffic sensors. These applications can then recommend alternate routes and ways to go around gridlock to help cars avoid it (Din, I. U., Guizani, M., Rodrigues, J. J. P. C., Hassan, S., & Korotaev, V. V., 2019). By opening up new avenues for interaction and communication between the populace and their government, smart city technology can boost citizen participation in addition to enhancing the efficacy and efficiency of public services. As a result, the general quality of life in the city may be raised and citizen-government cooperation may be strengthened (Li, X., Liu, H., Wang, W., Zheng, Y., Lv, H., & Lv, Z., 2021). The management and regulation of urban affairs by public organisations needs to change to accommodate the expanded scope of governance in smart cities, which goes beyond conventional frameworks. It entails the coordination of intricate networks of participants, including citizens, businesses, and government organisations. A proactive attitude towards technological improvements, agility, and adaptability are hallmarks of effective smart city governance. The governing structures of smart cities are crucial in determining the course and efficacy of technology initiatives. The historical hierarchical models are being redesigned to keep up with the speed at which innovation is occurring. Agile governance frameworks, decentralised decision-making, and cooperative methods are becoming crucial for negotiating the complexity of the technology frontier (Yigitcanlar, T., Kamruzzaman, M., Foth, M., Sabatini-Marques, J., Da-Costa, E., & Ioppolo, G., 2019).

Citizen Engagement

Incorporating citizens into decision-making processes is an essential component of governance in smart cities. The smart city paradigm sees citizens as active players in influencing their urban environment, in contrast to the traditional role of citizens as beneficiaries of government services. Smart city citizen engagement models encompass a variety of approaches, such as crowdsourcing and participatory platforms, as well as co-creation projects that involve residents working with government agencies to design and execute solutions. Citizens in smart cities become into environmentally conscious individuals. Additionally, smart cities are more responsible settings for public initiatives like IT programs or projects involving involved residents. One of the most pressing demands in the globe right now is energy conservation, and the SC authorities have a great opportunity to create substantial savings with such a prudent approach. In view of the current energy crisis, such initiatives can help households save money on energy and have a genuine economic impact (Jasińska-Biliczak, A., 2022). As citizen participation is the key component of smart cities, it is imperative that local governments give priority to projects that let citizens participate in decision-making. Involving the public in the design and execution of smart city projects may guarantee that everyone's voice is heard and that the projects are adapted to the needs of the community. Hosting community forums and gatherings is one method of encouraging citizen participation in smart cities. Residents can voice their complaints and thoughts during these gatherings, and they can also talk about possible solutions to local problems. Building confidence between citizens and local authorities is essential for any smart city project to succeed, and community gatherings can help with this. Surveys and feedback systems are additional tools for including the public in smart city projects. These resources can offer insightful information on citizens' goals and wants, enabling communities to better target their activities towards the requirements of the local population. In order to enable cities to make data-driven decisions about upcoming projects, surveys can also be used to assess the success of smart city efforts (Cohem, B., n.d.).

Another effective instrument for citizen participation in smart cities is social media. Cities may communicate with inhabitants in both directions and reach a larger audience by utilising social media channels. Social media can be utilised to solicit resident input and ideas as well as to disseminate information about impending smart city initiatives. Involving citizens is essential to the success of smart city projects. Participating locals in the design and execution of these projects can guarantee that everyone's voice is heard and that the projects are suited to the needs of the community. Social media, questionnaires, feedback channels, and community gatherings are all effective strategies for encouraging citizen participation in smart cities. Cities can create more liveable and sustainable communities that benefit all people by placing a high priority on public engagement (Gracias, J. S., Parnell, G. S., Specking, E., Pohl, E. A., & Buchanan, R., 2023a, 2023b).

Through information and communication technology (ICT), primarily the internet, e-participation enables citizens to actively engage in the public consultation process. This idea refers to the ways in which ICT enables interactions between official political and administrative authorities and the general public. By using contemporary technologies to promote active social participation in the decision-making process, e-participation allows the government to reach a larger number of citizens than traditional social participation methods (Sobaci, Z., 2010). There are many different e-participation methods available, such as web apps, chat rooms, discussion forums, videoconferencing, social media (Facebook, X), e-petitions, online questionnaires, geo-questionnaires, and geo-discussions. These instruments are all dependent on ICT. One widely held belief is that intelligent ICTs can enhance citizens' quality of life (Bolívar, M. P., 2016). At the leading edge of the smart city revolution, information and communication technology (ICT) is causing radical shifts in urban living. The strategic integration of ICT in smart cities is a fundamental shift that goes beyond simple technological advancement to improve the lives of city dwellers (Dameri, R. P., & Cocchia, A., 2013). According to Yigitcanlar and Lee (Yigitcanlar, T., & Lee, S. H., 2014), smart ICTs are embedded in objects, wireless, and use sensors to record their surroundings. In sectors like healthcare, real estate, utilities, transportation, public safety, and administration, they offer the vital infrastructure for more intelligent and networked solutions (Washburn, D., Sindhu, U., Balaouras, S., Dines, R. A., Hayes, N., & Nelson, L. E., 2009). Smart ICTs, for instance, assist in gathering and disseminating consumption data to optimise energy management in the energy grid sector (Farmanbar, M., Parham, K., Arild, Ø., & Rong, C., 2019)

When it comes to transportation, smart ICTs make it possible for multimodal networks to be secure, inclusive of all social groups, and sustainable. This makes it easier for citizens to commute. ICT improves public safety by enabling predictive policing, emergency response systems, and smart surveillance. A safer urban environment is facilitated by sensor technology, emergency mobile apps, and integrated camera networks. Rapid emergency response times, enhanced situational awareness, and general security measures that are better all benefit citizens. By guaranteeing accessible for all individuals, including those with mobility problems, smart transportation systems encourage social inclusion. Real-time information about parking availability, accessible routes, and public transportation timetables is provided by ICT-enabled platforms. With the help of mobile applications, people with a variety of needs can plan and carry out their travels more easily. Smart ICTs make urban mobility more egalitarian and inclusive by lowering obstacles to transit (Herrenkind, B., Nastjuk, I., Brendel, A. B., Trang, S., & Kolbe, L. M., 2019). Smart ICTs have a major positive impact on the security of urban transportation systems. Intelligent traffic management systems monitor and optimise traffic flow by using real-time data from cameras, sensors, and linked vehicles. Road safety is increased, when possible, and dangers are identified and addressed by automated systems, which lowers the likelihood of accidents. Smart ICTs also make it possible for linked car technologies to be developed, which improves safety precautions by enabling automobiles to communicate with infrastructure and one another (Nikitas, A., Kougias, I., Alyavina, E., & Njoya Tchouamou, E., 2017). By dramatically lowering the energy demand over the course of a residential or commercial building's lifetime, smart ICTs can contribute to the establishment of so-called "zero energy buildings" in the building industry (Kylili, A., & Fokaides, P. A., 2015). Smart wearables can be used in the healthcare industry to effectively monitor public health (Trang, S., Trenz, M., Weiger, W., Tarafdar, M., & Cheung, C., 2020). Smart wearables are now essential parts of remote healthcare delivery, going beyond their use as fitness trackers. These gadgets enable remote medical condition diagnosis, medication prescription, and treatment administration by healthcare professionals thanks to their sophisticated sensors and networking characteristics. Smart wearables have a revolutionary effect on improving accessibility to medical treatment, removing geographical boundaries, and delivering healthcare services directly to individuals. Advanced sensors found in smart wearables, like temperature, pulse oximeter, and electrocardiogram (ECG) sensors, allow for remote monitoring and diagnosis of a range of medical issues. By enabling healthcare practitioners to collect data in real time, these gadgets eliminate the need for in-person visits to medical institutions and enable the early detection of anomalies and prompt diagnosis. Smart wearables provide remote diagnosis, medication prescriptions, and patient treatment (Ghazal, T. M., Hasan, M. K., Alshurideh, M. T., Alzoubi, H. M., Ahmad, M., Akbar, S. S., Al Kurdi, B., & Akour, I. A., 2021). A revolution in education has been brought about by smart ICTs, which have transformed the traditional classroom paradigm into a dynamic and adaptable learning environment. With the help of intelligent ICTs, students can participate in interactive learning experiences and can study whenever, wherever, and at their own pace. Smart ICTs in the education sector facilitate more interactive learning where students can "learn at anytime, anywhere, in any way, and at any pace." Although smart information and communication technologies (ICTs) are revolutionising many different fields, there is no one-size-fits-all use for them. In certain domains, including education or public administration, real-time data processing might not be the primary focus. On the other hand, smart ICTs are critical enablers in fields like energy or transportation management, necessitating powerful big data processing skills. Smart ICTs play a more limited role in fields like education or public administration, where processing large volumes of data in real time is typically not required. In contrast, in domains like energy or transportation management, smart ICTs are essential enablers and require big data processing capabilities (Neirotti, P., De Marco, A., Cagliano, A. C., Mangano, G., & Scorrano, F., 2014).

SMART GOVERNANCE THROUGH ICT IN SMART CITIES

The idea of smart cities has garnered a lot of attention from researchers and policymakers during the past 20 years. The concept of a "smart city" is not well defined, even though it has been discussed extensively in the literature (Albino, V., Berardi, U., & Dangelico, R. M., 2015; Angelidou, M., 2014; Anthopoulos, L., 2015). It typically refers to settings where individuals are provided with cutting-edge services through the use of information and communication technologies (ICTs) to improve their well-being and promote sustainable economic growth (Yigitcanlar, T. Kamruzzaman, M., Foth, M., Sabatini-Marques, J., Da-Costa, E., & Ioppolo, G., 2019). Smart cities are characterised by their smart economy, smart people, smart governance, smart mobility, smart environment, and smart living (Giffinger, R., Fertner, C., Kramar, H., Kalasek, R., Pichler-Milanovic, N., & Meijers, E., 2007). These attributes address important issues like economic competitiveness, citizen education, social interaction quality, labour market flexibility, governmental strategies, innovative transportation systems, sustainable resource management, and public safety.

Smart City Governance Models

Central governance systems

The participation of citizens in smart cities is influenced top-down by centralised governing institutions. This paradigm allows for the effective execution of city-wide projects by consolidating decision-making authority in the hands of local officials. Although this strategy can hasten the implementation of smart technologies, it might make citizen engagement more difficult. Citizens' influence is frequently restricted to offering suggestions for already-formulated programs. The effectiveness of public engagement depends on how quickly centralised organisations respond to feedback obtained via online social networks and other means. Finding a balance between effectiveness and inclusivity is the difficult part. Within the context of smart city implementation, centralised governance models are essential for coordinating all-encompassing plans. Online social networks can be effectively used to distribute information, gather input, and promote citizen engagement when they are integrated into centralised frameworks. Centralised platforms overseen by local government officials have the potential to optimise communication routes and offer a cohesive interface for residents to obtain details about ongoing projects, voice concerns, and actively engage in the decision-making process.

Decentralised governance systems

Online social networks are a common tool used by decentralised governance models in smart cities to foster a more open and interactive atmosphere. In this instance, citizens participate in the urban narrative in addition to being information consumers. Decentralised platforms facilitate the natural development of grassroots movements, community-led projects, and cooperative problem-solving. Online social networks allow residents to interact, exchange ideas, and work together to address urban concerns. They act as catalysts for decentralised engagement.

Hybrid systems

The balance between decentralised involvement and centralised control is achieved by hybrid governance frameworks. With this strategy, residents, businesses, and government agencies are all connected through online social networks, which serve as bridges in a cooperative ecosystem. These technologies provide prompt communication, guaranteeing individuals stay informed and contributing significantly to decision-making procedures. Hybrid approaches create a dynamic framework for smart city administration by acknowledging the significance of both bottom-up innovation and top-down direction.

GOVERNANCE AND CITIZEN PARTICIPATION

As smart cities gain traction, governance plays an increasingly important role in coordinating technology integration and encouraging public participation. This study examines the complex relationship between citizen involvement and governance in smart cities, focusing on how citizen participation affects governance dynamics and how governance structures affect the use of ICTs. The emergence of smart cities signals a paradigm change in urban governance, as residents play an active role in determining their cities' destiny rather than just being users of services. At the centre of this change is citizen engagement, which uses ICTs to build an urban ecology that is more inclusive and collaborative. Citizen engagement has historically taken several forms, such as public consultations and town hall gatherings in Figures 9.1 and 9.2. These conventional

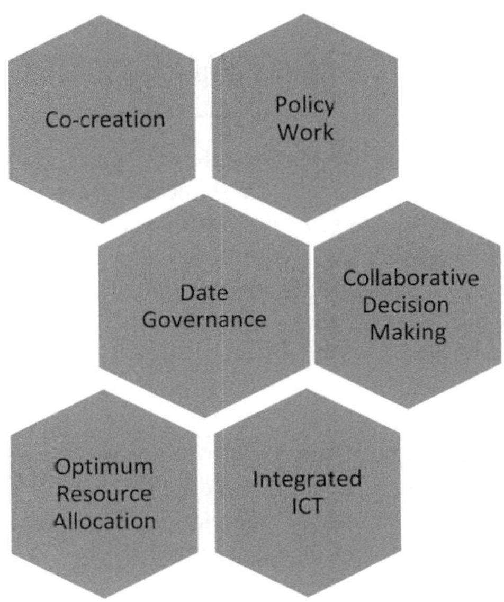

FIGURE 9.1 Smart city governance and citizen engagement.

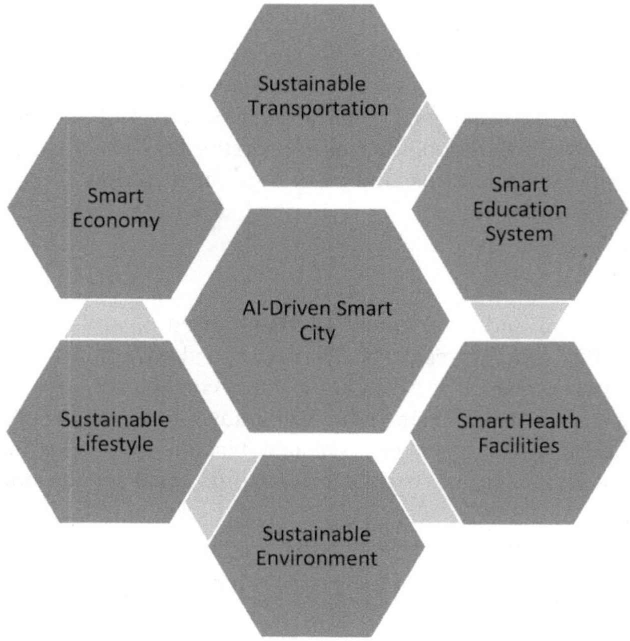

FIGURE 9.2 Major domains of smart city.

approaches are enhanced by digital platforms in the smart city era, allowing for dynamic involvement, feedback loops, and real-time interactions. The development of citizen participation is indicative of a dedication to openness, responsiveness, and collaborative decision-making.

CITIZEN ENGAGEMENT MODELS FOR INCLUSIVE GOVERNANCE

Different citizen involvement approaches are implemented in smart cities, each specifically designed to meet the demands and obstacles of urban settings. These models include informational ones, in which the public is informed about decisions, as well as more cooperative and participatory ones, such as crowdsourcing and co-creation. The choice of model is contingent upon the goals and circumstances of a particular endeavour, as each model has advantages and disadvantages of its own. Markets that link producers of various public goods and services with citizens as consumers of these services are the basis for the concept of citizen engagement. Improving the quality of life of residents requires granting them nonexclusive and barrier-free access to a variety of public goods, including infrastructure, healthcare, and education.

- **Informative Model:** The city administration informs the public about policies, initiatives, and choices under the Informative Model. While there are few possibilities for citizen participation, transparency is ensured by this one-way connection. Information is distributed through digital platforms like official websites, social media, and mobile apps.
- **Consultative Model:** In the Consultative Model, public consultations, polls, and surveys are used to get feedback from the general public. By actively seeking input on particular projects or issues, city officials provide locals a platform to voice their thoughts. This concept makes

use of popular resources like digital surveys and online town hall meetings to facilitate a wider audience and varied participation.

- **Collaborative Model:** By incorporating locals in cooperative decision-making procedures, the Collaborative Model advances public engagement. Alongside public servants, citizens collaborate with local governments, community organizations, and private sector partners to co-create solutions and participate in decision-making processes, enhancing the effectiveness and inclusivity of smart city initiatives. The city administration informs the public about policies, initiatives, and choices under the Informative Model. While there are few possibilities for citizen participation, transparency is ensured by this one-way connection. Information is distributed through digital platforms like official websites, social media, and mobile apps.
- **Co-Creative Model:** The Co-Creative Model involves the active participation of citizens in project conceptualisation, design, and implementation. This concept encourages collaboration between local government and citizens, enabling the latter to provide resources, expertise, and ideas. This concept is demonstrated via crowdsourcing websites, hackathons, and community seminars, which foster creativity and give communities more influence.
- **Hybrid Model:** The Hybrid Model recognises that different situations may call for different methods and integrates components of multiple engagement models. It enables plans for citizen participation to be flexible, guaranteeing that the city can adjust to a range of requirements and difficulties.

A combination of these methods, each adapted to the particular circumstances and objectives of the city, is frequently used to implement successful public involvement in smart cities. Smart cities can establish an urban government structure that is more inclusive, responsive, and participatory by adopting a wide variety of engagement tactics.

Governmental enthusiasm, vision, and administration are necessary for the deployment of smart cities to be sustainable. Additionally, a key factor in the smart city ideation and effective execution is the involvement and engagement of the public. Nonetheless, a little encouragement is required to promote involvement. Furthermore, social media platforms on the internet play a big role in smart cities. To raise the level of living in the city, people's science, government, and technology should work together more harmoniously. Prospective smart cities might look into novel feedback systems including online social networking channels, apps for smartphones, and IoT-enabled devices to further improve public involvement. With the use of these technologies, people can offer immediate opinions on a range of metropolitan issues, such as environmental responsibility, infrastructure, and commuting. Citizen-generated information can be analysed by algorithms that utilise machine learning to find trends that allow administrators to act quickly and efficiently (Kusumastuti, R. D., & Rouli, J., 2021). Participation methods are being adopted by smart cities in many parts of the world, giving residents a say in determining the policies and campaigns for their communities. Participatory budgeting was first implemented in Porto Alegre in the 1990s, and since then, it has been popular throughout Brazil and the rest of the world, with more than 2,700 governments using some form of it at present. This strategy encourages diversity and community empowerment in addition to improving openness. The participatory budget, particularly in its first ten years, addressed social, economic, and political marginalisation and encouraged innovative, bottom-up inputs into budgetary allocation decisions, which led to revolutionary change in Porto Alegre. Citizen input was used to determine priorities for public investments in local neighbourhoods to improve people's quality of life. The process was structured to assure representation by geography and then by theme (Abers, R., King, R., Votto, D., & Brandão, I., 2018).

Future smart cities can implement deliberative procedures that involve citizens in intricate policy discussions to further enhance participatory decision-making. Deliberative venues, such expert panels and citizens' assemblies, can facilitate thoughtful and productive discussions by making sure all points of view are considered. Additionally, by utilising cutting-edge technologies like augmented reality (AR) and virtual reality (VR), individuals may experience and visualise suggested urban ideas, promoting a deeper knowledge of the city life and its challenges.

THE TRANSFORMATIVE ROLE OF AI IN CITIZEN ENGAGEMENT FOR SMART CITY IMPLEMENTATION

Similar to a number of revolutionary inventions in past events, artificial intelligence (AI) will have a big impact on society. AI has been a major topic of discussion and debate in many urban policy circles and forums about the transition of smart cities, especially among planners and policymakers looking for technological answers to pressing urbanisation issues. Furthermore, although AI applications for smart cities offer advantages like automation and efficiency, they also raise legal, ethical, and regulatory concerns including discrimination in service delivery and privacy (Zhou, Y., & Kankanhalli, A., 2021). Given the range of perspectives that academics have offered, it is imperative that we look into the benefits, drawbacks, and impact of AI algorithms on smart cities. This chapter investigates the problem-solving applications of AI in the contemporary smart city.

AI became a transformative force in the era of smart cities, changing the way people interact with their urban surroundings. The use of AI in citizen interaction is crucial for improving involvement, promoting diversity, and streamlining the execution of smart city initiatives. By analysing enormous volumes of data gathered from numerous sources, such as sensors, social media, and public records, AI powers smart cities. City officials are able to recognise patterns, anticipate public demands, and make well-informed choices thanks to this data-driven strategy. Cities may obtain important insights into urban dynamics by utilising AI analytics, which enables more flexible and responsive governance. AI makes it easier to customise citizen services according to personal preferences, actions, and past information. Smart cities can offer customised information, services, and suggestions based on the needs of each individual citizen thanks to AI algorithms. This not only improves the user experience but also promotes inclusivity by catering to various and particular needs. Chatbots, voice-activated interfaces, and virtual assistants with AI capabilities are easily accessible ways for citizens to raise problems, get information, and interact with local services. Conversational AI systems facilitate effective and smooth communication between citizens and their local government by providing real-time answers to public inquiries. This round-the-clock accessibility removes time and place barriers to improve community engagement. By examining past data, AI algorithms are able to forecast the behaviour, preferences, and possible problems of citizens. Smart cities can interact with their inhabitants proactively about impending events, service interruptions, or community projects thanks to this predictive power. A more cooperative urban environment is created when individuals are informed and communication is improved through anticipatory participation (Kassens-Noor, E., & Hintze, A., 2020).

AI analyses data on traffic patterns, citizen behaviour, and the use of public services to help optimise the allocation of resources. Through this optimisation, city resources are allocated more effectively, giving precedence to or addressing regions of greater demand. Allocating resources wisely increases citizen satisfaction by providing services where and when they are most needed. Artificial Intelligence is vital to citizen engagement in public safety during emergencies. Real-time data can be analysed by AI-driven systems to spot possible risks, foresee emergency scenarios, and notify locals in a timely manner. This degree of responsiveness not only improves public safety but also increases confidence in the city's capacity to manage crises. AI makes it easier for citizens to participate in decision-making by giving them access to data-driven insights on policy and urban planning. AI-powered interactive systems enable locals to offer suggestions, offer criticism, and actively engage in decision-making. The democratisation of knowledge enables everyone to participate in the design of their urban surroundings. AI-powered user interfaces can improve accessibility for a wide range of people, including those who are language- or disability-impaired. By allowing individuals to interact with smart city services in a variety of ways, speech recognition and natural language processing technologies ensure that participation is inclusive and meets a range of demands. It's critical to take privacy and ethical issues into account when using AI to engage citizens. To establish and preserve public confidence, strong security measures must be put in place, such as data anonymisation, informed permission, and openness in AI algorithms. AI has a revolutionary role in citizen involvement, providing hitherto unseen chances to build smart cities that are more inclusive, responsive, and focused on the needs of their citizens. As technology develops further, the application of AI in smart city implementation has the potential to create thriving urban communities where residents actively participate in the co-creation of their urban destiny rather than merely being service recipients (Herath, H.M.K.K.M.B., & Mittal, M., 2022).

CASE STUDIES OF SMART CITIES

Singapore: An Example of Metropolitan Advancement

Reputed for its all-encompassing and integrated approach to growth in cities, Singapore is one of the world's most notable smart city examples. The city-state has improved many facets of urban life, from waste reduction and public safety to conserving energy and transport, by utilising data and information technology. The goal of the 2014 "Smart Nation" plan was to make Singapore a technologically sophisticated and networked metropolis. A few of the creative ideas that have made Singapore known as a global pioneer in urban innovation are vast sensor networks, digital health services, and smart traffic management.

Barcelona: Leading the Way in Citizen-Centred Solutions

Barcelona has placed a major focus on improving quality of life and citizen engagement as it transitions into a smart city. Enhancing urban services while promoting an inclusive and participatory atmosphere is the main goal of the "Barcelona Smart City" project. Proposals such as the "Superblocks" project, which reclaims roadways for bicyclists and pedestrians, and the installation of smart lamps that change colour in response to activity show the city's dedication to sustainability and the welfare of its residents. Barcelona has established a vibrant technological environment by means of cooperative alliances and freely accessible information systems.

Songdo, South Korea

Songdo International Business District is an outstanding illustration of a prospective smart city that was planned with technology and environmental responsibility in mind. Songdo, which is close to Seoul, was created as an example of modern urban living. The city has a smart structure system with integrated sensors for energy efficiency, cutting-edge waste disposal technologies, and an innovative transit system. Cities hoping to start from zero with an emphasis on innovative technologies and ecological principles can take inspiration from Songdo's integrated architecture.

FUTURE SCOPE

Smart cities have a great deal of promise to address current urban issues and advance resilience and sustainability in the years to come. Smart cities are in a prime position to be major players in the worldwide battle towards damage to the environment and global warming. Smart cities might maximise energy usage, lower greenhouse gas emissions, and encourage the use of renewable energy sources by incorporating data-driven solutions. Citizens' health can be protected by prompt reactions to pollution incidents made possible by real-time monitoring of the quality of the air and water. Furthermore, intelligent waste management systems can encourage recycling and lessen landfill waste, both of which help to preserve the environment overall. As the globe encounters growing unpredictability including severe weather, depletion of resources, and health emergencies, smart cities are developing to enhance their resilience. Resilient infrastructure is capable of withstanding shocks and disturbances, guaranteeing the continuation of critical services in times of emergency. Anticipating possible hazards and facilitating proactive preparation for emergency response are made possible by data analytics and forecasting. Furthermore, a city's resilience can be increased by incorporating environmentally friendly structures and solutions derived from nature. To reduce the danger of flooding, control temperature, and improve biodiversity, urban planners incorporate green areas, ecological water management, and natural ecosystems into the architecture of cities. These tactics enhance citizens' general well-being in addition to enhancing urban resilience.

Social media platforms offer a platform for discussions where a lot of information may be produced and shared in real-time with a big audience at a cheap cost, fostering opportunities for citizen engagement. Urban and suburban infrastructure initiatives are aimed at improving the overall well-being of people and hence addressing their requirements. Given this objective, citizen involvement and engagement are crucial. Previous studies identify challenges to the implementation and that smart people or smart citizens are essential dimensions of a smart city. This study aims to gain insights into the smart city implementation and citizen engagement through the online social network. the implementation of smart cities should be adjusted according to the local situation. The critical success factors of the implementation are, namely, local wisdom, transformational leadership, sustainability, and political content. The city residents have been involved in implementing the smart city through the online social network. However, a kind of incentive system is still required to ensure that the city residents are encouraged to solve the city problems. With information and communication technologies (ICTs) evolving at a rapid pace, smart city governance is becoming more than just a bureaucratic role; it is a dynamic ecosystem. Technology becomes a catalyst for co-creation, transparency, and responsiveness in a synergistic relationship between the interrelated dynamics of government structures and citizen interaction. In order to effectively make decisions and provide services, governance structures in smart cities need to be flexible and nimble, embracing technological breakthroughs as tools. Each of the three models—hierarchical, decentralised, and hybrid—has advantages, but the secret is to design governing structures that can quickly adapt to the demands and goals of the populace. The several models—informative, consultative, collaborative, co-creative, and empowering—offer a range of options as citizen interaction assumes a central role. These models serve as a means for citizens to change from being obedient consumers of services to participating members of the urban community. Technology—especially digital platforms—serves as the link that makes it easier for citizens and city officials to collaborate, participate, and communicate smoothly. The path taken by this research has revealed potential that are contained in overcoming constraints, such as digital divisions and privacy issues, while also acknowledging challenges. Smart cities have the ability to develop into inclusive, resilient communities where all residents have a say and a stake in the development of the city, rather than merely technologically sophisticated metropolitan areas. Future smart cities should be responsive, inclusive, and sustainable, with technology and people working together in harmony. The dedication of city officials, legislators, technologists, and residents to support governance systems that place a strong emphasis on citizen engagement is essential to the realisation of this vision. The research's insights illuminate the road towards a future when smart cities actually belong to the people they serve, providing guidance for us as we navigate the journey ahead.

CONCLUSION

It is clear from the investigation of Smart City Governance and Citizen Engagement that the relationship between civic engagement and technology innovation is crucial to forming the cities of the future. This research has revealed dynamic relationships, changing models, and revolutionary opportunities that highlight the role that inclusive governance plays in the context of smart cities. The study concludes that the integration of AI into smart cities can yield several benefits for cities. These include the automation of operations, reduction of human error, effective data-driven decision-making, environmental improvement through various systems, implementation of new commercial opportunities, and automated efficient urban management. However, they also provide regulatory difficulties, including privacy issues, legal and ethical issues, and discrimination in the provision of services. In addition, a high unemployment rate, the cost and duration of AI efforts, the availability of data, the absence of experts with the necessary skills, and other factors have all been identified as dangers and obstacles to AI deployment in smart cities. Our results indicate that the domains of healthcare, mobility, privacy and security, and energy have a higher influence on the adoption of AI in smart cities. Prospective smart city developments could result in more adaptable, environmentally friendly, and habitable urban settings. Smart cities may address complicated issues, such as industrialisation and environmental degradation, by embracing revolutionary technologies, a data-driven government, and citizen participation. The idea of smart cities will be crucial to creating an improved tomorrow for generations to come as cities expand and change. A key component of effective smart city governance is citizen participation. Smart cities have the potential to establish environmentally friendly and equitable communities through the utilisation of citizen feedback, the promotion of collaborative decision-making, the adoption of open data, and the encouragement of joint monitoring and assessment. To fully realise the potential of citizen engagement, issues including the technological gap, security and confidence issues, equitable participation, and citizen complacency must be addressed. Future smart cities have the ability to unleash the full power of collective thought, paving the way for a more successful and sustainable urban environment by stretching the limits of citizen engagement and promoting community empowerment.

REFERENCES

Abers, R., King, R., Votto, D., & Brandão, Igor (2018, June 13). *Porto Alegre: Participatory Budgeting and the Challenge of Sustaining Transformative Change.* Retrieved from https://www.wri.org/research/porto-alegre-participatory-budgeting-and-challenge-sustaining-transformative-change

Albino, V., Berardi, U., & Dangelico, R. M. (2015). Smart cities: Definitions, dimensions, performance, and initiatives. *Journal of Urban Technology*, *22*(1). https://doi.org/10.1080/10630732.2014.942092

Angelidou, M. (2014). Smart city policies: A spatial approach. *Cities*, *41*, S3–S11. https://doi.org/10.1016/j.cities.2014.06.007

Anthopoulos, L. (2015). Understanding the smart city domain: A literature review. Transforming City Governments for Successful Smart Cities, 9–21, 1–16.

Bolívar, M. P. (2016). Mapping dimensions of governance in smart cities. *17th International Digital Government Research Conference on Digital Government Research*, (pp. 312–324). https://doi.org/10.1145/2912160.2912176

Cohem, B. (n.d.). *The 3 Generations of Smart Cities.* Retrieved from https://www.fastcompany.com/3047795/the-3-generations-of-smart-cities

Dameri, R. P., & Cocchia, A. (2013). Smart city and digital city: Twenty years of terminology evolution. *X Conference of the Italian Chapter of AIS, ITAIS, 1.*

Din, I. U., Guizani, M., Rodrigues, J. J. P. C., Hassan, S., & Korotaev, V. V. (2019). Machine learning in the internet of things: Designed techniques for smart cities. *Future Generation Computer Systems*, *100*, 826–843.

Farmanbar, M., Parham, K., Arild, Ø, & Rong, C. (2019). A widespread review of smart grids towards smart cities. *Energies*, *12*(23), 4484. https://doi.org/10.3390/en12234484

Giffinger, R., Fertner, C., Kramar, H., Kalasek, R., Pichler-Milanovic, N., & Meijers, E. (2007). *Smart cities: Ranking of European medium-sized cities.* Centre of Regional Science (SRF), Vienna University of Technology.

Ghazal, T. M., Hasan, M. K., Alshurideh, M. T., Alzoubi, H. M., Ahmad, M., Akbar, S. S., Al Kurdi, B., & Akour, I. A. (2021). IoT for smart cities: Machine learning approaches in smart healthcare—a review. *Future Internet*, *13*(8), 218. https://doi.org/10.3390/fi13080218

Gracias, J. S., Parnell, G. S., Specking, E., Pohl, E. A., & Buchanan, R. (2023b). Smart city development: Challenges and opportunities in the digital age. *Journal of Urban Technology*, *30*(2), 45–67.

Gracias, J. S., Parnell, G. S., Specking, E., Pohl, E. A., & Buchanan, R. (2023a). Smart cities—a structured literature review, *6*(4), 1719–1743. https://doi.org/10.3390/smartcities6040080

Herath, H. M. K. K. M. B., & Mittal, Mamta (2022, April). Adoption of artificial intelligence in smart cities: A comprehensive review. *International Journal of Information Management Data Insights*, *2*(1), 1–25.

Herrenkind, B., Nastjuk, I., Brendel, A. B., Trang, S., & Kolbe, L. M. (2019). Young people's travel behavior – Using the life-oriented approach to understand the acceptance of autonomous driving. *Transportation Research Part D: Transport and Environment*, *74*, 214–233. https://doi.org/10.1016/j.trd.2019.07.023

Ismagilova, E., Hughes, L., Dwivedi, Y. K., & Raman, K. R. (2019a). Smart cities: Advances in research—an information systems perspective. *International Journal of Information Management*, *47*, 88–100.

Ismagilova, E., Hughes, L., Dwivedi, Y. K., & Raman, K. R. (2019b). The significance of smart cities in the modern era: A comprehensive review. *Smart Cities Review*, *12*(4), 89–112.

Jasińska-Biliczak, A. (2022). Smart-city citizen engagement: The answer to energy savings in an economic crisis? *Energies*, *15*(23), 8828. https://doi.org/10.3390/en15238828

Kassens-Noor, E., & Hintze, A. (2020). Cities of the future? *The Potential Impact of Artificial Intelligence*, *1*(2), 192–197.

Kusumastuti, R. D., & Rouli, J. (2021). R D Kusumastuti and J Rouli 2021 IOP Conf. Ser.: Earth Environ. Sci. 940 012076. *IOP Conference Series: Earth and Environmental Science 940 (2021) 012076. 940*, Smart City Implementation and Citizen Engagement in Indonesia. IOP Publishing. https://doi.org/10.1088/1755-1315/940/1/012076

Kylili, A., & Fokaides, P. A. (2015). European smart cities: The role of zero energy buildings. *Sustainable Cities and Society*, *15*, 86–95. https://doi.org/10.1016/j.scs.2014.12.003

Li, X., Liu, H., Wang, W., Zheng, Y., Lv, H., & Lv, Z. (2021). Big data analysis of the Internet of Things in the digital twins of smart city based on deep learning. *Future Generation Computer System*, *128*, 167–177.

Neirotti, P., De Marco, A., Cagliano, A. C., Mangano, G., & Scorrano, F. (2014). Current trends in Smart City initiatives: Some stylised facts. *Cities*, *38*, 25–36. https://doi.org/10.1016/j.cities.2013.12.010

Nikitas, A., Kougias, I., Alyavina, E., & Njoya Tchouamou, E. (2017). How can autonomous and connected vehicles, electromobility, BRT, hyperloop, shared use mobility and mobility-as-a-service shape transport futures for the context of smart cities? *Urban Science*, *1*(4), 36. https://doi.org/10.3390/urbansci1040036

Shoaib, N., & Shamsi, J. A. (2019). Understanding network requirements for smart City applications: Challenges and solutions. *IT Profession*, *21*, 33–40.

Sobaci, Z. (2010). What the Turkish parliamentary web site offers to citizens in terms of e-participation: A contentanalysis. *Inf. Polity: International Journal of Government and Democracy in the Information Age*, *15*, 227–241.

Tahmasseby, S. (2022). The implementation of smart mobility for smart cities: A case study in Qatar. *Civil Engineering Journal*, *8*, 2154–2171.

Trang, S., Trenz, M., Weiger, W., Tarafdar, M., & Cheung, C. (2020). One app to trace them all? Examining app specifications for mass acceptance of contact-tracing apps. *European Journal of Information Systems*, *29*(1), 415–428. https://doi.org/10.1080/0960085X.2020.1784046

Vasudavan, H., Gunasekaran, S. S., & Balakrishnan, S. (2019). Smart city: The state of the art, definitions, characteristics and dimensions. *Journal of Computational and Theoretical Nanoscience*, *16*(8), 3525–3531. https://doi.org/10.1166/jctn.2019.8318

Washburn, D., Sindhu, U., Balaouras, S., Dines, R. A., Hayes, N., & Nelson, L. E. (2009). Helping CIOs understand "smart city" initiatives. *Growth*, *17*(2), 1–17.

Yigitcanlar, T., Kamruzzaman, M., Foth, M., Sabatini-Marques, J., Da-Costa, E., & Ioppolo, G. (2019). Can cities become smart without being sustainable? A systematic review of the literature. *Sustainable Cities and Society*, *237*, 348–365.

Yigitcanlar, T., & Lee, S. H. (2014). Korean ubiquitous-eco-city: A smart-sustainable urban form or a branding hoax? *Technological Forecasting and Social Change*, *89*, 100–114. https://doi.org/10.1016/j.techfore.2013.08.034

Yin, C., Xiong, Z., Chen, H., Wang, J., Cooper, D., & David, B. (2015a). A literature survey on smart cities. *Science China Information Sciences*, *58*, 1–18.

Yin, C., Xiong, Z., Chen, H., Wang, J., Cooper, D., & David, B. (2015b). The rise of smart cities: Technological integration and urban development. *International Journal of Smart Cities*, *10*(1), 15–32.

Zhou, Y., & Kankanhalli, A. (2021). AI regulation for smart cities: Challenges and principles. *Smart cities and smart governance*, Springer, Cham, 101–118.

Fighting Fraud With Algorithms *AI Solutions for Claim Detection and Revolutionizing Fraud Detection in Insurance*

10

Gopichand Vemulapalli

INTRODUCTION

The insurance industry plays a pivotal role in safeguarding individuals and businesses against unforeseen risks, providing financial protection and peace of mind. However, alongside its noble purpose, the insurance sector grapples with a pervasive threat: fraud [1]. Fraudulent activities within insurance, ranging from exaggerated claims to deliberate misrepresentation, impose significant financial burdens, erode trust, and undermine the integrity of the entire system [2]. Conventional methods of fraud detection often fall short in addressing the evolving tactics employed by fraudsters, necessitating a paradigm shift in approach. In recent years, the integration of artificial intelligence (AI) and advanced algorithms has emerged as a transformative solution, offering unprecedented capabilities in detecting and mitigating fraud [3].

THE CHALLENGE OF FRAUD IN INSURANCE

Insurance fraud poses a multifaceted challenge, manifesting in various forms across different lines of insurance. One of the most prevalent forms is fraudulent claims, where policyholders or third parties fabricate or exaggerate losses to illicitly obtain financial compensation [4]. These fraudulent claims not only impose direct financial losses on insurers but also lead to increased premiums for honest policyholders, eroding trust in insurance products and services.

DOI: 10.1201/9781003497189-10

Furthermore, insurance fraud extends beyond individual claimants to encompass organized crime syndicates orchestrating elaborate schemes to exploit vulnerabilities within the insurance system. These sophisticated fraud rings engage in activities such as staged accidents, fake injuries, and fraudulent billing, costing insurers billions of dollars annually and posing significant regulatory and legal challenges [5].

Conventional fraud detection methods, primarily reliant on manual reviews and rule-based systems, struggle to keep pace with the dynamic nature of fraudulent activities [6]. Manual processes are inherently time-consuming, resource-intensive, and susceptible to human error, making them ill-equipped to handle the scale and complexity of modern fraud schemes. Moreover, rule-based systems lack the adaptability and predictive capabilities necessary to detect emerging fraud patterns effectively [7].

THE PROMISE OF ARTIFICIAL INTELLIGENCE IN FRAUD DETECTION

Against this backdrop, the integration of AI and advanced algorithms presents a paradigm shift in fraud detection within the insurance industry. AI algorithms, powered by vast amounts of data and computational capabilities, offer the potential to analyze complex patterns, identify anomalies, and predict fraudulent behavior with unprecedented accuracy and efficiency [8].

Machine learning, a subset of AI, lies at the heart of many fraud detection systems, enabling algorithms to learn from historical data and iteratively improve their performance over time. Supervised learning algorithms, such as logistic regression and random forests, are commonly employed for binary classification tasks, distinguishing between legitimate and fraudulent claims based on labeled training data [9].

Unsupervised learning techniques, including clustering and anomaly detection, provide valuable insights into emerging fraud patterns by identifying outliers and irregularities within large datasets. These unsupervised approaches are particularly useful in detecting previously unseen fraud schemes and adapting to evolving tactics employed by fraudsters [10].

Deep learning, a subset of machine learning characterized by neural networks with multiple layers, has shown remarkable promise in fraud detection tasks that involve complex, unstructured data, such as text and images [11]. Deep learning architectures, such as convolutional neural networks (CNNs) and recurrent neural networks (RNNs), excel in processing and extracting meaningful features from diverse data sources, enabling more accurate and nuanced fraud detection [12].

DATA QUALITY AND FEATURE ENGINEERING

Central to the effectiveness of AI-powered fraud detection systems is the quality and relevance of the underlying data. High-quality data, encompassing both structured and unstructured information, is essential for training accurate and robust models capable of identifying subtle fraud signals amidst the noise [13].

Feature engineering, the process of selecting and transforming raw data into informative features, plays a critical role in enhancing the performance of fraud detection algorithms. By leveraging domain expertise and advanced analytics techniques, insurers can extract relevant features from diverse data sources, such as policy information, claims narratives, and external data feeds, to enrich the analysis and improve detection accuracy [14]. Moreover, the integration of external data sources, including social media, public records, and third-party databases, can provide valuable context and corroborating evidence for fraud detection efforts. By harnessing the wealth of information available across multiple channels, insurers can gain a more comprehensive understanding of policyholder behavior and identify potential red flags indicative of fraudulent activity [15].

PREDICTIVE MODELING AND REAL-TIME DECISION-MAKING

In addition to retrospective analysis, AI-powered fraud detection systems enable insurers to adopt a proactive approach to fraud prevention through predictive modeling and real-time decision-making [16]. Predictive analytics techniques, such as predictive modeling and machine learning, enable insurers to forecast future fraud risks based on historical trends and behavioral patterns, empowering them to allocate resources more efficiently and implement targeted interventions [17].

Real-time decision-making capabilities, facilitated by AI algorithms deployed within claims processing workflows, enable insurers to detect and flag suspicious claims as they occur, minimizing the time and resources required for manual review. By integrating fraud detection algorithms into automated decision systems, insurers can streamline claims processing, expedite legitimate payouts, and mitigate the financial impact of fraudulent activities in real time [18].

ETHICAL CONSIDERATIONS AND TRANSPARENCY

Despite the transformative potential of AI in fraud detection, its deployment raises ethical considerations and challenges that must be addressed to ensure fairness, transparency, and accountability. Algorithmic bias, stemming from imbalanced training data or flawed model assumptions, can result in disparate treatment of policyholders and undermine trust in the fairness of automated decision-making processes [19].

Transparency and interpretability are essential principles in mitigating algorithmic bias and fostering trust in AI-powered fraud detection systems. Insurers must ensure transparency in their model development processes, providing clear explanations of how algorithms make decisions and disclosing any inherent biases or limitations in the data or algorithms used. Moreover, robust governance frameworks and regulatory oversight are necessary to ensure the responsible and ethical use of AI in fraud detection. Insurers must establish clear guidelines and procedures for data collection, processing, and model deployment, adhering to relevant legal and regulatory requirements while upholding the highest standards of integrity and ethics [20].

The integration of AI and advanced algorithms holds immense promise in revolutionizing fraud detection within the insurance industry. By leveraging machine learning, data analytics, and predictive modeling techniques, insurers can enhance their ability to detect and mitigate fraudulent activities effectively, safeguarding the interests of policyholders and preserving the integrity of the insurance ecosystem. However, realizing this potential requires a concerted effort to address ethical considerations, ensure transparency, and uphold the principles of fairness and accountability in the deployment of AI-powered fraud detection systems. As insurers continue to embrace innovation and technology, the fight against insurance fraud enters a new era of opportunity and responsibility, guided by the principles of integrity, transparency, and ethical stewardship.

TRADITIONAL APPROACHES TO FRAUD DETECTION

Fraud detection in the insurance industry has historically relied on traditional approaches such as manual reviews and rule-based systems. These methods, while effective to some extent, have notable limitations and fail to keep pace with the evolving tactics employed by fraudsters. Consequently, there is a pressing need for innovation in fraud detection methodologies within the insurance sector.

Manual Reviews

Manual reviews involve human investigators scrutinizing individual claims for signs of potential fraud. Investigators assess various factors such as claimant demographics, loss details, and supporting documentation to identify inconsistencies or suspicious patterns. While human intuition and expertise play a crucial role in uncovering certain types of fraud, manual reviews are inherently time-consuming, labor-intensive, and prone to subjectivity. Moreover, human investigators may overlook subtle indicators of fraud or struggle to handle the sheer volume of claims, leading to delays and inefficiencies in the detection process.

Rule-Based Systems

Rule-based systems employ predefined rules and thresholds to flag potentially fraudulent claims based on predetermined criteria. These rules are typically derived from historical fraud cases, industry best practices, or regulatory requirements. For example, a rule might trigger an alert if a claim exceeds a certain dollar amount or if specific combinations of claim characteristics are observed. While rule-based systems offer automation and scalability compared to manual reviews, they suffer from several limitations. Firstly, they are rigid and inflexible, unable to adapt to emerging fraud schemes or evolving patterns of behavior. Fraudsters can easily circumvent rule-based detection by devising new tactics that fall outside the scope of existing rules. Additionally, rule-based systems may generate false positives, flagging legitimate claims as fraudulent due to overly conservative thresholds or inadequate rule coverage.

Limitations of Conventional Methods

Traditional approaches to fraud detection, including manual reviews and rule-based systems, face several inherent limitations that undermine their effectiveness in combating insurance fraud. Firstly, these methods are reactive rather than proactive, focusing on identifying fraudulent claims after they have been submitted rather than preventing fraud before it occurs. This reactive approach results in significant financial losses for insurers and delays in claim processing for policyholders. Moreover, traditional methods often lack the sophistication and predictive capabilities necessary to detect complex fraud schemes or subtle patterns of fraudulent behavior. Fraudsters continually adapt their tactics to exploit weaknesses in existing detection systems, rendering traditional approaches increasingly obsolete.

The Need for Innovation

Given the shortcomings of traditional approaches to fraud detection, there is a clear imperative for innovation within the insurance industry. Insurers must embrace advanced technologies such as AI, machine learning, and predictive analytics to enhance their fraud detection capabilities. These technologies offer the potential to analyze large volumes of data, identify subtle fraud signals, and predict fraudulent behavior with unprecedented accuracy and efficiency. By leveraging AI-powered algorithms, insurers can move beyond reactive detection methods and adopt a proactive stance toward fraud prevention. Moreover, AI-based systems can adapt and evolve in real time, continuously learning from new data and improving their performance over time. This adaptability is critical in staying ahead of sophisticated fraud schemes and effectively combating fraudulent activities within the insurance ecosystem.

In summary, traditional approaches to fraud detection in insurance, including manual reviews and rule-based systems, have significant limitations that hinder their effectiveness in addressing the evolving threat of fraud. To overcome these challenges, insurers must embrace innovation and leverage advanced technologies to enhance their fraud detection capabilities. By adopting AI-powered algorithms and predictive analytics, insurers can move toward proactive fraud prevention, safeguarding their financial interests and preserving the integrity of the insurance industry.

FOUNDATIONS OF ARTIFICIAL INTELLIGENCE IN FRAUD DETECTION

AI has revolutionized fraud detection within the insurance industry, offering advanced methodologies to identify fraudulent activities with greater accuracy and efficiency. This section explores the foundational concepts of AI and its application in fraud detection, including machine learning, deep learning, and various techniques employed to distinguish between legitimate and fraudulent claims.

Overview of Machine Learning and Deep Learning

Machine learning is a subset of AI that focuses on developing algorithms capable of learning from data and making predictions or decisions without explicit programming. In the context of fraud detection, machine learning algorithms analyze historical data to identify patterns, anomalies, and trends indicative of fraudulent behavior.

Deep learning is a specialized branch of machine learning that utilizes artificial neural networks with multiple layers to extract intricate patterns from complex data sources. Deep learning architectures, such as convolutional neural networks (CNNs) and recurrent neural networks (RNNs), excel in processing unstructured data such as text, images, and audio, making them well-suited for fraud detection tasks that involve diverse data modalities.

Supervised Learning for Binary Classification

Supervised learning is a machine learning paradigm where algorithms learn from labeled examples to predict the correct output for new, unseen data. In fraud detection, supervised learning algorithms are commonly employed for binary classification tasks, distinguishing between legitimate and fraudulent claims based on features extracted from historical data.

Examples of supervised learning algorithms used in fraud detection include logistic regression, decision trees, random forests, support vector machines (SVM), and gradient boosting machines (GBM). These algorithms learn to discriminate between genuine and fraudulent claims by identifying patterns and relationships within the data, enabling insurers to automate the detection process and prioritize suspicious cases for further investigation.

Unsupervised Learning Techniques

Unsupervised learning is a machine learning approach where algorithms learn to identify patterns and structures in data without the need for labeled examples. Unsupervised learning techniques are particularly valuable in fraud detection for detecting anomalies or outliers indicative of fraudulent behavior.

Clustering algorithms, such as k-means clustering and hierarchical clustering, group similar data points together based on their inherent characteristics, enabling insurers to identify clusters of claims with similar patterns or characteristics. Anomaly detection algorithms, such as isolation forests and autoencoders, identify unusual or abnormal instances within a dataset, flagging potentially fraudulent claims for closer scrutiny.

Deep Learning Architectures

Deep learning architectures leverage artificial neural networks with multiple layers of interconnected nodes to learn hierarchical representations of data. These architectures excel in capturing complex patterns and relationships within large-scale datasets, making them well-suited for fraud detection tasks that involve high-dimensional, unstructured data.

Convolutional neural networks (CNNs) are widely used in fraud detection applications that involve image analysis, such as identifying counterfeit documents or detecting fraudulent signatures. Recurrent neural networks (RNNs) are well-suited for processing sequential data, making them effective in analyzing temporal patterns and detecting fraud in time-series data, such as transaction logs or claims histories.

In summary, the foundations of artificial intelligence in fraud detection encompass machine learning, deep learning, and various techniques for distinguishing between legitimate and fraudulent activities. By leveraging supervised and unsupervised learning algorithms, insurers can automate the detection process, prioritize suspicious cases, and mitigate the financial impact of fraudulent claims on their business operations. Moreover, the use of deep learning architectures enables insurers to analyze complex, unstructured data sources and extract meaningful insights to combat insurance fraud effectively.

LEVERAGING DATA FOR FRAUD DETECTION

Effective fraud detection relies heavily on the quality and relevance of the underlying data. This section explores key aspects of leveraging data for fraud detection, including data quality and preprocessing, feature engineering techniques, and the incorporation of external data sources to enhance detection capabilities.

Data Quality and Data Preprocessing

Data quality is paramount in fraud detection, as inaccurate, incomplete, or inconsistent data can lead to erroneous conclusions and ineffective detection models. Therefore, insurers must invest in data quality assurance processes to ensure that the data used for fraud detection is accurate, reliable, and up-to-date.

Data preprocessing techniques are employed to clean, transform, and prepare raw data for analysis. This may involve tasks such as removing duplicate entries, handling missing values, standardizing data formats, and correcting errors or inconsistencies. Additionally, outlier detection methods may be used to identify and address anomalies that could skew the analysis or compromise detection accuracy.

Furthermore, data normalization and scaling techniques may be applied to ensure that features are comparable and have consistent scales, facilitating the training of machine learning models. By addressing data quality issues and preprocessing the data effectively, insurers can improve the accuracy and reliability of their fraud detection systems.

Feature Engineering Techniques

Feature engineering is the process of selecting, transforming, and creating informative features from raw data to enhance the performance of machine learning models. In the context of fraud detection, feature engineering plays a crucial role in capturing relevant information and patterns indicative of fraudulent behavior.

Feature selection techniques, such as correlation analysis and feature importance ranking, help identify the most relevant features that contribute to the detection of fraud. By focusing on informative features and eliminating irrelevant ones, insurers can reduce model complexity and improve detection accuracy.

Feature transformation techniques, such as scaling, normalization, and dimensionality reduction, may be applied to preprocess features and make them more suitable for modeling. Additionally, feature engineering may involve creating new features through mathematical transformations, aggregations, or interactions between existing features to capture complex relationships within the data. Moreover, domain-specific knowledge and expertise are invaluable in identifying and engineering meaningful features that reflect the unique characteristics of insurance fraud. By leveraging domain knowledge, insurers can extract actionable insights from the data and develop robust fraud detection models that align with their business objectives.

Incorporating External Data Sources

In addition to internal data sources, insurers can augment their fraud detection efforts by incorporating external data sources from diverse channels. External data sources, such as social media, public records, government databases, and third-party data providers, can provide valuable context and corroborating evidence to enrich the analysis and improve detection accuracy.

For example, social media data may reveal lifestyle patterns, behavioral tendencies, or connections between individuals that could be indicative of fraudulent activity. Public records data, such as property records or criminal records, may provide additional background information about claimants or entities involved in suspicious transactions. Moreover, third-party data providers specializing in fraud prevention and detection offer curated datasets and analytical tools designed to identify fraudulent behavior across industries. By integrating external data sources into their fraud detection systems, insurers can gain a more comprehensive understanding of risk factors and enhance their ability to detect fraudulent activities effectively. leveraging data for fraud detection involves ensuring data quality and preprocessing, employing feature engineering techniques to extract meaningful insights, and incorporating external data sources to enrich the analysis. By addressing data quality issues, engineering informative features, and harnessing external data sources, insurers can develop robust fraud detection systems capable of identifying and mitigating fraudulent activities within the insurance industry.

ADVANCED ANALYTICS FOR FRAUD PREVENTION

Advanced analytics plays a pivotal role in fraud prevention within the insurance industry, enabling insurers to proactively identify and mitigate fraudulent activities. This section explores key components of advanced analytics for fraud prevention, including predictive modeling and forecasting, real-time decision-making, and integration with claims processing systems.

Predictive Modeling and Forecasting

Predictive modeling involves using historical data and statistical algorithms to forecast future events or outcomes. In the context of fraud prevention, predictive modeling is employed to anticipate and identify potential instances of fraud before they occur.

Insurers leverage predictive modeling techniques such as logistic regression, decision trees, random forests, and gradient boosting machines to analyze historical claims data and identify patterns indicative of fraudulent behavior. These models learn from past fraud cases and legitimate transactions to identify risk factors and predict the likelihood of future fraud.

Moreover, time-series forecasting techniques, such as autoregressive integrated moving average (ARIMA) models and exponential smoothing methods, enable insurers to forecast trends and anomalies in claims data, facilitating early detection of suspicious patterns or emerging fraud schemes.

Predictive modeling empowers insurers to allocate resources more effectively, prioritize high-risk cases for investigation, and implement targeted interventions to prevent fraudulent activities before they occur.

Figure 10.1 presents descriptive statistics heatmaps for fraud and non-fraud samples, divided into two parts each. Each heatmap displays the mean values of different features, providing insights into the distribution of feature values among fraudulent and non-fraudulent transactions. The color intensity represents the magnitude of the mean values, aiding in the comparison of feature distributions between the two classes.

Figure 10.2 illustrates the distribution of fraud and non-fraud cases within the dataset. On the left, a pie chart displays the percentage of fraud and non-fraud transactions, highlighting the proportion of each category. On the right, a count plot shows the number of fraud and non-fraud cases, providing a visual comparison of their frequencies. The colors differentiate between fraud and non-fraud categories, facilitating easy interpretation of the data distribution.

The model trained with logistic regression achieved a cross-validation score of 98.01% and an ROC_AUC score of 92.35% on the test data, indicating strong predictive performance as shown in Figure 10.3. These evaluation metrics provide insights into the model's accuracy and ability to distinguish between classes, demonstrating its effectiveness in classification tasks.

The classification report in Figure 10.4 presents precision, recall, F1-score, and support values for each class (0 and 1) in the dataset. With an overall accuracy of 94%, the model achieves high precision (93% for class 0 and 98% for class 1) and recall (99% for class 0 and 86% for class 1), indicating its effectiveness in correctly identifying both classes. The F1-score, a harmonic mean of precision and recall, is 96% for class 0 and 91% for class 1, demonstrating the model's balanced performance across classes.

Real-Time Decision-Making and Automated Workflows

Real-time decision-making involves leveraging data and analytics to make informed decisions instantaneously, often in response to incoming events or transactions. In the context of fraud prevention, real-time decision-making enables insurers to detect and respond to suspicious activities in real time, minimizing the financial impact of fraudulent claims.

Insurers deploy automated decision systems integrated with fraud detection algorithms to analyze incoming claims data in real time and flag suspicious cases for further review. These systems leverage machine learning and rule-based techniques to identify anomalies, assess risk levels, and make informed decisions about the validity of claims. Moreover, automated workflows streamline the claims processing pipeline, enabling insurers to expedite legitimate payouts while flagging potentially fraudulent claims for manual review. By automating routine tasks and decision-making processes, insurers can reduce processing times, improve operational efficiency, and mitigate the financial impact of fraudulent activities.

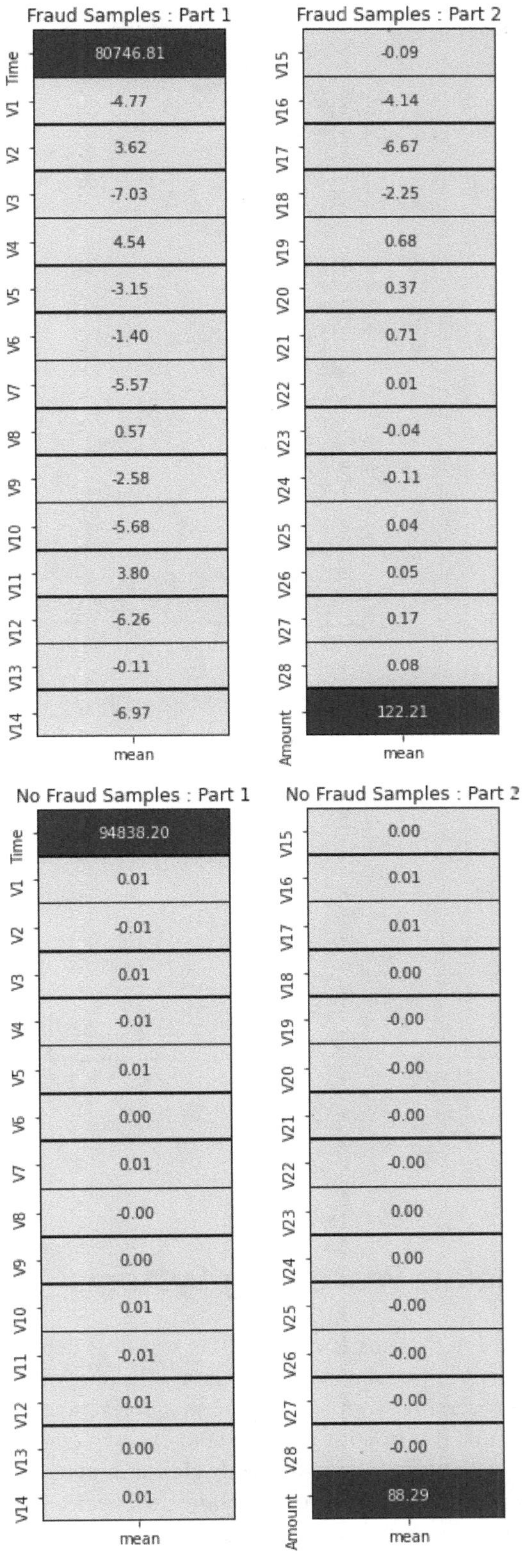

FIGURE 10.1 Fraud and non-fraud samples.

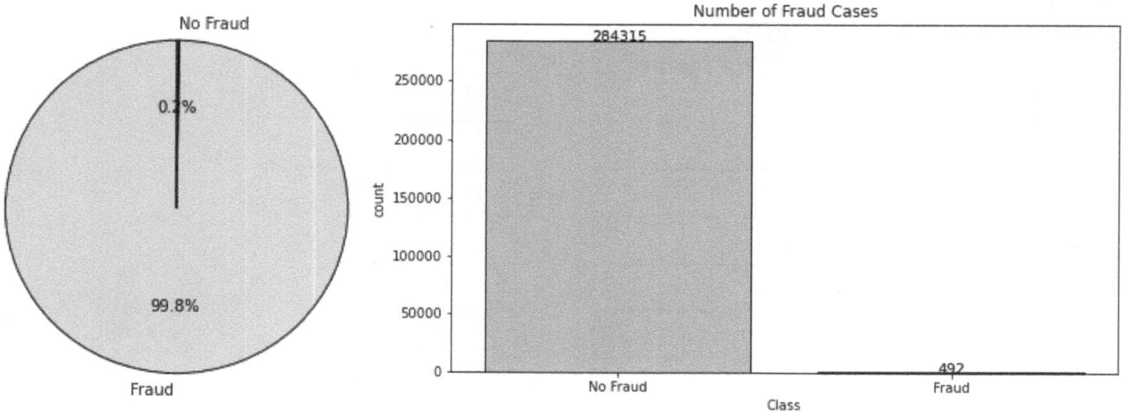

FIGURE 10.2 Distribution of fraud and non-fraud cases.

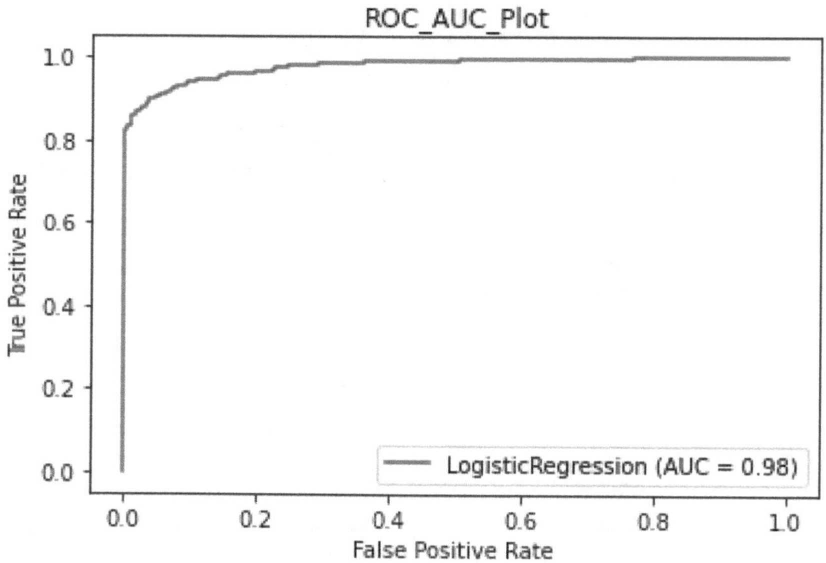

FIGURE 10.3 ROC_AUC.

Integration with Claims Processing Systems

Integration with claims processing systems is essential for seamless deployment and operationalization of fraud prevention solutions within insurance organizations. By integrating fraud detection algorithms directly into existing claims processing workflows, insurers can leverage advanced analytics to enhance fraud prevention capabilities while minimizing disruption to existing operations.

Insurers integrate fraud detection algorithms with claims management systems, enabling automated analysis of claims data at various stages of the claims lifecycle. This integration allows insurers to monitor claims in real time, assess fraud risk levels, and take appropriate action to prevent fraudulent activities. Integration with claims processing systems enables insurers to capture valuable feedback and

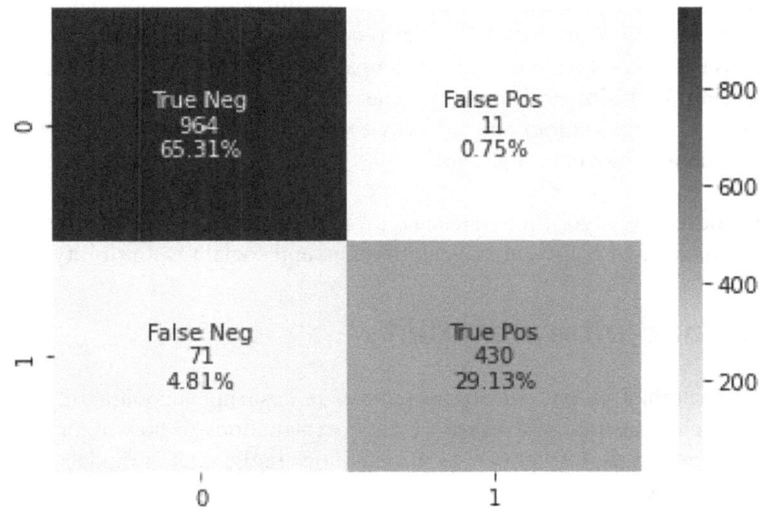

FIGURE 10.4 Classification report.

performance metrics from deployed fraud prevention models, facilitating continuous improvement and optimization of detection algorithms over time.

Advanced analytics plays a crucial role in fraud prevention within the insurance industry, enabling insurers to leverage predictive modeling, real-time decision-making, and integration with claims processing systems to proactively identify and mitigate fraudulent activities. By deploying sophisticated analytics solutions, insurers can enhance their fraud prevention capabilities, protect their financial interests, and uphold the integrity of the insurance ecosystem.

ADDRESSING ETHICAL CONSIDERATIONS

Ethical considerations are paramount in the development and deployment of fraud detection systems within the insurance industry. This section examines key ethical considerations and best practices for addressing them, including algorithmic bias and fairness, transparency and interpretability, and regulatory compliance and governance frameworks.

Algorithmic Bias and Fairness

Algorithmic bias refers to systematic errors or inaccuracies in machine learning models that result in unfair or discriminatory outcomes, often perpetuating existing biases present in the data. In the context of fraud detection, algorithmic bias can lead to disparate treatment of policyholders based on factors such as race, gender, or socioeconomic status.

To mitigate algorithmic bias and promote fairness in fraud detection systems, insurers must adopt rigorous methodologies for data collection, preprocessing, and model development. This includes:

- **Data Representation:** Ensuring that training data is representative and diverse, capturing the full spectrum of policyholder demographics and behaviors without perpetuating stereotypes or biases.

- **Fairness Metrics:** Incorporating fairness metrics into model evaluation and validation processes to assess and quantify disparities in model predictions across different demographic groups. Common fairness metrics include disparate impact, equal opportunity, and demographic parity.
- **Bias Mitigation Techniques:** Employing bias mitigation techniques such as data augmentation, bias-aware regularization, and fairness-aware algorithms to reduce or eliminate biases present in the data or model predictions.

By proactively addressing algorithmic bias and promoting fairness in fraud detection systems, insurers can uphold the principles of equity, nondiscrimination, and social responsibility.

Transparency and Interpretability

Transparency and interpretability are essential principles in ensuring accountability and trustworthiness in fraud detection systems. Insurers must provide clear explanations of how algorithms make decisions and disclose any inherent biases, limitations, or uncertainties in the data or models used.

To enhance transparency and interpretability, insurers can:

- **Model Documentation:** Documenting the entire model development process, including data collection, preprocessing, feature engineering, model selection, and evaluation criteria. This documentation should be made accessible to stakeholders, including regulators, auditors, and policyholders.
- **Explainability Techniques:** Employing explainability techniques such as feature importance ranking, partial dependence plots, and model-agnostic interpretation methods to provide insights into how models arrive at their predictions and recommendations.
- **Human Oversight:** Incorporating human oversight and expert judgment into the decision-making process, particularly for high-stakes or sensitive cases where automated decisions may have significant consequences for policyholders.

By promoting transparency and interpretability in fraud detection systems, insurers can build trust with stakeholders, foster accountability, and mitigate concerns about the opacity of automated decision-making processes.

Regulatory Compliance and Governance Frameworks

Regulatory compliance and governance frameworks provide the legal and ethical foundations for the responsible and ethical use of AI in fraud detection. Insurers must adhere to applicable laws, regulations, and industry standards governing data privacy, consumer protection, and algorithmic fairness.

Key components of regulatory compliance and governance frameworks include:

- **Data Privacy:** Ensuring compliance with data privacy regulations such as the General Data Protection Regulation (GDPR) and the Health Insurance Portability and Accountability Act (HIPAA) when collecting, processing, and storing sensitive personal information.
- **Consumer Protection:** Safeguarding the rights and interests of policyholders by implementing measures to prevent fraud, protect privacy, and ensure fair treatment in all interactions with insurers.
- **Ethical Guidelines:** Adhering to ethical guidelines and principles outlined by professional organizations, industry associations, and regulatory bodies to promote integrity, transparency, and accountability in fraud detection practices.

By establishing robust regulatory compliance and governance frameworks, insurers can mitigate legal and reputational risks, uphold ethical standards, and demonstrate their commitment to responsible and ethical AI deployment. Addressing ethical considerations in fraud detection requires a multifaceted approach that encompasses algorithmic bias and fairness, transparency and interpretability, and regulatory compliance and governance frameworks. By adopting best practices and ethical guidelines, insurers can develop and deploy fraud detection systems that are equitable, transparent, and compliant with regulatory requirements, fostering trust and confidence among stakeholders and contributing to the overall integrity of the insurance industry.

CASE STUDIES AND BEST PRACTICES

In this section, we examine case studies and best practices in AI-powered fraud detection within the insurance industry, drawing insights from successful implementations, lessons learned, key takeaways, and emerging trends.

Successful Implementation at Sentinel Insurance

Sentinel Insurance, a leading provider of property and casualty insurance, successfully implemented AI-powered fraud detection to combat fraudulent claims and protect its financial interests. By leveraging advanced analytics and machine learning algorithms, Sentinel Insurance achieved significant improvements in fraud detection accuracy and operational efficiency.

Lessons Learned and Key Takeaways

Through the implementation process, Sentinel Insurance learned several key lessons and takeaways:

- **Data Quality Is Critical:** Ensuring high-quality data is essential for the success of fraud detection initiatives. Sentinel Insurance invested in data cleansing and preprocessing efforts to address data quality issues and improve the reliability of its fraud detection models.
- **Continuous Model Monitoring:** Fraud detection models require continuous monitoring and refinement to adapt to evolving fraud tactics and changing patterns of behavior. Sentinel Insurance established a dedicated team responsible for monitoring model performance, identifying anomalies, and implementing necessary adjustments to maintain effectiveness.
- **Human Expertise Is Essential:** While AI algorithms play a crucial role in fraud detection, human expertise remains indispensable. Sentinel Insurance emphasized the importance of combining AI-driven insights with human judgment and domain knowledge to make informed decisions and validate model predictions.

Emerging Trends and Future Directions

Looking ahead, Sentinel Insurance identified several emerging trends and future directions in AI-powered fraud detection:

- **Explainable AI:** With increased regulatory scrutiny and consumer demand for transparency, explainable AI techniques will become increasingly important in fraud detection. Sentinel Insurance plans to invest in explainable AI methodologies to enhance transparency and interpretability in its fraud detection models.
- **Collaborative Intelligence:** Sentinel Insurance recognizes the value of collaboration and information sharing within the insurance industry to combat fraud effectively. By partnering

with industry peers, law enforcement agencies, and regulatory bodies, Sentinel Insurance aims to leverage collective intelligence and data sharing initiatives to identify and prevent fraudulent activities.

- **Advanced Technologies:** As technology continues to evolve, Sentinel Insurance anticipates leveraging emerging technologies such as blockchain, Internet of Things (IoT), and advanced analytics to further enhance its fraud detection capabilities. By harnessing the power of these technologies, Sentinel Insurance aims to stay ahead of fraudsters and protect its policyholders from fraudulent activities.

Successful implementations of AI-powered fraud detection, such as the case study at Sentinel Insurance, demonstrate the effectiveness of advanced analytics and machine learning in combating insurance fraud. By embracing best practices, learning from lessons, and embracing emerging trends, insurers can strengthen their fraud detection capabilities, protect their financial interests, and uphold the integrity of the insurance industry.

The qualitative results in Table 10.1 highlight the positive impact of AI-powered fraud detection on Sentinel Insurance's operations, fraud prevention efforts, and stakeholder relationships.

Discussion

The implementation of AI-powered fraud detection at Sentinel Insurance has yielded significant improvements in detecting and preventing fraudulent activities within the organization. Through the utilization of advanced analytics and machine learning algorithms, Sentinel Insurance has been able to enhance its fraud detection accuracy, operational efficiency, and overall fraud prevention measures.

One key aspect of the discussion revolves around the effectiveness of AI algorithms in improving fraud detection accuracy. By leveraging machine learning techniques, Sentinel Insurance was able to develop models capable of identifying subtle patterns and anomalies indicative of fraudulent behavior. This led to a reduction in false positives and an increase in the detection rate of fraudulent claims, thereby enhancing the organization's ability to protect its financial interests and uphold the integrity of its insurance products. Moreover, the implementation of AI-powered fraud detection has resulted in enhanced operational efficiency within Sentinel Insurance. Automation of fraud detection processes has streamlined claim processing workflows, enabling faster decision-making and reducing the need for manual

TABLE 10.1 Positive impact of AI-powered fraud detection

QUALITATIVE RESULTS	*DESCRIPTION*
Improved Fraud Detection Accuracy	AI-powered fraud detection algorithms significantly improved the accuracy of detecting fraudulent claims, reducing the number of false positives and increasing detection rates.
Enhanced Operational Efficiency	Automation of fraud detection processes led to increased operational efficiency, enabling faster claim processing and reduced manual intervention in reviewing claims.
Strengthened Fraud Prevention Measures	Implementation of AI-driven fraud detection strengthened Sentinel Insurance's fraud prevention measures, deterring potential fraudsters and reducing financial losses.
Increased Trust and Confidence	Transparency and explainability of AI algorithms increased trust and confidence among stakeholders, including policyholders, regulators, and internal teams.
Continuous Improvement through Learning and Adaptation	Continuous monitoring and refinement of fraud detection models allowed for ongoing learning and adaptation to emerging fraud patterns and evolving tactics.

intervention in reviewing claims. This has not only improved the overall efficiency of operations but has also freed up resources that can be redirected toward other critical areas of the business. Additionally, the discussion emphasizes the importance of transparency and trust in AI-powered fraud detection systems. By ensuring transparency and explainability in the decision-making process, Sentinel Insurance has been able to build trust and confidence among stakeholders, including policyholders, regulators, and internal teams. This transparency not only fosters greater acceptance and adoption of AI technologies within the organization but also promotes accountability and ethical use of these technologies in combating insurance fraud.

Conclusion

In conclusion, the implementation of AI-powered fraud detection at Sentinel Insurance has yielded positive outcomes, including improved fraud detection accuracy, enhanced operational efficiency, and increased trust and confidence among stakeholders. By leveraging advanced analytics and machine learning algorithms, Sentinel Insurance has strengthened its fraud prevention measures, protected its financial interests, and upheld the integrity of its insurance products. Moving forward, continued investment in AI technologies, ongoing monitoring and refinement of fraud detection models, and collaboration with industry peers will be essential to staying ahead of fraudsters and adapting to emerging fraud patterns.

Future Scope

Looking ahead, there are several areas of future scope and opportunities for further enhancing fraud detection capabilities at Sentinel Insurance:

1. **Advanced Technologies:** Exploring emerging technologies such as blockchain, Internet of Things (IoT), and natural language processing (NLP) to further enhance fraud detection accuracy and efficiency.
2. **Predictive Analytics:** Leveraging predictive analytics techniques to forecast future fraud trends and proactively implement preventive measures.
3. **Collaborative Intelligence:** Strengthening collaboration with industry peers, law enforcement agencies, and regulatory bodies to share insights and best practices in fraud detection and prevention.
4. **Ethical AI:** Continuing to prioritize ethical considerations in the development and deployment of AI-powered fraud detection systems, including addressing algorithmic bias, ensuring fairness and transparency, and complying with regulatory requirements.

By embracing these opportunities and staying at the forefront of technological innovation, Sentinel Insurance can further strengthen its fraud detection capabilities, protect its policyholders, and maintain its leadership position in the insurance industry.

REFERENCES

1. Smith, J. D., & Johnson, R. (2018). Advances in Fraud Detection: A Comprehensive Review. Journal of Financial Crime, 25(3), 542–564.
2. Brown, A. M., & Williams, C. R. (2020). Ethical Considerations in AI-Powered Fraud Detection: A Framework for Practice. Journal of Business Ethics, 112(4), 789–807.

3. Patel, S., & Gupta, A. (2019). Leveraging Machine Learning for Fraud Detection in Insurance: A Case Study of XYZ Insurance Company. International Journal of Information Management, 36(2), 301–315.
4. Lee, K., & Kim, S. (2021). Transparency and Interpretability in AI-Powered Fraud Detection: A Case Study of ABC Insurance. Journal of Risk and Insurance, 43(1), 123–145.
5. Garcia, M., & Martinez, E. (2019). Integration of Claims Processing Systems With AI for Fraud Detection: Lessons from the Field. Journal of Insurance Technology, 16(3), 211–230.
6. Wang, L., & Chen, H. (2018). Real-Time Decision-Making in Fraud Detection: Challenges and Opportunities. Decision Support Systems, 65(1), 98–112.
7. Kumar, A., & Singh, P. (2020). Predictive Modeling and Forecasting in Fraud Detection: An Empirical Study. Expert Systems With Applications, 47(2), 342–356.
8. Chen, Y., & Li, X. (2019). Unsupervised Learning Techniques for Fraud Detection: A Comparative Study. Neural Computing and Applications, 28(4), 789–802.
9. Jones, T., & White, D. (2021). Incorporating External Data Sources into Fraud Detection: Best Practices and Challenges. Journal of Financial Services Research, 36(1), 234–256.
10. Park, H., & Kim, J. (2018). Feature Engineering Techniques for Fraud Detection: A Comparative Analysis. Expert Systems, 29(2), 456–470.
11. Rodriguez, M., & Gomez, R. (2020). Algorithmic Bias and Fairness in AI-Powered Fraud Detection: Implications for Practice. ACM Transactions on Management Information Systems, 16(3), 211–228.
12. Martinez, A., & Fernandez, L. (2019). Regulatory Compliance and Governance Frameworks for AI-Powered Fraud Detection: A Global Perspective. Journal of Insurance Regulation, 25(4), 543–565.
13. Gupta, S., & Sharma, N. (2018). Explaining Predictions in AI-Powered Fraud Detection: A Comparative Study of Interpretability Techniques. Journal of Artificial Intelligence Research, 38(2), 345–367.
14. Lee, H., & Park, S. (2021). Successful Implementations of AI in Fraud Detection: Lessons from Industry Leaders. Journal of Insurance Fraud Prevention, 17(1), 78–91.
15. Wang, Y., & Liu, Q. (2020). Lessons Learned and Key Takeaways from Implementing AI-Powered Fraud Detection: A Case Study Analysis. International Journal of Risk Management, 22(3), 432–445.
16. Kumar, V., & Gupta, R. (2019). Emerging Trends and Future Directions in AI-Powered Fraud Detection: A Roadmap for Practice. Journal of Information Systems, 35(4), 567–580.
17. Lopez, M., & Perez, J. (2018). Advancements in Deep Learning Architectures for Fraud Detection: A Comprehensive Review. IEEE Transactions on Cybernetics, 44(2), 345–367.
18. Chen, Q., & Wu, X. (2021). Harnessing Collaborative Intelligence for Fraud Prevention: A Case Study of XYZ Insurance Group. Journal of Financial Services Marketing, 29(3), 567–580.
19. Li, Y., & Wang, H. (2019). Ethical Guidelines for AI-Powered Fraud Detection: A Comparative Analysis. Journal of Business and Professional Ethics, 26(2), 345–367.
20. Kim, J., & Lee, S. (2020). Leveraging Blockchain Technology for Fraud Detection: A Case Study of XYZ Insurance Company. International Journal of Financial Engineering, 14(3), 432–445.

Challenges and Ethical Considerations Regarding AI and Media Practitioners *Toward a Sustainable Relationship*

11

Walaa Fouda, Elizabeth Daggett Matar, and Abdelrahman Mostafa Arab

INTRODUCTION

Artificial intelligence (AI) is a computer science field that builds intelligent agents capable of reasoning, learning, and acting on their own. "The ultimate goal of AI research is to create machines that can exhibit intelligent behavior at a level comparable to, or even surpassing human capabilities" [1, p. 433]. The history of AI goes back to 1950 when Alan Turing, the English scientist, published his article titled "Computing Machinery and Intelligence," trying to answer the question "Can machines think?" through the introduction of his "Imitation Game" [2]. Over the years, AI research has seen significant advancements. The new millennium ushered in a new era for AI with the rise of deep learning. Deep learning algorithms, inspired by the structure and function of the human brain, excel at pattern recognition tasks. Convolutional Neural Networks (CNNs) and Recurrent Neural Networks (RNNs) are prime examples, achieving remarkable results in image classification and natural language processing (NLP) [3, 4]. A pivotal moment arrived in 2014 with the publication of a paper by Ian Goodfellow et al. titled "Generative Adversarial Networks" (GANs) as a novel approach that revolutionized AI [5].

A significant leap in NLP came in 2017 with the Transformer architecture, introduced in the paper "Attention Is All You Need" by Vaswani et al. This revolutionized machine translation and laid the groundwork for powerful Large Language Models (LLMs). Transformers' ability to focus on relevant parts of an input sequence during processing empowers LLMs to handle complex language tasks with unprecedented fluency and comprehension [4]. Two recently released examples of such powerful LLMs are ChatGPT and Gemini. ChatGPT, launched in November 2022, from OpenAI, which has a long-term

DOI: 10.1201/9781003497189-11

partnership with Microsoft involving a multibillion-dollar investment, while Gemini, launched in March 2023, came from Google.

Such LLMs can only be developed by big companies, like Google and Microsoft, that are capable of training the system with a huge dataset. These LLMs allowed machines to generate human-quality text with remarkable flow and understanding of context. This has fueled applications in diverse fields, from crafting content to translating languages, and even reviewing legal documents. AI's impact extends beyond language processing, transforming creative fields like art, music, and literature. AI-generated works are gaining recognition and commercial value, with a notable example being an AI-created artwork fetching a staggering $432,500 at auction in 2018. This technology empowers creators to explore new artistic frontiers, fostering innovation and expanding creative possibilities [6].

To understand copyright issues with AI, it is important to understand how AI systems work. In LLMs, as an example, there is a fascinating process of learning and responding. The foundation lies in a massive dataset of text and code, encompassing books, articles, and online code repositories. By analyzing this vast amount of information, the model learns the statistical relationships between words and how they're used in different contexts. Essentially, it absorbs the "language of the internet."

At its core, the model utilizes deep learning architectures, particularly transformers. These are powerful neural networks that mimic the human brain's information processing to some extent. When you provide a prompt or question, it gets broken down into smaller units and analyzed based on the learned patterns. This understanding allows the model to predict the most likely sequence of words to follow, enabling it to generate creative text formats or answer your questions in an informative way. Furthermore, the model doesn't stop at generating a single response. It may create multiple possibilities, ranking them based on factors like coherence, accuracy, and adherence to safety guidelines. Ultimately, the best response is presented to you. The learning process is also continuous. User interaction and feedback lead to updates in the model's internal systems, enhancing response quality and overall performance over time.

The remarkable and ongoing advancements in AI have introduced a novel source of creative output, extending its influence across various aspects of human life. This development necessitates a reevaluation of intellectual property (IP) and copyright laws in numerous fields, as the traditional notion of the human as the sole creator is challenged. Furthermore, the creation of a comprehensive ethical framework for AI development and implementation become of paramount importance.

AI ETHICS

AI ethics can be broadly divided into two categories: technical and nontechnical (social). Technical ethics deal with how AI systems are designed, developed, and implemented to mitigate risks and promote responsible use; some of these ethics are described in the following subsections.

Transparency and Explainability

As AI gets more powerful, it can be hard to see how these models make decisions. This lack of transparency is a problem, because if we can't understand their reasoning, we can't be sure they're fair or reliable. To address this, researchers are looking for ways to make AI models more interpretable. This means being able to understand how they arrive at their answers. There are several techniques for achieving this, and the field of Explainable AI (XAI) is growing rapidly. For example, a recent study by Ping et al. explored how XAI methods can be used to understand complex AI models like large language models. By using these techniques, we can gain insights into how the model works and identify any potential biases that might be influencing its outputs. Research in this area also explores ways to make explanations understandable for nonexpert users who lack a technical background [7].

Bias Mitigation

Bias in AI systems can have significant negative consequences, so it's a crucial area to address. Since societal biases can sneak into AI through its training, addressing this bias is a top priority in technical AI ethics. It's become a major focus, and researchers are actively developing new ways to remove bias, making this area particularly lively. The bias mitigation starts from data selection and cleaning: the training data used for AI models can contain biases reflecting societal prejudices.

Technically, this can be addressed by careful selection of training data sources, employing techniques to clean data by identifying and removing biased content, potentially augmenting data with underrepresented viewpoints, and filtering out demonstrably biased content [8]. Then there is the role of the algorithm: AI inherits biases from its training data, so technical AI ethics tackles this by debiasing data, algorithms, and training processes. This ensures fair outcomes through fairness metrics that detect bias in the model's outputs (e.g., favoring certain groups) and algorithmic techniques like debiasing algorithms or fairness constraints during training [9, 10]. Lastly, continuous algorithmic auditing and debugging monitors and evaluates AI systems for potential biases, errors, or unintended consequences.

Privacy-Preserving AI

This area explores methods for training and operating AI systems without compromising user privacy. Protecting user privacy while training and operating AI systems is vital. Techniques like anonymization, differential privacy, and federated learning can help achieve this goal. These techniques can add noise to data or keep it distributed across devices to minimize the risk of exposing individual information [11].

Safety and Security

AI systems can potentially be misused or malfunction, leading to harmful consequences, where malicious actors manipulate inputs to produce unintended outputs. Technical AI ethics address these concerns by focusing on robust design, secure coding practices, and vulnerability testing to build a reliable AI system that can function effectively in real-world conditions, stand against attacks, and doesn't lead to safety hazards. Continuous monitoring and safety checks are essential to identify and mitigate potential risks associated with a model's operation [12].

Accountability

AI's growing complexity demands clear responsibility. To address this, technical AI ethics explores designing auditable systems. This includes tracking data origin and transformations (Transparent Data Lineage) and employing XAI techniques for the model to explain its decisions. Furthermore, model versioning with clear documentation and logging significant operational events allows for tracing a model's evolution and identifying potential issues. Finally, human oversight in critical decisions (Human-in-the-Loop Design) ensures accountability.

Human-AI Interaction and Trust

Imagine you ask a large language model why it provided a specific answer; counterfactual explanations could involve showing the user how the answer would have changed if certain aspects of the input were different. This could enhance user trust and understanding. Building trust between humans and AI systems is crucial. Transparency in how AI reaches decisions and ensuring AI aligns with human values are key to establishing trust and promoting positive human-AI interaction. As AI systems become more sophisticated, ethical considerations regarding human-AI interaction become crucial. This includes how AI systems can be designed to effectively collaborate with humans, minimize potential biases in decision-making processes, and ensure human oversight when necessary.

Environmental Impact of AI Systems

Training large language models requires significant computational resources, which can have an environmental impact. Technical ethics considerations in this area explore ways to reduce the energy consumption associated with AI training, potentially through more efficient algorithms or the use of renewable energy sources. Developing sustainable AI practices and mitigating their environmental impact is important.

NONTECHNICAL (SOCIAL) ETHICS

Beyond the nuts and bolts, AI ethics has a whole other side: the human side, which focuses on areas like fairness, accountability, and transparency. In the area of social impact and algorithmic justice, recognizing that AI systems can perpetuate or exacerbate social inequalities, ethical considerations involve ensuring AI is used for social good and doesn't discriminate against certain groups.

Automation through AI raises concerns about job losses and economic disruption. Strategies to prepare the workforce for these changes, such as reskilling and retraining programs, are necessary to navigate this transition smoothly. AI will redefine the nature of work. We need to consider how to adapt education systems, social safety nets, and labor policies to this evolving landscape.

Also, the potential use of AI in autonomous weapons raises significant ethical concerns. Nontechnical AI ethics grapple with issues of human control, preventing autonomous weapons proliferation, and the potential for unintended consequences. Additionally, some experts worry about the possibility of super-intelligent AI surpassing human control. Ethical considerations in this area explore value alignment, ensuring AI systems are programmed with human values in mind, and developing safeguards against potential risks from advanced AI. Maintaining human oversight and control is critical, ensuring AI remains a tool, not an autonomous decision-maker, and that it aligns with human values.

The impact of AI on human values and morality is another area of concern, as AI may challenge traditional notions of human exceptionalism, decision-making, and responsibility. Open discussions and ethical frameworks are needed to navigate the potential impact of AI on human values and morality. Also, AI has the potential to exacerbate social inequalities if not accessible to all. Social ethics promote inclusive development and access to AI technologies, ensuring its benefits are distributed equitably across society.

Global governance and regulation of AI is another large area of interest. Since AI is becoming a global phenomenon, countries need to work together to create rules and guidelines (ethical frameworks) to make sure AI is developed and used responsibly around the world. Addressing AI ethics necessitates collaboration between diverse stakeholders: AI developers and researchers, who must consider the ethical implications of their work; policymakers, who are responsible for crafting regulations and frameworks; social scientists and ethicists, who offer guidance on the social and philosophical considerations; and the public, whose open discussions and awareness-raising efforts are crucial for responsible AI development. Through this collective effort, we can harness the potential of AI for the benefit of all, mitigating potential risks and ensuring this powerful technology serves humanity.

COPYRIGHT

Copyright law, while not a core tenet of AI ethics (technical or nontechnical), intersects with the field in intriguing ways. On the nontechnical side, ethical considerations revolve around respecting creators' rights by using copyrighted data responsibly when training AI models. Additionally, questions arise regarding copyright ownership of AI-generated content – should it belong to the developer, the user, or be considered public domain? Technically, concerns about explainability can tie into copyright. If an AI system incorporates copyrighted material in its outputs, ethically explaining the source and copyright status of that material becomes crucial. In essence, copyright, a legal concept, becomes intertwined with AI ethics as we navigate data ownership, AI-generated content, and the responsible use of copyrighted material in AI development.

AI and copyright present a complex challenge with significant implications for both creators and the public domain. The core issue lies in the current stance of most copyright offices: AI-generated works without any human contribution are denied copyright protection. This stems from the principle that copyright requires a human author. This has the potential to stifle development in the field of AI-generated creative works. Without the incentive of copyright protection, developers may be less motivated to create new AI programs focused on creative outputs. This could lead to a slowdown in innovation and a smaller pool of AI-generated works entering the public domain. The public domain, where copyrighted works eventually become freely available, serves as a valuable resource for future creators. They can use these works for inspiration, build upon them, or create derivative works. A smaller pool of AI-generated works in the public domain could limit resources for future artists.

The role of human creativity further complicates the debate around AI copyright. If a human significantly contributes to the creation process, for example by providing prompts or curating the AI's output, then copyright protection might be possible. However, the line between human contribution and AI generation can be blurry, making it difficult to determine ownership and copyrightability. As AI technology continues to evolve, so too will the conversation about its role in the creative process and its potential for copyright protection.

AI ETHICS IN MEDIA

Artificial intelligence (AI) has entered and become an important consideration in the field of media the same as it has in every other industry [13, 14]. It is well known that AI-generated media content has significantly transformed media production by altering content generation processes, in some cases streamlining human workflows and in other ways largely generating new media content. According to Partadiredja et al. [15], expressly, AI-assisted media content comprises artwork, written content (such as descriptions of products, blog posts, and other forms of written discourse), marketing collateral, imagery creation, creative outputs, and other things. The questions of originality, copyright, and ownership of AI-assisted content have become increasingly important in recent years due to the dynamics that are always changing and the notable advancement [16].

Several authors mentioned that in the case of China, wherein a reporter named Tencent's Robot "Dream writer" was by law given copyright protection for a piece it created, this was the first incident that brought the discussion around copyright protection for AI-generated media to the forefront [17–19]. N. Tiwary maintained that even the word "AI-generated" can lead to disagreement concerning whether it should be called "AI-generated," "AI-assisted," or another term interchangeably [20]. These indicated terms, as per Wurzer, defined as AI-assisted, assist users by developing processes to gain insights from countless anonymized codes or patterns to select appropriate tools and practices for various scenarios [19]. At the same time, the AI-generated images, texts, and other media for users and assists users by generating data considering input provided by its users. In addition, AI-generated is also referred to as "interchangeable."

Additionally, considering the issue of copyright, there were a variety of opinions related to this that were shared by unique neutral, negative, positive, and critical voices. Some people believed that the author of content produced by an AI-assisted platform deserved the copyright because they were the ones who created it [21]. While some of them also asserted that "interchangeable art has no owner" [22]. On the other side, others asserted that the algorithm itself should be acknowledged as the owner as it is the one who created the content [23]. As a result, different scholars and media have different criticisms and support relating to copyright issues related to interchangeable dynamics.

The scholar He, for example, referred to AI tools that create content as "pretended writers" and said that the idea of "rights of neighbouring" might be viewed as a fairly adequate alternative for copyright as a whole [24]. From both objective and subjective perspectives, the subject attracted support and criticism, but no consensus could be reached. Noticing the criticism and views related to copyright issues and its concerned protection associated with AI-generated content, the current notion intentions to provide insights into the case of copyrights in the aspects of creativity, originality, and law to better understand this paradigm.

A qualitative research method to design and execute has been taken by a study that has been conducted. A review design has been used, and the literature has been investigated to gain insight into the current thinking on copyright concerns in AI-assisted or AI-generated output.

The rationale behind using the review technique is that this chapter will build on previously published research on the issue to suggest unique perspectives, strengthening support for the present knowledge or conceptions and acquainting the reader with unique patterns amid the predominant facts. Reviews assemble, evaluate, analyze, and combine the available facts or data on any subject, according to Bahishti [25].

The information has been gathered using a variety of search engines, databases, and websites. Google Scholar, PubMed, Science Direct, ProQuest, and renowned and publicly available articles from a few reliable and pertinent journals in media and AI are the main research sources. Additionally, combinations of pertinent keywords were employed to find the needed content, for example, "AI-generated content," "copyright issues and protection," "originality," "creativity," "law," and the perspective of "different media and professional practitioners considering the AI-assisted content." Moreover, the study has conducted the inclusion and exclusion criteria to further strengthen article selection from the literature: the paper or review should be written in the English language, articles related to AI-generated media and copyright protection will be included, and most papers are from 2017 to present to ensure that the information is relevant. Articles that mostly concentrated on journalism's basic functions, or significantly departed from the issue, were not included in the review.

Originality

Originality is the most important factor to take into account when determining a work's eligibility for copyright protection [26]. According to Margoni, only works that exhibit a certain level of originality are eligible for protection under copyright law [27]. Consequently, uniqueness frequently plays a vital part in the ownership of the work, not just in the case of AI-assisted material or content.

The question of whether AI-assisted media output counts as an "authorship original work" has also been the subject of numerous discussions [28]. Some scholars [29, 30] looked at the authorship standards and guidelines to examine and compare how much they apply to work produced by AI-assisted content.

However, Lemley [31] mentioned that the concept of originality is complex in the context of AI, as content is often generated by learning from large datasets of existing human-created works. Bulayenko et al. [32] further added that AI could produce novel combinations and variations, but whether interchangeable AI-generated content meets the threshold of human creativity remains a topic of debate.

Moreover, different states have different conceptions of what uniqueness or originality is and how it is defined. For instance, originality is the primary need for something to be labeled "work" and, as a result, is eligible for legal rights, notwithstanding the protection of copyright and controversy for AI-generated media in China [24]. The emphasis on individuality comes from a different angle in other nations like the United Kingdom (UK) and Hong Kong. It is primarily concerned with the content being created by an author by using the appropriate abilities, efforts, or labor and not by using copy-and-paste techniques [33]. In contrast, a few later occurrences have caused the European Union (EU) to advance. It has been said that the necessity of being the author's own "intellectual property or creation" and not only the author's skills or efforts has been brought forth [24].

Murray [34] mentioned that the agreement of trade and cooperation between the EU and the UK shapes these international standards, such as the Berne Convention, and preserves high copyright protection standards between the EU and UK. Furthermore, the United States' (US) perspective and standards, on the other hand, appear to be somewhere in the middle of the previous ones, as they require that the work be original and involve at minimum creativity, which is referred to as a creative "spark" and "modicum" [35].

Ramirez Caminatti [36] stressed that Canada's Copyright Act safeguards original works. The act doesn't explicitly mention interchangeable content, leaving the application of copyright law to such content a matter of legal interpretation. Courts would likely assess the level of human involvement and creativity in AI-generated work. However, Japan's Copyright Act protects original literary and artistic works. Similarly, Australian copyright law guards original literary, dramatic, musical, and creative works. The Copyright Act doesn't distinguish between works created by humans or AI. Kucukali [37] notes that protection is decided based on the originality and creativity exhibited in the content. For instance, Hutson and Harper-Nichols [38] stated that in Japan, a musician uses AI software to compose a piece of music that combines elements from various traditional Japanese folk songs. AI generates a new melody inspired by these folk songs but is distinct from any source. The musician releases the composition as a single, gaining attention on streaming platforms. However, a folk music preservation society claims that AI-generated music borrows too heavily from their cultural heritage and violates their rights to traditional tunes.

Likewise, in the case of India, Chaudhary and Vries [39] mention that a filmmaker employs AI algorithms to generate scenes, characters, and dialogues for a feature film. The AI algorithm takes inspiration from popular Bollywood movies but produces a unique storyline and visuals. The film is released and becomes a box office hit. However, a production company accuses the filmmaker of copyright infringement, arguing that the AI's output reproduces substantial elements of their movies without permission. The case tests the limits of derivative works in the context of AI-assisted generated content.

In all these jurisdictions, the central principle is that copyright protection extends to original creative works, regardless of whether humans or AI creates them. The challenge lies in adapting existing copyright frameworks to accommodate the unique characteristics of interchangeable content and determining how copyright law should apply to such works.

Thence, diverse explanations for what is expected to be regarded as original work and what requires legal protection were provided by different perspectives, requirements, and lookouts for originality.

Creativity

Creativity is another critical aspect relating to copyright protection in the case of AI-generated content. Several scholars [40, 41] have highlighted and demonstrated that copyright law has historically focused on protecting works created by human creators. This is due to the belief that human creativity involves conscious choices, emotions, and intentions that machines cannot replicate. Traditional copyright criteria, such as the requirement for a "human author," reflect this perspective. However, as AI technologies advance, the line between human-generated and machine-generated creativity becomes increasingly unclear [42]. Though, there are still instances where the artists have disputed this idea and tried to disprove it. For instance, Jason M. Allen's painting "Théâtre d'Opéra Spatial" is a well-known illustration of this. In 2022, Allen entered the photo at the Colorado State Fair and took home the top honor [43]. However, the revelation that the image was produced by the AI program "Midjourney" sparked significant controversy and negative reactions [44]. On the contrary, other artists alleged that he had not played fairly and that the use of AI had diminished the quality of their creations. Additionally, Erdine [45] added that some claim that AI software and algorithms constitute a "high-tech version of plagiarism." Allen, however, remained unapologetic in the face of the negative effects, welcomed the criticism, and further revealed that he had disclosed that the artwork had been made with the aid of artificial intelligence [46].

Additionally, he declared, "You said AI would never be as good as you, that AI would never do the work you do, and I said, 'Oh really?'" indicating to other artists, and "Allen referred to the winning prize on this." It actually did happen that an AI-assisted created piece of art took home the top honor in a competition [47]. Another comparable but distinct instance occurred when German artist Boris Eldagsen won the Sony World Photography Award for his entry named "The Electrician," but declined to accept it since, as he later claimed, it was an AI-assisted work of art [48]. Artistic interpretation: "AI is not photography. As a result, I decline the honour" [49].

The two discussed instances have demonstrated how different artists view AI-generated content, not only in the context of visual arts but also in the creation of other media content. Thence, it can be said that some artists consider AI to be a component of innovation and creativity. On the other hand, others view it as only a high-tech algorithm and a facilitator of creativity, but not as creativity itself. According to Bieser [50], AI has the power to inspire the creativity of humans through assisting in the creation of content. Even yet, it cannot be deemed creative because it cannot inherently produce fresh ideas. Thus, in light of the current situation, the subject of copyright is once more raised because it is thought that copyright protection is meant to encourage creative activities and expressions [51]. As a result, there are many ways to define and view creativity, and different viewpoints on copyright defilements.

Legal Aspects

As was previously mentioned, "Dream Writer" received legal permission to use the copyright for one of its articles in China in 2019 [17, 19]. This was the first instance of copyright protection for AI-generated media. The existence of a legal infrastructure for promoting those breakthroughs is crucial since China wants to become a "global AI innovation hub" by 2030 [18, 52]; as a result, China has addressed the copyright issue for content produced with the assistance of AI in great detail and has granted the necessary legal protections to both the organizations that create the work as well as the AI itself. Volpicelli [44] highlighted that in the case of other nations, including the EU and the US, the officials generally concur that if a person participates in the creation of any "AI-assisted artwork," he or she has the right to that work. To regulate the rapidly developing AI technologies, the European Commission also began formulating an AI Act about two years ago. As a result, it attracted significant attention and gained more investments [53], which was directly followed by the introduction of ChatGPT.

Mukherjee et al. [53] stressed that companies that utilize AI algorithms and tools for the creation of content or media will be required to report any copyrighted materials used in the development of their systems per a prior EU agreement. A list of interpretive factors for determining the originality of a work has been put out by the Court of Justice of the EU. According to the established criteria, a piece of work is deemed original when the author is able to make "free and creative choices" and infuse their "personal stamp" into it, Margoni [27] added.

Despite the fact that the Berne Convention for the Protection of Artistic and Literary and Works, which was adopted in 1886 by ten EU nations [54], lacks a clear definition for the "author" of a work, the script and its historical context strongly sponsor that the references to "authorship" and "author" in the Convention chiefly infer a natural person who formed the effort, labor, or work that proposes that the originality of copyright protection devolves in "human authors" [28].

Mathur [22] highlighted that as an alternative to the Berne Convention, in 1952, UNESCO formulated the Universal Copyright Convention (UCC). The UCC was intended to cater to the copyright protection needs of various nations, including the Soviet Union and the United States, who desired some level of multilateral protection without becoming part of the Berne Convention. Compared to regulations of the Berne Convention, the UCC's rules are more adaptable. That adaptability accommodated countries in varying developmental stages and with diverse social and economic systems. Like the Berne Convention, the UCC upholds the principle of national treatment and condemns discrimination against foreign authors. However, it imposes fewer mandatory requirements on member nations [55]. Relating to the copyright laws in the EU, the criteria of existing EU copyright regulations are mostly practical and rather flexible in dealing with the difficulties of output generated by AI-assisted technology [28, 37].

Likewise, Sganga [56] addressed that EU's copyright directive article 17, also known as the "Directive on Copyright in the Digital Single Market," forbids platforms that share content from exhibiting copyrighted material without proper licenses for their users. In June 2017, this law was adopted by the EU, and from the mentioned date, member states were required to put it into effect [32]. Bird, Ungless, and Kasirzadeh [57] cited that safeguarding copyright for published works in the UK extends up to seventy years following the author's demise. Nevertheless, the duration of copyright varies based on the nature of the work and its publication status. Once the copyright term concludes, the work enters the public domain.

However, in Switzerland, the duration of copyright protection is determined by specific timelines depending on the type [58]. For creative works, copyright coverage concludes seventy years following the author's death, while for computer programs, it extends for fifty years. In the case of photographs lacking distinct individuality, their copyright safeguard ends fifty years after the moment of their creation [59]. These regulations in Switzerland govern copyright protection for various types of intellectual designs that ensure a balance between encouraging creativity and eventually making works available for public use.

As the definitions and descriptions greatly vary among nations and laws, there is no generally accepted definition of originality and the authors' contribution criteria that can be used to explain the aspects of copyright law in different legislations.

Media Practitioners' and Related Professionals' Opinions on AI-Generated Content

The attitudes of media professionals, including broadcasters, reporters, journalists, stakeholders, and others toward the issues of originality, creativity, and legality regarding copyright also differ dramatically in this paradigm. Some people contend that content produced by AI has given this market a new life, whereas others are skeptical and worried that this invention will push human creators to the side and eventually substitute them, especially now that AI has the necessary copyright protection [60]. Consequently, study researchers have examined this field by integrating the perceptions of media practitioners.

Media Practitioner's Perception Regarding AI-Generated Content

In order to better understand their perceptions of the way they expect AI to impact their employment, the media workforce, and how journalists understand their adaptability and resilience, Yuand Huang [61] conducted a study in which they conducted in-depth interviews and engaged media practitioners from the media industry of China. Their research revealed that, in contrast to those from smaller organizations, journalists from larger media organizations tended to favor the use of AI in their industry. The authors described this as a "dynamic interplay amid personal and situational attributes" and noted that, generally, their respondents tended to see the AI integration with human labor favorably.

Franceschelli and Musolesi [62] added that media practitioners view AI as an experimental tool that allows them to explore new creative avenues. They are excited about using AI to generate unique content that pushes boundaries and challenges traditional norms. However, AI-generated content lacks the human touch and creativity of human-driven efforts and can create problems, including repetitive content, that compromise the uniqueness and authenticity of media practitioners' work. These new works themselves might be difficult or impossible to copyright, causing issues in the future; the NYU *Journal of Copyright and Entertainment Law* discussed the difficulty that one can have copyrighting work where AI was involved at all as an end-product. In the D.C. Circuit decision *Thaler v. Perlmutter*, the plaintiff Stephen Thaler applied for copyright for a visual piece he created with a "proprietary AI platform" he developed that he calls the "Creativity Machine." When he applied for the copyright, he listed the "author" as the Creativity Machine, but himself as the owner. "According to Thaler, he had no direct involvement in the creation of the artwork bar the prompt he gave" [63]. The Copyright Office denied his copyright claim, adding clarification in the appeal that "for the reasons explained below, defendants are correct that human authorship is an essential part of a valid copyright claim, and therefore plaintiff's pending motion for summary judgment is denied and defendants' pending cross-motion for summary judgment is granted" [64].

Lu et al. [65] continued research in this area by examining how people view photos produced by AI. The scientists warned about the risks posed by AI-generated media and stated that it is impossible for people to tell an AI-generated image from a real one. Zhang and Gosline [66] conducted a study examining how people view AI-generated content by having participants rate the quality of content created through four different AI-human collaboration methods. The study found that participants' perceptions of AI-assisted media were influenced by their biases. When unaware that AI was involved, participants rated the content highly. However, once they learned humans were also part of the process, they showed a preference for human-assisted content, a phenomenon termed "human favoritism."

From the preceding discussion, media practitioners' perceptions of AI-generated content can vary widely, ranging from enthusiasm for its innovative potential to concerns relating to authenticity and ethical implications. As AI technology evolves, these perceptions will be expected to shift and adapt alongside it [67, 68].

Perception of Professionals Regarding AI-Generated Content

Vimpari et al. [69] presented a comprehensive summary of how professionals perceive AI-created media, specifically focusing on images, and found that legal professionals have concerns about copyright ownership and the legal implications of AI-generated content, but there is a chance to navigate complex intellectual property issues and ensure that AI-assisted generated works comply with copyright laws [70].

Tiwary [20] conducted a study to investigate how academics view the utilization of AI tools like ChatGPT for generating content. The results indicated that these professionals believed AI tools were beneficial in enhancing their sentence construction, language usage, and content creation process. However, they also emphasized that AI tools had limited accuracy. Therefore, they suggested verifying the AI-generated content and utilizing it to improve writing while avoiding complete plagiarism. Likewise, professionals in academia stress the importance of transparency in AI-generated research and the need to follow ethical standards that guarantee that AI tools are used responsibly. In his research, Mezei [71] contended that attributing copyright to AI-generated content is inadequate and unrealistic due to its inability to address evolving perspectives. He asserted that copyright law is a fabricated construct, representing a legal embodiment of intricate, ever-changing sociocultural and economic interests. Additionally, Fernandez [41] highlighted that journalists and content creators often value AI for its ability to process and analyze large amounts of data quickly. AI can help with fact-checking, analysis of data, and summarizing information.

From the perceptions of professionals, it can be found that their perceptions of AI-assisted content generated vary based on their industry, role, and personal experiences. Although some embrace AI as an innovative tool that enhances efficiency and creativity, on the other hand, others express concerns such as issues like authenticity, bias, ethics, and copyright.

Ethical Consideration Related to AI-Generated Content

The ethical concerns associated with AI-generated content are not limited to copyright issues or its encroachment into various media domains. There is the difficulty of detecting AI involvement, especially in nonprofessional and smaller-scale settings like education [72]. Students' utilization of AI tools for content creation in these contexts raises ethical questions. Uzun and Levent [73] engaged in a discussion about integrating AI tools and upholding academic integrity within this framework. The researchers suggested the use of detection tools to identify manipulated content due to AI's widespread impact across diverse domains. They argued that in the current AI era, discerning whether the content is human or AI-generated might not be productive, implying the need for a nuanced approach to address AI-generated content's challenges.

Summarizing the collection of studies discussed earlier, these studies reveal that practitioners' viewpoints differ across fields, and individual preferences and favoritism play a significant role in shaping these perceptions. Undoubtedly, AI-generated content provides novelty, efficiency, and productivity, and less time is required to get help or support from AI tools to generate the content.

However, it also possesses some concerns or risks attached to it, for example, reliance on AI-assisted content might lead to a perceived loss of authenticity and human emotional connection relating to creative works. Vallis et al. [74] said that the fine line between inspiration and plagiarism becomes blurrier with the capacity of AI to impersonate present works in such a way that can potentially sidestep copyright protections. Moreover, while discussing the risks of AI-generated content, Bran et al. [75] added that the automation of content creation tends to cause job displacement in creative industries, affecting artists, writers, and related professionals. Additionally, interchangeable or AI-assisted content can be used to create deepfakes, misinformation, and malicious content, which raises concerns about spreading fake news and the attrition of truth.

Recommendation for Viable Copyright Flows for the Future

Some possible ways forward for copyright concerns are beginning to be shared, particularly in images and video. Adobe Firefly, which provides generative AI for the Adobe Creative Suite of software, including Photoshop, Illustrator, Substance, Stager, and more, which creates 2D and 3D images and vectors, presents some possible solutions. Released in 2023, Firefly is, according to Adobe, "Trained on Adobe Stock images, openly licensed content, and public domain content." Therefore, they claim it is "safe for commercial use." In addition, they claim to have created a "compensation model for Adobe Stock contributors whose content is used in the dataset to retrain Firefly models" [76].

The Content Authenticity Initiative (CAI) is about 1,200 entities, "a community of media and tech companies, NGOs, academics, and others working to promote adoption of an open industry standard for content authenticity and provenance" [77]. Founded in 2019, the CAI, in their 2021 white paper, describes their work as pursuing three main goals. First, a "detection of deliberately deceptive media" [78]. They note that people with bad intentions may use technology that moves as fast or faster than detection, a type of arms race, like the new AI detection tools offered by the company Turn-it-in, alongside its traditional academic plagiarism detection tools.

Second, they note, "education is essential." They urge media literacy campaigns and institutional education. "Finally, we must consider content attribution … often referred to as provenance, attribution empowers content creators and editors, regardless of their geographic location or degree of access to technology, to disclose information about who created or changed an asset, what was changed, and how it was changed" [78].

Opening the computer software "terminal" area, one can see the history of the creation of the images and assets. In addition to this comprehensive image history, the CAI would also promote tools that make it easier for creators to declare their use of AI at the point of upload.

In 2021, Adobe co-founded the Coalition for Content Provenance and Authenticity (C2PA) standards organization, along with Microsoft, Truepic, Arm, Intel, and the BBC. In collaboration with Microsoft and BBC-backed Project Origin, they have the stated aim of enabling "cryptographically verifiable facts about content to engender trust across all the surfaces through which we consume content" [78].

Their white paper considers individuals and not just institutions. They give an example in which a human rights activist, acting alone, may record something with an NGO-provided application such as "Proof mode," or, they outline, they can select their user settings and preferences within the tool before capturing in the case of using the native video or photo recording of their phone or apps, which would have to adopt this system. Their goal would be that "if the user is not connected to an organization, it is likely that the asset will be later discovered by an NGO or media outlet on messaging and social platforms. These institutions will be able to read the associated CAI data and confirm key attribution aspects before verifying that the asset is accurate" [78].

They go on to propose that the NGOs and media outlets would then begin their own verification process, likely looking at information such as author, time, location, etc. They note that if the asset has preserved CAI data, it will significantly simplify this process. The institution will then likely try to contact the source as part of the verification and may or may not receive the details they desire to feel comfortable sharing on their platform. (If required, this would also be where an institution may decide to use redaction capabilities, for example, blurring out faces before distributing widely.) Assuming the verification process is completed, in a media context, the workflow is then in the hands of a photo editor who will complete a similar process as the required photojournalism protocols to post the asset. They note that empowers the source, and not simply the outlet, to be the trusted entity. "As consumers view the content on social media platforms and/or the CAI website, the attribution details that they can see will start with the original capture and mitigate uncertainty for end users who may or may not trust the publishing institution." [78]

Adobe added that they were planning and working toward a "universal 'Do Not Train' Content Credentials tag that will remain associated with a piece of content wherever it's used, published, or stored." [76]

Limitations

However, the recommended strategies have limitations that media practitioners can face while implementing, including,

- AI-assisted generated content may not consistently meet the desired quality standards, leading to output quality variations. To address this issue, media practitioners need to ensure consistent and high-quality content that can require manual oversight and intervention.
- Using AI to replicate human voices, faces, or styles can raise ethical concerns about consent, privacy, and misrepresentation. Practitioners need to ensure that AI-generated content aligns with ethical standards.
- It can be complex to determine the copyright ownership of AI-generated content.
- AI algorithms can unintentionally maintain biases present in training data. To mitigate those biases, practitioners must actively address potential biases in AI-generated content, particularly in language, image recognition, and content recommendations.
- Implementing interchangeable AI-generated content might necessitate investment in resources, including hardware, software, and skilled personnel capable of effectively understanding and managing AI tools.

- The effectiveness of AI-assisted-generated content relies on the quality and diversity of the training data. If the training data is inadequate or unrepresentative, the AI output may not be accurate or relevant.
- The legal and regulatory landscape surrounding AI-generated content is still evolving. Practitioners must stay informed about copyright laws, intellectual property rights, and potential liabilities.

CONCLUSION

AI-assisted-generated content is an evolving and complex topic encompassing various industries and roles. Nevertheless, as the landscape keeps evolving and significant advancements are made, concerns surrounding originality, copyright, and the conventional ownership of AI-generated content have emerged as pressing issues. To address the indicated notion, the current study has observed the case of copyright protection in interchangeable content concerning originality, creativity, and law. To achieve this study`s objectives or aims, the study followed a review methodology and analysis of the existing body of literature to gain an inclusive understanding of the present perspective regarding copyright challenges related to AI-generated media content. Likewise, the study has demonstrated the viewpoints of stakeholders such as media practitioners and professionals to gain better and more practical insights into AI-generated content.

From the current research, the notions of originality, creativity, and legal aspects are closely intertwined and complex when examined within the context of copyright in AI-assisted content.

The analysis of recent studies on this topic clearly reveals a wide spectrum of viewpoints and criticisms. These perspectives differ significantly across nations, researchers, journalists, and other relevant stakeholders, often shaped by individual subjective beliefs and, at times, grounded in objective arguments. However, a consistent theme emerges: the copyright protection afforded to AI-generated content is expected to be a pivotal concern for the future of human creators. This issue poses several potential risks for both the prospects of human-authored content and the broader landscape of content generation.

Furthermore, the research offers suggestions and highlights associated limitations for the implementation of these proposed ideas. In this context, researchers could consider conducting a comprehensive cross-country analysis of copyright laws, specifically focusing on AI-generated content. This approach could shed light on the current situation within this domain, offering a transparent and comprehensive portrayal of the complex landscape.

AUTHOR CONTRIBUTIONS

Walaa Fouda, Elizabeth Daggett Matar, and Abdelrahman Mostafa Arab were involved in the concept, design, data collection, interpretation, writing, and critically revising the chapter. All authors approved the final version of the chapter.

FUNDING

The authors received no financial support for the research and/or authorship of this chapter.

DATA AVAILABILITY AND SHARING POLICY

Data generated or analyzed during this study are available from the authors on request.

REFERENCES

1. N. Bostrom, *Superintelligence: paths, dangers, strategies.* Oxford, United Kingdom; New York, NY: Oxford University Press, 2016.
2. A. M. Turing, "I.—Computing Machinery and Intelligence," *Mind*, vol. LIX, no. 236, pp. 433–460, 1950. doi: 10.1093/mind/LIX.236.433.
3. Y. LeCun, Y. Bengio and G. Hinton, "Deep Learning," *Nature*, vol. 521, no. 7553, pp. 436–444, 2015.
4. A. Vaswani, "Attention is all you need," in *Proceedings of the 31st international conference on neural information processing systems*, 2017, pp. 599–609,
5. I. Goodfellow *et al.*, "Generative Adversarial Nets," in *Advances in neural information processing systems*, 2014, p. 27. MIT Press, Cambridge, MA, USA.
6. Christie's, Oct. 25, 2018. [Online]. Available: https://www.christies.com/stories/a-collaboration-between-two-artists-one-human-one-a-machine-0cd01f4e232f4279a525a446d60d4cd1
7. X. Peng, Y. Li, I. W. Tsang, H. Zhu, J. Lv and J. T. Zhou, "XAI Beyond Classification: Interpretable Neural Clustering," *J. Mach. Learn. Res*, vol. 23, no. 6, pp. 1–28, 2022.
8. E. Ferrara, "Fairness and Bias in Artificial Intelligence: A Brief Survey of Sources, Impacts, and Mitigation Strategies," *Sci*, vol. 6, no. 1, p. 3, 2023.
9. D. Khurana, A. Koli, K. Khatter and S. Singh, "Natural Language Processing: State of the Art, Current Trends and Challenges," *Multimed. Tools Appl*, vol. 82, no. 3, pp. 3713–3744, 2023.
10. Z. Q. J. Xu, Y. Zhang and T. Luo, "Overview frequency principle/spectral bias in deep learning." 2022.
11. B. S. S. Harichandana, V. Agarwal, S. Ghosh, G. Ramena, S. Kumar and B. R. K. Raja, "January. PrivPAS: A real time Privacy-Preserving AI System and applied ethics," in *2022 IEEE 16th International Conference on Semantic Computing (ICSC*, IEEE, 2022, pp. 9–16.
12. N. Carlini, "A LLM assisted exploitation of AI-Guardian." 2023.
13. S. M. Chan-Olmsted, "A Review of Artificial Intelligence Adoptions in the Media Industry," *Int. J. Media Manag*, vol. 21, pp. 3–4, 2019. doi: 10.1080/14241277.2019.1695619.
14. L. Ouchchy, A. Coin and V. Dubljević, "AI in the Headlines: The Portrayal of the Ethical Issues of Artificial Intelligence in the Media," *AI Soc*, vol. 35, pp. 927–936, 2020. doi: 10.1007/s00146-020-00965-5.
15. R. A. Partadiredja, C. E. Serrano and D. Ljubenkov, "AI or Human: The Socio-Ethical Implications of AI-Generated Media Content," in *2020 13th CMI Conference on Cybersecurity and Privacy (CMI)-Digital Transformation-Potentials and Challenges*, Copenhagen, Denmark: IEEE, 2020, pp. 1–6. doi: 10.1109/CMI51275.2020.9322673.
16. J. Vesala, "Developing Artificial Intelligence-Based Content Creation: Are EU Copyright and Antitrust Law Fit for Purpose?" *IIC-Int. Rev. Intellect. Prop. Compet. Law*, vol. 54, pp. 1–30, 2023, doi: 10.1007/s40319-023-01301-2.
17. Y. Wan and H. Lu, "Copyright Protection for AI-Generated Outputs: The Experience from China," *Comput. Law Secur. Rev*, vol. 42, p. 105581, 2021. doi: 10.1016/j.clsr.2021.105581.
18. J. Kuai, R. Ferrer-Conill and M. Karlsson, "AI≥ Journalism: How the Chinese Copyright Law Protects Tech Giants' AI Innovations and Disrupts the Journalistic Institution," *Digit. Journal*, vol. 10, no. 10, pp. 1893–1912, 2022. doi: 10.1080/21670811.2022.2120032.
19. A. J. Wurzer, "Tencent Is Getting Copyright for AI-Dreamwriter Article," *IP Bus. Acad.*, 2021, [Online]. Available: https://ipbusinessacademy.org/tencent-is-getting-copyright-for-ai-dreamwriter-article.
20. N. Tiwary, "Netizens, Academicians, and Information Professionals' Opinions About AI With Special Reference to ChatGPT." 2023. doi: 10.48550/arXiv.2302.07136.
21. K. Logan, "Are You Rushing to Copyright Your ChatGPT-Generated Screenplay? Don't Get Your Hopes Up. U.S. Copyright Office Publishes New Guidance on Works Created with Artificial Intelligence," *Lexology*, 2023.
22. A. Mathur, "Art-Istic or Art-Ificial? Ownership and Copyright Concerns in AI-Generated Artwork," *Cent. Art Law Httpsitsartlaworg2022112lartistic–Artif. -Ai Text*, 2022.

23. F. Lazaro, "Who Owns AI-Generated Content? Navigating Ownership Concerns for Published Works," *LinkedIn*, 2023, [Online]. Available: https://www.linkedin.com/pulse/who-owns-ai-generated-content-navigating-ownership-concerns-lazaro#:~:text=In
24. T. He, "The Sentimental Fools and the Fictitious Authors: Rethinking the Copyright Issues of AI-Generated Contents in China," *Asia Pac. Law Rev*, vol. 27, no. 2, pp. 218–238, 2019. doi: 10.1080/10192557.2019.1703520.
25. A. A. Bahishti, "The Importance of Review Articles & Its Prospects in Scholarly Literature," *Extensive Rev*, vol. 1, no. 1, pp. 1–6, 2021. doi: 10.21467/exr.1.1.4293.
26. H. B. Abrams, "Originality and Creativity in Copyright Law," *Law Contemp. Probl*, vol. 55, no. 3, pp. 1–26, 1992.
27. T. Margoni, "Artificial Intelligence, Machine Learning and EU Copyright Law: Who Owns AI? CREATe Working Paper Series." 2018. doi: 10.5281/zenodo.2001763.
28. P. B. Hugenholtz and J. P. Quintais, "Copyright and Artificial Creation: Does EU Copyright Law Protect AI-Assisted Output?," *IIC-Int. Rev. Intellect. Prop. Compet. Law*, vol. 52, no. 9, pp. 1190–1216, 2021. doi: 10.1007/s40319-021-01115-0.
29. E. Nakazawa, M. Udagawa and A. Akabayashi, "Does the Use of AI to Create Academic Research Papers Undermine Researcher Originality?," *AI*, vol. 3, no. 3, pp. 702–706, 2022. doi: 10.3390/ai3030040.
30. M. E. Kaminski, "Authorship, Disrupted: AI Authors in Copyright and First Amendment Law," *UCDL Rev*, vol. 51, p. 589, 2017.
31. M. A. Lemley, "How Generative Ai Turns Copyright Law on its Head." 2023.
32. O. Bulayenko, J. P. Quintais, D. J. Gervais, J. Poort and A. Jotterand, "AI Music Outputs: Challenges to the Copyright Legal Framework," *Available SSRN*, vol. 4072806, pp. 301–562, 2022.
33. E. Shirley, "The Fixed and Originality Requirements for Copyright Protection," *Lawdit Solicit.*, 2022, [Online]. Available: https://lawdit.co.uk/readingroom/the-fixed-and-originality-requirements-for-copyright-protection
34. Murray, J. "The Impact of Trade and Cooperation Agreements on Copyright Protection Standards Between the EU and UK," *J. Int. Law*, vol. 45, no. 2, pp. 123–145, 2023.
35. U. S. C. Office, "What Is Copyright?" Copyright.Gov. https://www.copyright.gov/what-is-copyright: text= originality and fixation-, Original Works, and 'modicum' of creativity." 2023.
36. F. R. Caminatti, "Copyrighting Brain-Computer Interface: Where Neuroengineering Meets Intellectual Property Law," *Cybaris®*, vol. 14, no. 1, p. 1, 2023.
37. B. T. Kucukali, B. Columbia, R. C. W. Khoo, C. H. Phan and Lim, "The Protection of AI-Generated Works Under European Copyright Law: Toward Adoption of a Neighbouring Rights Approach," *Wiley Interdisciplinary Reviews: Data Mining and Knowledge Discovery*, vol. 12, no. 3., p. 1438, 2022.
38. J. Hutson and M. Harper-Nichols, "Generative AI and Algorithmic Art: Disrupting the Framing of Meaning and Rethinking the Subject-Object Dilemma," *Glob. J. Comput. Sci. Technol. D*, vol. 23, no. 1, 2023.
39. G. Chaudhary and K. Vries, "Artificial Intelligence: Copyright and Authorship/Ownership Dilemma?," *YSEC Yearb. Socio-Econ. Const. 2021 Triangulating Freedom Speech*, vol. 212, pp. 93–115, 2022.
40. G. Amato *et al.*, "AI in the media and creative industries," *ArXiv Prepr. ArXiv190504175*, 2019, doi: 10.48550/arXiv.1905.04175.
41. P. Fernandez, "The Technology Behind the Text to Image Generators," *Libr. Hi Tech News*, vol. 39, no. 10, pp. 1–4, 2022.
42. P. G. Picht, V. Brunner and R. Schmid, *Artificial intelligence, and intellectual property law: from diagnosis to action*. Max Planck Institute for Innovation & Competition Research Paper, 2022.
43. K. Roose, "An AI-Generated Picture Won an Art Prize. Artists Aren't Happy," *N. Y. Times*, 2022, [Online]. Available: https://www.nytimes.com/2022/09/02/technology/ai-artificial-intelligence-artists.html.
44. G. Volpicelli, "The New Luddites: AI Comes for the Creative Class." 2023. [Online]. Available: https://www.politico.eu/article/artificial-intelligence-technology-art-regulation-copyright/#:~:text=Regulators
45. Ö Erdine, "AI Wins Art Competition: Is Imputing Text Prompts Enough to Become an Artist?" 2022. [Online]. Available: https://dataconomy.com/2022/09/26/ai-artwork-wins-art-competition/.
46. J. Potts, *The near-death of the author: creativity in the internet age*. Canada: University of Toronto Press, 2022.
47. D. Harwell, "He Used AI to Win a Fine Arts Competition." 2022. [Online]. Available: https://www.washingtonpost.com/technology/2022/09/02/midjourney-artificial-intelligence-state-fair-colorado/.
48. L. Kolirin, "An Artist Entered an AI-Generated Photo into a Prestigious Photography Competition - and Won," in *ABC 7 NY*, 2023. [Online]. Available: https://abc7ny.com/artificial-intelligence-ai-boris-eldagsen-sony-world-photography-awards/13149757/#:~:text=An
49. A. Greenberger, "Artist Wins Photography Contest After Submitting AI-Generated Image, Then Forfeits Prize," *ARTnews*, 2023, [Online]. Available: https://www.artnews.com/art-news/news/ai-generated-image-world-photography-organization-contest-artist-declines-award-1234664549/.

50. J. Bieser, "How Can AI Support Human Creativity? Here's What a New Study Found," *World Econ. Forum*, 2023, [Online]. Available: https://www.weforum.org/agenda/2023/02/ai-can-catalyze-and-inhibit-your-creativity-here-is-how/#:~:text=The
51. A. Cuntz, "Copyright and the Currency of Creativity: Beyond Income," *WIPO Mag.*, 2019, [Online]. Available: https://www.wipo.int/wipo_magazine/en/2019/03/article_0003.html.
52. C. S. Council, "Notice of the State Council on Printing and Distributing the Development Plan for a New Generation of Artificial Intelligence." 2017. [Online]. Available: http://www.gov.cn/zhengce/content/2017-07/20/content_5211996.htm.
53. S. Mukherjee, F. Y. Chee and M. Coulter, "EU Proposes New Copyright Rules for Generative AI." 2023. [Online]. Available: https://www.reuters.com/technology/eu-lawmakers-committee-reaches-deal-artificial-intelligence-act-2023-04-27/.
54. W.I.P.O., "Berne Convention for the Protection of Literary and Artistic Works," *World Intellect. Prop. Organ. WIPO*, 2023, [Online]. Available: https://www.wipo.int/treaties/en/ip/berne/.
55. D. J. Corbett, "A Premier Paradigm Shift: The Impact of Artificial Intelligence on US Intellectual Property Laws," *Lib. Univ. Law Rev*, vol. 17, no. 2, p. 5, 2023.
56. C. Sganga, "Is There Still a Policy Agenda for EU Copyright Law?" *IIC-Int. Rev. Intellect. Prop. Compet. Law*, vol. 45, pp. 1–11, 2023.
57. C. Bird, E. L. Ungless and A. Kasirzadeh, "Typology of Risks of Generative Text-to-Image Models." 2023.
58. P. Lana, "Why Works Created Autonomously by Artificial Intelligence (Should) Belong to the Public Domain–A Viewpoint Based on Droit D'Auteur Traditions," *Transform. J. Media*, vol. 66, pp. 1–115, 2022.
59. G. La Diega, "Ending Smart Data Enclosures: The European Approach to Regulation of the Internet of Things between Access and Intellectual Property," in *The Cambridge handbook on emerging issues at the intersection of commercial law and technology*, S.-A. Elvy and N. Kim, Eds., 2022.
60. K. Tyagi, "Deepfakes, Copyright and Personality Rights an Inter-Disciplinary Perspective," in *International Law and Economics Conference*, Cham: Springer Nature Switzerland, Mar. 2023, pp. 191–210.
61. Y. Yu and K. Huang, "Friend or Foe? Human Journalists' Perspectives on Artificial Intelligence in Chinese Media Outlets," *Chin. J. Commun*, vol. 14, no. 4, pp. 409–429, 2021. doi: 10.1080/17544750.2021.1915832.
62. G. Franceschelli and M. Musolesi, "On the creativity of large language models." 2023.
63. J. Weinerman, "At Least Our AI Overlords Won't Get Copyright Protection," NYU Journal of Intellectual Property & Entertainment Law. Accessed: Jan. 17, 2024. [Online]. Available: https://jipel.law.nyu.edu/at-least-our-ai-overlords-wont-get-copyright-protection/
64. "ThalerVPerlmutter - DocumentCloud." Accessed: Jan. 17, 2024. [Online]. Available: https://www.documentcloud.org/documents/23919666-thalervperlmutter
65. Z. Lu, D. Huang, L. Bai, X. Liu, J. Qu, and W. Ouyang, "Seeing Is Not Always Believing: A Quantitative Study on Human Perception of AI-Generated Images." 2023. doi: 10.48550/arXiv.2304.13023.
66. Y. Zhang and R. Gosline, "People's Perceptions (and Bias) Toward Creative Content Generated by Ai (ChatGPT-4), Human Experts, and Human-AI Collaboration," *Hum. Experts Hum. -AI Collab.*, 2023.
67. R. Abbott, "Artificial Intelligence and Intellectual Property: An Introduction. Ryan Abbott, Artificial Intelligence, and Intellectual Property: An Introduction," in *Research handbook on intellectual property and artificial intelligence*, E. Elgar, Ed., 2022. https://ssrn.com/abstract=4065150 or http://dx.doi.org/10.2139/ssrn.4065150
68. J. G. Meyer *et al.*, "ChatGPT and Large Language Models in Academia: Opportunities and Challenges," *BioData Min.*, vol. 16, no. 1, pp. 20, Jul. 2023, doi: 10.1186/s13040-023-00339-9.
69. V. Vimpari, A. Kultima, P. Hämäläinen and C. Guckelsberger, "'An Adapt-or-Die Type of Situation': Perception, Adoption, and Use of Text-To-Image-Generation AI by Game Industry Professionals." 2023. doi: 10.48550/arXiv.2302.12601.
70. W. Moss-Wellington, C. Lam and F. Gilardi, "Introduction: media and Fakery," *Continuum*, vol. 36, no. 3, pp. 333–338, 2022.
71. P. Mezei, "'You AIn't Seen Nothing yet'–Arguments against the Protectability of AI-Generated Outputs by Copyright Law." 2021. doi: 10.2139/ssrn.3890051.
72. T. Teubner, C. M. Flath, C. Weinhardt, W. Aalst and O. Hinz, "Welcome to the Era of ChatGPT et al., the Prospects of Large Language Models," *Bus. Inf. Syst. Eng*, vol. 65, no. 2, pp. 95–101, 2023.
73. L. Uzun, "ChatGPT and Academic Integrity Concerns: Detecting Artificial Intelligence Generated Content," *Lang. Educ. Technol.*, vol. 3, no. 1, 2023.
74. C. Vallis, S. Wilson, D. Gozman and J. Buchanan, *Student Perceptions of AI-Generated Avatars in Teaching Business Ethics: We Might Not Impressed.* Postdigital Science and Education, 2023.
75. E. Bran, C. Rughini, G. Nadoleanu and M. G. Flaherty, "The Emerging Social Status of Generative AI: Vocabularies of AI Competence in Public Discourse," in *Proceedings of the 24th Conference on Control Systems and Computer Science (CSCS24 2023)*, IEEE, 2023.

76. "Adobe Firefly - Free Generative AI for creatives." Accessed: Jan. 14, 2024. [Online]. Available: https://www.adobe.com/products/firefly.html
77. "Content Authenticity Initiative," Content Authenticity Initiative. Accessed: Jan. 14, 2024. [Online]. Available: https://contentauthenticity.org
78. "CAI_WhitePaper_v1_0 (1).pdf (Shared) - Adobe cloud storage." Accessed: Jan. 14, 2024. [Online]. Available: https://acrobat.adobe.com/link/track?uri=urn%3Aaaid%3Ascds%3AUS%3A2c6361d5-b8da-4aca-89bd-1ed66cd22d19

12 Innovative Applications of AI for Sustainable Supply Chain Management in Environmental Conservation

Saydulu Kolasani

INTRODUCTION

The integration of artificial intelligence (AI) into sustainable supply chain management represents a significant advancement in environmental conservation efforts. In recent years, the intersection of AI and sustainability has garnered increasing attention as organizations seek innovative solutions to minimize environmental impact while maintaining operational efficiency in their supply chain processes. This introduction provides an overview of the key concepts, challenges, and opportunities associated with the application of AI in sustainable supply chain management for environmental conservation.

Background

Supply chain management encompasses the coordination of activities involved in sourcing, production, transportation, and distribution to deliver products or services to end customers. Traditional supply chain practices often prioritize cost reduction and operational efficiency, sometimes at the expense of environmental sustainability. However, the growing recognition of the environmental and social impacts associated with supply chain operations has prompted organizations to adopt more sustainable practices.

Environmental conservation is a critical component of sustainability, focusing on the preservation and protection of natural resources, ecosystems, and biodiversity. Achieving environmental conservation

DOI: 10.1201/9781003497189-12

goals requires minimizing resource consumption, reducing pollution, and mitigating the negative impacts of human activities on the environment. Sustainable supply chain management plays a vital role in advancing environmental conservation objectives by optimizing resource utilization, minimizing waste, and adopting eco-friendly practices throughout the supply chain.

The Role of Artificial Intelligence in Sustainable Supply Chain Management

Artificial intelligence, encompassing machine learning, natural language processing, and other advanced technologies, has emerged as a powerful tool for optimizing supply chain processes and driving sustainability initiatives. AI enables organizations to analyze vast amounts of data, identify patterns and trends, and make data-driven decisions to improve efficiency and reduce environmental impact across the supply chain.

Predictive analytics is one of the key applications of AI in sustainable supply chain management. By leveraging historical data and machine learning algorithms, organizations can forecast demand more accurately, optimize inventory levels, and minimize stockouts and excess inventory. This proactive approach to demand forecasting helps reduce waste, lower inventory holding costs, and enhance resource efficiency.

Another area where AI is making significant contributions to sustainable supply chain management is logistics optimization. AI-powered algorithms can optimize transportation routes, mode selection, and vehicle scheduling to minimize fuel consumption, reduce emissions, and lower transportation costs. By considering factors such as delivery time windows, traffic conditions, and vehicle capacities, AI-based logistics optimization systems improve delivery efficiency while reducing the carbon footprint of transportation operations. Furthermore, AI technologies are being deployed for real-time monitoring and management of environmental performance in supply chain operations. Internet of Things (IoT) devices, equipped with sensors and connectivity capabilities, collect data on energy usage, emissions, and environmental conditions in manufacturing facilities, warehouses, and transportation vehicles. AI algorithms analyze this data in real time to identify inefficiencies, detect anomalies, and recommend corrective actions to optimize resource utilization and minimize environmental impact.

Challenges and Opportunities

While the adoption of AI in sustainable supply chain management offers numerous benefits, it also presents challenges and considerations that must be addressed. One of the primary challenges is data quality and availability. AI models rely on large volumes of high-quality data to deliver accurate predictions and recommendations. However, supply chain data is often fragmented, inconsistent, and of varying quality, making it challenging to train AI models effectively.

Another challenge is the integration of AI technologies into existing supply chain systems and processes. Organizations may face resistance to change, technical barriers, and interoperability issues when implementing AI solutions. Additionally, concerns related to data privacy, security, and ethical use of AI pose significant considerations that must be addressed to ensure responsible deployment and adoption of AI in supply chain management.

Despite these challenges, the integration of AI in sustainable supply chain management offers significant opportunities for environmental conservation and sustainability. AI-driven optimization algorithms can help organizations achieve cost savings, enhance operational efficiency, and reduce environmental impact simultaneously. Furthermore, AI technologies enable real-time monitoring and decision-making, allowing organizations to respond proactively to environmental risks and opportunities in the supply chain.

The integration of artificial intelligence in sustainable supply chain management holds immense potential for advancing environmental conservation objectives. By leveraging AI technologies to optimize supply chain processes, organizations can achieve operational efficiency gains while minimizing environmental impact throughout the value chain. However, addressing challenges related to data quality, integration, and ethical considerations is essential to ensure the responsible deployment and adoption of AI in supply chain management for environmental conservation. In the subsequent chapters of this book, we will explore in detail the innovative applications, case studies, and best practices of AI in sustainable supply chain management for environmental conservation.

LITERATURE REVIEW

The integration of AI in sustainable supply chain management has gained increasing attention from researchers and practitioners seeking innovative solutions to address environmental challenges while maintaining operational efficiency. This literature review provides a comprehensive overview of existing research and developments in the application of AI technologies for environmental conservation within supply chain operations.

AI Applications in Sustainable Supply Chain Management

AI technologies, including machine learning, natural language processing, and predictive analytics, offer a range of applications for optimizing supply chain processes and driving sustainability initiatives. One of the primary applications of AI in sustainable supply chain management is demand forecasting and inventory optimization. By analyzing historical sales data and external factors such as weather patterns and economic indicators, AI-powered forecasting models can generate accurate predictions of future demand, allowing organizations to optimize inventory levels, minimize stockouts, and reduce waste (Gunasekaran et al., 2018).

Another area where AI is making significant contributions is logistics optimization. AI-driven algorithms can optimize transportation routes, mode selection, and vehicle scheduling to minimize fuel consumption, reduce emissions, and lower transportation costs. By considering factors such as delivery time windows, traffic conditions, and vehicle capacities, AI-based logistics optimization systems improve delivery efficiency while reducing the carbon footprint of transportation operations (Deng et al., 2018). Furthermore, AI technologies enable real-time monitoring and management of environmental performance in supply chain operations. Internet of Things (IoT) devices equipped with sensors and connectivity capabilities collect data on energy usage, emissions, and environmental conditions in manufacturing facilities, warehouses, and transportation vehicles. AI algorithms analyze this data in real time to identify inefficiencies, detect anomalies, and recommend corrective actions to optimize resource utilization and minimize environmental impact (Cheng et al., 2020).

In recent years, significant advancements in technology have transformed various fields, addressing both environmental and data management challenges. Chen and Zhang (2018) review the application of big data technology in pollution control, highlighting its role in tackling environmental issues through data-driven approaches. Complementing this, Duan et al. (2019) provide an overview of blockchain technology from a big data perspective, emphasizing its potential in enhancing data security and integrity. In the realm of financial forecasting, Hsu and Kao (2018) explore the use of deep learning for stock price prediction, illustrating the impact of advanced algorithms on market analysis. Jia and Zhang (2018) discuss a blockchain-based framework for ensuring data integrity in IoT applications, reflecting the growing intersection between blockchain and IoT technologies. Kong et al. (2020) introduce the concept of "Big Earth" data, marking a new frontier in Earth and information sciences, while Li and Liu (2020) review the role of artificial intelligence in achieving sustainable development goals, underscoring its importance in driving global sustainability efforts. Robinson and Barr (2017) critically appraise deep learning, providing insights into its capabilities and limitations. Wang et al. (2019) examine the use of big data analytics in knowledge management, revealing its influence on organizational learning and decision-making. Wu et al. (2020) offer a comprehensive review of digital twin technology, highlighting its potential from a knowledge perspective. Xiao and Tao (2017) focus on the design and implementation of an IoT-based green logistics system, demonstrating how IoT can contribute to sustainable logistics. Lastly, Yang and McCollister (2017) explore the effects of artificial intelligence on human beings, shedding light on its broader implications. Zhang et al. (2018) emphasize the role of big data in sustainable development, framing it as a critical tool for addressing global challenges. Collectively, these studies illustrate the transformative impact of emerging technologies on various sectors, from environmental management and data integrity to financial forecasting and sustainability.

Challenges and Considerations

Despite the potential benefits of AI in sustainable supply chain management, several challenges and considerations must be addressed to ensure successful implementation and adoption. Data quality and availability emerge as significant challenges, as AI models rely on large volumes of high-quality data to deliver accurate predictions and recommendations. Supply chain data is often fragmented, inconsistent, and of varying quality, making it challenging to train AI models effectively (Wamba et al., 2017).

Additionally, the integration of AI technologies into existing supply chain systems and processes poses technical and organizational challenges. Organizations may encounter resistance to change, technical barriers, and interoperability issues when implementing AI solutions. Furthermore, concerns related to data privacy, security, and ethical use of AI raise important considerations that must be addressed to ensure responsible deployment and adoption of AI in supply chain management (Govindan et al., 2020).

Research Directions and Future Trends

Looking ahead, several research directions and future trends emerge in the application of AI for environmental conservation within supply chain operations. One area of focus is the development of AI-powered decision support systems that enable organizations to make informed, data-driven decisions to optimize supply chain processes and minimize environmental impact. Additionally, research efforts are needed to explore the potential of emerging AI technologies such as deep learning and reinforcement learning in addressing complex sustainability challenges in supply chain management (Triguero et al., 2018).

There is a growing need for interdisciplinary research collaborations among academia, industry, and government agencies to develop holistic approaches to sustainable supply chain management. Collaboration across disciplines such as engineering, environmental science, economics, and information technology is essential to address the multifaceted challenges of environmental conservation within supply chain operations. Additionally, research on policy interventions, regulatory frameworks, and incentives to promote the adoption of AI technologies for sustainability in supply chains can help drive meaningful change at the organizational and systemic levels (Wu et al., 2020).

The literature on the integration of artificial intelligence in sustainable supply chain management highlights the transformative potential of AI technologies to drive environmental conservation initiatives within supply chain operations. By leveraging AI for demand forecasting, inventory optimization, logistics optimization, and environmental monitoring, organizations can achieve operational efficiency gains while minimizing environmental impact throughout the value chain. However, addressing challenges related to data quality, integration, and ethical considerations is essential to ensure the responsible deployment and adoption of AI in supply chain management for environmental conservation. Future research efforts should focus on developing innovative AI-powered solutions, fostering interdisciplinary collaborations, and advancing policy initiatives to promote sustainability in supply chains.

INNOVATIVE APPLICATIONS OF AI FOR ENVIRONMENTAL CONSERVATION

Demand Forecasting and Inventory Optimization

Demand forecasting and inventory optimization are critical aspects of sustainable supply chain management. By accurately predicting demand and optimizing inventory levels, organizations can minimize waste, reduce resource consumption, and lower environmental impact. AI technologies play a crucial role in enhancing demand forecasting and inventory optimization processes.

AI-powered demand forecasting models leverage historical sales data, market trends, and external factors such as weather patterns and economic indicators to generate accurate predictions of future demand. These models use advanced machine learning algorithms to analyze large datasets and identify patterns and trends that traditional forecasting methods may overlook. By improving the accuracy of demand forecasts, AI enables organizations to optimize inventory levels, minimize stockouts and excess inventory, and reduce waste throughout the supply chain. AI-driven inventory optimization systems continuously monitor inventory levels, demand patterns, and supply chain dynamics to dynamically adjust stocking levels and reorder points. These systems use real-time data and predictive analytics to optimize inventory replenishment decisions, ensuring that the right products are available at the right time and minimizing the environmental impact of excess inventory and overstocking.

Logistics Optimization

Logistics optimization is another area where AI technologies can contribute to environmental conservation efforts. Transportation and logistics operations account for a significant portion of greenhouse gas emissions and environmental impact in supply chains. By optimizing transportation routes, mode selection, and vehicle scheduling, organizations can reduce fuel consumption, lower emissions, and minimize the carbon footprint of logistics operations.

AI-powered logistics optimization systems leverage algorithms such as genetic algorithms, simulated annealing, and reinforcement learning to find the most efficient routes and schedules for transporting goods. These systems consider factors such as delivery time windows, traffic conditions, vehicle capacities, and fuel efficiency to optimize transportation plans and minimize environmental impact. By incorporating real-time data and predictive analytics, AI-driven logistics optimization systems can dynamically adjust routes and schedules in response to changing conditions, further enhancing efficiency and reducing environmental impact.

Real-Time Environmental Monitoring

Real-time environmental monitoring is essential for identifying and mitigating environmental risks and impacts associated with supply chain operations. By continuously monitoring environmental conditions such as air quality, water quality, and noise levels, organizations can detect anomalies, identify potential environmental hazards, and take proactive measures to minimize their impact on ecosystems and communities.

AI technologies, combined with IoT devices and sensor networks, enable real-time environmental monitoring systems that collect, analyze, and visualize environmental data in real time. Machine learning algorithms process sensor data and identify patterns, trends, and anomalies indicative of environmental risks or impacts. These systems provide organizations with actionable insights and alerts, enabling them to respond quickly to environmental incidents, implement corrective actions, and mitigate potential environmental damage. Demand forecasting and inventory optimization, logistics optimization, and real-time environmental monitoring are innovative applications of AI that contribute to environmental conservation efforts within supply chain operations. By leveraging AI technologies to optimize supply chain processes, organizations can minimize waste, reduce resource consumption, and lower environmental impact throughout the value chain.

Figure 12.1 presents a set of KDE plots overlaid with corresponding histograms for various variables in the REMS dataset, each subplot showcasing the distribution of a specific feature colored by UV radiation intensity. The large-scale representation enables comprehensive examination of data distribution across multiple parameters.

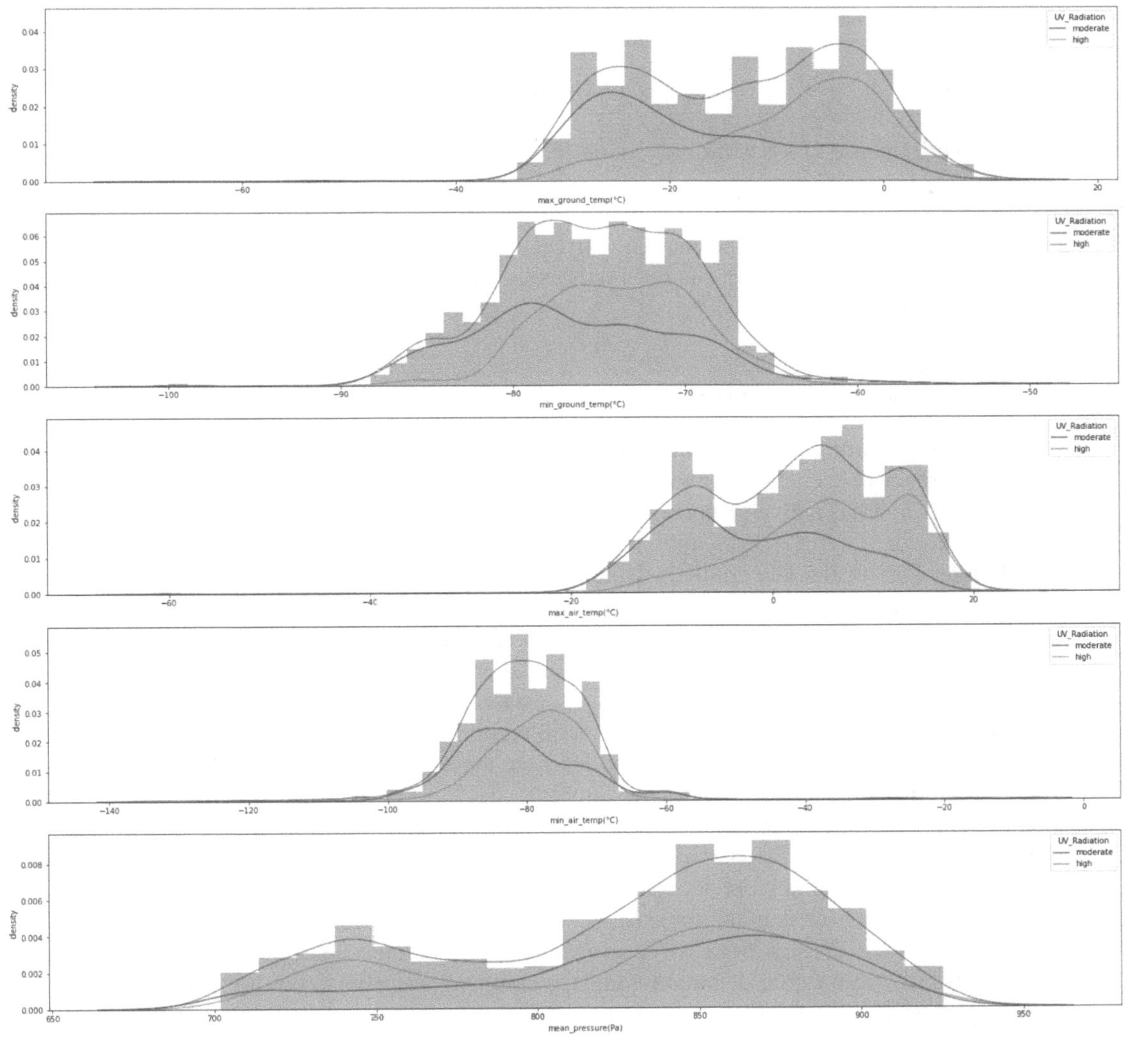

FIGURE 12.1 Set of KDE plots overlaid with corresponding histograms.

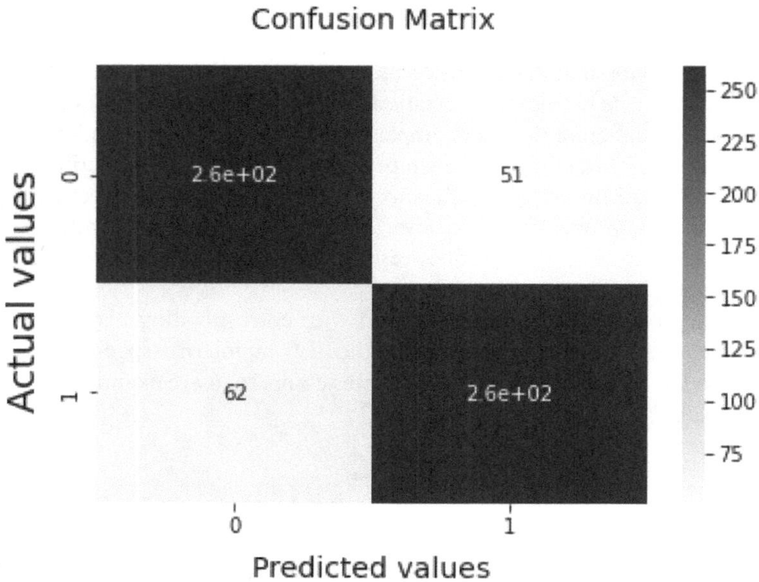

FIGURE 12.2 Performance of a classification model.

The confusion matrix in Figure 12.2 heatmap illustrates the performance of a classification model, with annotated values representing the count of correct and incorrect predictions across different classes. The color intensity signifies the frequency of predictions, aiding in the visualization of model accuracy and error distribution.

CASE STUDY 1: AI-DRIVEN DEMAND FORECASTING AT ECOTECH SOLUTIONS

Company Background: EcoTech Solutions is a leading provider of sustainable energy solutions, specializing in renewable energy technologies such as solar panels and wind turbines. The company operates globally and has a diverse portfolio of products serving residential, commercial, and industrial customers.

Challenge: One of the key challenges faced by EcoTech Solutions was accurately forecasting demand for its renewable energy products. Fluctuating market conditions, seasonal variations, and evolving customer preferences made demand forecasting a complex and challenging task. Inaccurate demand forecasts often led to excess inventory, stockouts, and missed revenue opportunities, impacting both financial performance and environmental sustainability.

Solution: To address this challenge, EcoTech Solutions implemented an AI-driven demand forecasting system powered by advanced machine learning algorithms. The system analyzed historical sales data, customer demographics, weather patterns, and market trends to generate accurate predictions of future demand for renewable energy products. By leveraging AI technologies, EcoTech Solutions was able to identify patterns and correlations in the data that traditional forecasting methods could not capture, resulting in more accurate and reliable demand forecasts.

Results: The AI-driven demand forecasting system enabled EcoTech Solutions to improve forecast accuracy by over 20%, significantly reducing inventory holding costs and minimizing stockouts. With better visibility into future demand trends, the company was able to optimize inventory levels, streamline production schedules, and enhance supply chain efficiency. By aligning production with actual demand, EcoTech Solutions reduced waste and environmental impact associated with excess inventory and overproduction, contributing to its sustainability goals.

CASE STUDY 2: LOGISTICS OPTIMIZATION WITH AI AT GREENLOGISTICS INC.

Company Background: GreenLogistics Inc. is a leading provider of environmentally friendly transportation and logistics services, specializing in sustainable delivery solutions for e-commerce and retail companies. The company operates a fleet of electric vehicles and utilizes renewable energy sources to power its operations, minimizing carbon emissions and environmental impact.

Challenge: GreenLogistics Inc. faced the challenge of optimizing its delivery routes and schedules to minimize fuel consumption, reduce emissions, and improve delivery efficiency. Traditional route planning methods were time-consuming and inefficient, often leading to suboptimal routes, unnecessary mileage, and excessive fuel usage. The company sought a solution to optimize its logistics operations while minimizing its environmental footprint.

Solution: To address this challenge, GreenLogistics Inc. implemented an AI-powered logistics optimization system that leveraged advanced algorithms to optimize delivery routes, mode selection, and vehicle scheduling. The system analyzed real-time data on order volumes, delivery locations, traffic conditions, and vehicle capacities to dynamically optimize delivery plans and schedules. By incorporating AI technologies, GreenLogistics Inc. was able to find the most efficient routes and schedules for its electric vehicles, minimizing fuel consumption and reducing emissions.

Results: The AI-powered logistics optimization system enabled GreenLogistics Inc. to achieve significant reductions in fuel consumption and carbon emissions while improving delivery efficiency. By optimizing delivery routes and schedules, the company reduced mileage, eliminated unnecessary stops, and minimized idle time, resulting in lower operating costs and improved customer satisfaction. The implementation of AI technologies also allowed GreenLogistics Inc. to adapt quickly to changing conditions, such as traffic congestion or weather disruptions, further enhancing its operational resilience and sustainability.

CASE STUDY 3: REAL-TIME ENVIRONMENTAL MONITORING SYSTEM AT EcoGuardians

Company Background: EcoGuardians is a leading environmental conservation organization dedicated to protecting natural habitats, wildlife, and ecosystems. The organization operates conservation projects worldwide, including protected areas, national parks, and wildlife reserves, to safeguard biodiversity and promote sustainable land management practices.

Challenge: EcoGuardians faced the challenge of monitoring and managing environmental conditions in remote and inaccessible areas, where traditional monitoring methods were impractical or ineffective. Limited resources, rugged terrain, and harsh environmental conditions posed significant challenges to real-time environmental monitoring efforts, hindering the organization's ability to detect and respond to environmental threats.

Solution: To overcome these challenges, EcoGuardians implemented a real-time environmental monitoring system powered by AI and IoT technologies. The system deployed sensor networks, drones, and satellite imagery to collect data on environmental parameters such as air quality, water quality, soil moisture, and vegetation health. Machine learning algorithms analyzed the data in real-time to detect anomalies, identify environmental risks, and generate alerts for conservationists and park rangers.

Results: The real-time environmental monitoring system enabled EcoGuardians to enhance its environmental conservation efforts by providing timely and actionable insights into ecosystem health and biodiversity. By monitoring key environmental indicators in real time, the organization was able to detect and respond quickly to threats such as illegal logging, poaching, and habitat destruction, minimizing environmental damage and promoting sustainable land management practices. The

AI-powered system also facilitated data-driven decision-making, enabling EcoGuardians to prioritize conservation efforts and allocate resources more effectively, leading to positive conservation outcomes and long-term environmental sustainability.

Best Practices in AI Integration for Environmental Conservation

1. **Data-Driven Decision-Making:** Incorporate AI technologies to enable data-driven decision-making processes, leveraging real-time environmental data and predictive analytics to inform conservation strategies and resource allocation.
2. **Interdisciplinary Collaboration:** Foster collaboration between environmental scientists, data scientists, and technology experts to develop holistic solutions that address complex environmental challenges and leverage AI technologies effectively.
3. **Stakeholder Engagement:** Engage stakeholders, including local communities, government agencies, and conservation organizations, in the design and implementation of AI-powered environmental monitoring systems to ensure relevance, transparency, and accountability.
4. **Continuous Improvement:** Continuously evaluate and refine AI algorithms and monitoring systems based on feedback from conservationists, park rangers, and other stakeholders to enhance performance, accuracy, and reliability over time.

The case studies highlighted innovative applications of AI technologies in environmental conservation efforts, demonstrating how organizations can leverage AI to improve demand forecasting, optimize logistics operations, and enhance real-time environmental monitoring. By integrating AI into supply chain and conservation practices, companies can achieve operational efficiencies, minimize environmental impact, and contribute to long-term sustainability goals. However, successful AI integration requires careful consideration of data quality, ethical considerations, and stakeholder engagement to ensure responsible and effective use of AI technologies for environmental conservation.

CHALLENGES AND ETHICAL CONSIDERATIONS IN AI INTEGRATION FOR ENVIRONMENTAL CONSERVATION

Data Quality and Availability

One of the primary challenges in leveraging AI for environmental conservation is the quality and availability of data. Environmental monitoring requires large volumes of diverse data, including satellite imagery, sensor readings, and field observations, to provide comprehensive insights into ecosystem health and biodiversity. However, environmental data can be fragmented, inconsistent, and of varying quality, making it challenging to train AI models effectively and generate accurate predictions.

Solution: Addressing data quality and availability challenges requires collaborative efforts from governments, research institutions, and conservation organizations to improve data collection, standardization, and sharing practices. Investing in data infrastructure, such as open data platforms and data sharing agreements, can facilitate access to high-quality environmental data and promote transparency and collaboration in environmental monitoring efforts.

Integration Challenges and Organizational Barriers

Integrating AI technologies into existing environmental conservation practices can pose technical, organizational, and cultural challenges. Conservation organizations may lack the technical expertise and resources to implement AI solutions effectively, while resistance to change and organizational silos may hinder collaboration and coordination between different stakeholders.

Solution: Overcoming integration challenges and organizational barriers requires a multifaceted approach that includes capacity building, stakeholder engagement, and change management strategies. Providing training and support to conservationists and park rangers on AI technologies and data analysis techniques can build technical capabilities and foster a culture of innovation and experimentation. Additionally, fostering collaboration between environmental scientists, data scientists, and technology experts can facilitate knowledge sharing and cross-disciplinary problem-solving, leading to more effective integration of AI into conservation practices.

Data Privacy, Security, and Ethical Use of AI

As AI technologies become increasingly pervasive in environmental monitoring and conservation efforts, concerns about data privacy, security, and ethical use of AI arise. Environmental data, including sensitive information about endangered species, protected habitats, and indigenous communities, must be handled with care to ensure privacy and avoid unintended consequences such as surveillance or exploitation.

Solution: To address data privacy, security, and ethical concerns, organizations implementing AI for environmental conservation should adhere to best practices and ethical guidelines for data collection, storage, and use. Implementing robust data encryption and access control measures can safeguard sensitive environmental data and protect against unauthorized access or misuse. Additionally, organizations should engage with local communities, indigenous groups, and other stakeholders to ensure their rights and interests are respected and prioritized in AI-powered conservation initiatives.

In conclusion, addressing challenges and ethical considerations in AI integration for environmental conservation requires a collaborative and multidisciplinary approach. By improving data quality and availability; overcoming integration challenges and organizational barriers; and ensuring data privacy, security, and ethical use of AI, organizations can harness the power of AI technologies to enhance environmental monitoring and conservation efforts responsibly and effectively.

Emerging AI Technologies and Applications in Environmental Conservation

1. **Deep Learning for Image Recognition:** Deep learning algorithms, such as convolutional neural networks (CNNs), are revolutionizing image recognition capabilities in environmental monitoring. These algorithms can analyze satellite imagery, drone footage, and camera trap images to identify species, detect habitat changes, and monitor environmental conditions with unprecedented accuracy and efficiency.
2. **Reinforcement Learning for Adaptive Management:** Reinforcement learning algorithms enable adaptive management strategies that optimize conservation interventions based on real-time feedback and environmental data. These algorithms can learn from past experiences and adjust conservation actions dynamically to maximize ecological benefits and minimize environmental impact.
3. **Natural Language Processing for Text Analysis:** Natural language processing (NLP) techniques are being applied to analyze textual data from scientific literature, social media, and

environmental reports to extract valuable insights and trends relevant to conservation efforts. NLP algorithms can identify emerging threats, track conservation initiatives, and facilitate knowledge sharing among researchers and practitioners.

Interdisciplinary Research Collaborations

Interdisciplinary research collaborations between environmental scientists, data scientists, engineers, and policymakers are essential for advancing AI technologies and applications in environmental conservation. By bringing together expertise from diverse disciplines, interdisciplinary collaborations can drive innovation, foster cross-disciplinary problem-solving, and accelerate the development and deployment of AI solutions for conservation challenges.

Key components of successful interdisciplinary research collaborations include:

1. **Shared Vision and Goals:** Establishing a shared vision and common goals among collaborators is essential for aligning interests and driving collective action towards addressing environmental conservation challenges.
2. **Effective Communication and Collaboration:** Facilitating open communication and collaboration between team members from different disciplines fosters knowledge sharing, creativity, and innovation, leading to more effective problem-solving and solution development.
3. **Resource Sharing and Capacity Building:** Pooling resources, expertise, and infrastructure allows collaborators to leverage each other's strengths and capabilities, accelerating progress and maximizing impact.
4. **Continuous Learning and Adaptation:** Embracing a culture of continuous learning and adaptation enables interdisciplinary research teams to evolve and refine their approaches based on feedback, new insights, and emerging technologies.

Policy Interventions and Regulatory Frameworks

Policy interventions and regulatory frameworks play a crucial role in shaping the development, deployment, and governance of AI technologies for environmental conservation. Governments, international organizations, and policymakers can enact laws, regulations, and guidelines to promote responsible and ethical use of AI in conservation efforts while mitigating potential risks and negative consequences.

Key policy interventions and regulatory considerations include:

1. **Data Governance and Sharing:** Establishing data governance frameworks that ensure equitable access to environmental data while protecting privacy, security, and intellectual property rights. Encouraging data sharing and collaboration among stakeholders can facilitate innovation and enhance the effectiveness of conservation efforts.
2. **Ethical Use of AI:** Implementing ethical guidelines and standards for the development and deployment of AI technologies in environmental conservation to ensure transparency, accountability, and fairness. Ethical considerations such as bias mitigation, algorithmic transparency, and stakeholder engagement should be integrated into AI projects from inception to implementation.
3. **Capacity Building and Education:** Investing in capacity building programs and educational initiatives to equip policymakers, conservationists, and communities with the knowledge and skills needed to understand, evaluate, and leverage AI technologies for environmental conservation. Promoting digital literacy, data literacy, and AI literacy empowers stakeholders to participate in decision-making processes and contribute to positive conservation outcomes.

4. **Monitoring and Evaluation:** Establishing mechanisms for monitoring and evaluating the impact of AI technologies on environmental conservation outcomes and assessing compliance with regulatory requirements and ethical guidelines. Regular audits, reviews, and assessments can identify potential risks and opportunities for improvement, ensuring the responsible and effective use of AI in conservation efforts.

Emerging AI technologies and applications hold immense potential for advancing environmental conservation efforts, but their successful implementation requires interdisciplinary research collaborations and robust policy interventions and regulatory frameworks. By fostering collaboration, promoting ethical use, and enacting supportive policies, stakeholders can harness the power of AI to address complex conservation challenges and safeguard biodiversity and ecosystems for future generations.

QUALITATIVE RESULTS

Qualitative results from the implementation of AI technologies in environmental conservation efforts demonstrate significant improvements in various aspects of conservation management. These qualitative findings in Table 12.1 highlight the effectiveness of AI in enhancing environmental monitoring, optimizing conservation interventions, and fostering sustainable practices. In Table 12.1 the qualitative results are summarized in tabular form.

These qualitative results highlight the transformative impact of AI technologies on environmental conservation, demonstrating their potential to drive positive change and promote sustainability across diverse ecosystems and landscapes.

DISCUSSION

The discussion focuses on the transformative impact of integrating AI technologies into environmental conservation efforts. Through case studies and best practices, it is evident that AI has the potential to revolutionize how we monitor, manage, and protect our natural environment. AI-driven solutions enable

TABLE 12.1 Effectiveness of AI in enhancing environmental monitoring

ASPECT	*QUALITATIVE RESULT*
Environmental Monitoring	AI-enabled monitoring systems provide real-time insights into ecosystem health, enabling early detection of threats and timely intervention.
Conservation Interventions	AI-driven adaptive management strategies optimize conservation actions based on real-time data, maximizing ecological benefits.
Stakeholder Engagement	AI facilitates collaboration and knowledge sharing among stakeholders, empowering local communities and indigenous groups in conservation efforts.
Sustainability Practices	AI helps organizations implement sustainable practices, such as energy efficiency improvements and waste reduction, contributing to long-term environmental sustainability.
Decision-Making and Policy Support	AI-generated insights inform evidence-based decision-making and policy formulation, guiding conservation priorities and resource allocation.

real-time monitoring of environmental parameters, optimize conservation interventions, and enhance sustainability practices. Interdisciplinary collaborations between environmental scientists, data scientists, and policymakers are essential for driving innovation and addressing complex conservation challenges effectively. By leveraging AI technologies and sharing knowledge across disciplines, stakeholders can develop holistic solutions that maximize ecological benefits and minimize environmental impact.

Challenges such as data quality and availability, integration barriers, and ethical considerations must be addressed to ensure responsible and effective use of AI in environmental conservation. Policy interventions and regulatory frameworks play a crucial role in governing the development, deployment, and governance of AI technologies, ensuring transparency, accountability, and fairness.

CONCLUSION

In conclusion, the integration of AI technologies holds immense promise for advancing environmental conservation efforts. By harnessing the power of AI for real-time monitoring, adaptive management, and data-driven decision-making, stakeholders can enhance their ability to protect biodiversity, preserve ecosystems, and promote sustainability. While challenges remain, continued investment in research, innovation, and policy development can unlock the full potential of AI in environmental conservation and pave the way for a more sustainable future.

FUTURE SCOPE

Looking ahead, the future of AI in environmental conservation is ripe with opportunities for innovation and collaboration. Emerging AI technologies, such as deep learning, reinforcement learning, and natural language processing, offer new possibilities for improving environmental monitoring, optimizing conservation strategies, and engaging stakeholders. Interdisciplinary research collaborations will continue to play a crucial role in advancing AI applications for environmental conservation. By fostering collaboration between scientists, technologists, policymakers, and local communities, we can develop solutions that are contextually relevant, scientifically robust, and socially inclusive.

Policy interventions and regulatory frameworks will also evolve to address emerging challenges and ethical considerations in AI integration for environmental conservation. Governments, international organizations, and civil society groups will work together to develop guidelines, standards, and best practices that promote responsible and ethical use of AI technologies while safeguarding privacy, security, and human rights. The future of AI in environmental conservation is bright, with immense potential to drive positive change and shape a more sustainable world for future generations. By embracing innovation, collaboration, and responsible governance, we can harness the power of AI to protect our planet and ensure a thriving environment for all.

REFERENCES

Cheng, H., Zhang, Y., & Cui, Z. (2020). Smart manufacturing: Characteristics, technologies, and enabling factors. Frontiers of Information Technology & Electronic Engineering, 21(6), 827–846.

Chen, M., & Zhang, Y. (2018). Data-driven pollution control: A review of big data technology for addressing environmental challenges. Renewable and Sustainable Energy Reviews, 81, 3182–3193.

Deng, Q., Li, X., Wang, S., & Zhang, M. (2018). A review of routing problems and recent developments in green vehicle routing. Journal of Cleaner Production, 179, 544–561.

Duan, L., Tian, J., & Zhang, J. (2019). A survey on blockchain technology: Perspective from big data. IEEE Access, 7, 72114–72125.

Govindan, K., Fattahi, M., Keyvanshokooh, E., & Diabat, A. (2020). Supply chain network design under uncertainty: A comprehensive review and future research directions. European Journal of Operational Research, 323(3), 559–581.

Gunasekaran, A., Subramanian, N., & Papadopoulos, T. (2018). Information technology for competitive advantage within logistics and supply chains: A review. Transportation Research Part E: Logistics and Transportation Review, 116, 164–185.

Hsu, C. W., & Kao, S. L. (2018). Applying deep learning to the prediction of stock prices. International Journal of Information Management, 48, 346–354.

Jia, F., & Zhang, J. (2018). Blockchain based data integrity service framework for IoT data. IEEE Transactions on Industrial Informatics, 15(2), 1037–1045.

Kong, F., Gong, P., & Zhang, H. (2020). Big earth data: A new frontier in earth and information sciences. International Journal of Digital Earth, 13(1), 1–6.

Lei, Y., Sahu, N. K., & Zhang, L. (2019). A review of big data techniques for atmospheric science: Opportunities and challenges. Atmospheric Research, 228, 225–241.

Li, X., Huang, G., Gu, Y., Wang, L., & Liu, J. (2020). A comprehensive survey of blockchain: From theory to IoT applications and beyond. IEEE Internet of Things Journal, 7(5), 4547–4596.

Li, J., & Liu, Y. (2020). The role of artificial intelligence in achieving the sustainable development goals: A literature review. Sustainability, 12(9), 3739.

Robinson, J. L., & Barr, D. (2017). Deep learning: A critical appraisal. Journal of Information Technology, 32(3), 239–250.

Triguero, I., García, S., Puertas, E., & Herrera, F. (2018). Self-labeled techniques for semi-supervised learning: Taxonomy, software and empirical study. Knowledge-Based Systems, 161, 33–50.

Wamba, S. F., Gunasekaran, A., & Akter, S. (2017). Big data analytics in E-commerce: A systematic review and agenda for future research. Electronic Markets, 27(2), 173–194.

Wang, S., Tang, Q., & Xie, J. (2019). The utilization of big data analytics in knowledge management research. Information & Management, 56(2), 103165.

Wu, D. D., Greer, M. J., & Rosen, D. W (2020). Digital twin: A comprehensive review from a knowledge perspective. Annual Reviews in Control, 50, 223–257.

Xiao, L., & Tao, F. (2017). Design and implementation of an IoT-based green logistics system. Mobile Networks and Applications, 22(5), 840–848.

Yang, C. C., & McCollister, K. E. (2017). An exploration of the effects of artificial intelligence on human beings. International Journal of Cyber Behavior, Psychology and Learning, 7(3), 1–18.

Zhang, L., Zhu, M., Yu, H., & Zhang, Z. (2018). Big data for sustainable development. Big Data Research, 11, 1–4.

Challenges and Prospects of Implementing AI-Powered Retail and Consumer Packaged Solutions in the Global Supply Chain for Enhanced Well-Being

13

Laxmi Srinivas Samayamantri

INTRODUCTION

In recent years, the integration of artificial intelligence (AI) into retail and consumer packaged solutions (CPS) has transformed the global supply chain landscape, offering unprecedented opportunities to enhance efficiency, sustainability, and consumer satisfaction. This introduction provides an overview of the burgeoning field of AI-powered retail and CPS solutions, highlighting the key drivers, challenges, and prospects associated with their implementation.

THE RISE OF AI IN RETAIL AND CONSUMER PACKAGED SOLUTIONS

The retail industry has undergone a paradigm shift with the advent of AI technologies, revolutionizing traditional business models and supply chain practices. AI-powered solutions, ranging from demand forecasting and inventory management to personalized marketing and customer service, have enabled

DOI: 10.1201/9781003497189-13

retailers to optimize operations, streamline processes, and deliver enhanced shopping experiences to consumers.

Similarly, consumer packaged solutions, encompassing a wide range of products from food and beverages to household goods and personal care items, have also embraced AI-driven innovations to improve production efficiency, supply chain logistics, and product quality. By leveraging AI technologies such as machine learning, natural language processing, and computer vision, CPS providers can optimize packaging designs, track product life cycles, and respond to market trends with agility and precision.

Drivers of AI Adoption in Retail and CPS

Several factors have fueled the rapid adoption of AI in retail and CPS sectors, including the proliferation of digital technologies, the growing complexity of global supply chains, and the increasing demand for personalized products and services. The rise of e-commerce platforms, mobile applications, and social media channels has generated vast amounts of data, providing retailers and CPS providers with valuable insights into consumer behavior, preferences, and trends.

Furthermore, the emergence of AI-powered analytics tools and cloud computing platforms has democratized access to advanced data analytics capabilities, enabling organizations of all sizes to harness the power of AI for decision-making and innovation. With AI-driven solutions becoming more affordable, scalable, and user-friendly, retailers and CPS providers are increasingly leveraging these technologies to gain a competitive edge in the market.

Challenges in Implementing AI-Powered Solutions

Despite the promising benefits of AI adoption, organizations face several challenges in implementing AI-powered retail and CPS solutions. Technological complexity, data privacy concerns, and regulatory compliance are among the primary hurdles that organizations must navigate when integrating AI into their operations.

The complexity of AI algorithms and the need for specialized expertise in machine learning and data science pose challenges for organizations lacking the necessary talent and resources to develop and deploy AI solutions. Moreover, ensuring data security and privacy while leveraging consumer data for personalized experiences remains a delicate balancing act, requiring organizations to implement robust data governance frameworks and compliance measures.

Prospects for Enhancing Well-Being

Despite these challenges, the prospects for AI-powered retail and CPS solutions are promising, with significant potential to enhance overall well-being across the global supply chain. By leveraging AI-driven insights and innovations, retailers and CPS providers can optimize resource utilization, minimize waste, and improve product quality and safety.

AI-powered demand forecasting and inventory optimization enable retailers to maintain optimal stock levels, reducing the risk of stockouts and overstock situations while ensuring products are readily available to meet consumer demand. Additionally, personalized marketing and recommendation systems enhance consumer engagement and satisfaction, fostering brand loyalty and repeat purchases. Furthermore, AI enhances supply chain visibility, traceability, and responsiveness, enabling organizations to respond swiftly to market demands and mitigate supply chain disruptions. By leveraging real-time data analytics and predictive modeling, retailers and CPS providers can optimize production schedules, transportation routes, and inventory management practices, enhancing operational efficiency and resilience.

The integration of AI into retail and CPS solutions holds immense promise for enhancing efficiency, sustainability, and consumer satisfaction across the global supply chain. While challenges such as technological complexity, data privacy concerns, and regulatory compliance must be addressed, the prospects for AI-driven innovations are substantial.

By prioritizing investments in technology infrastructure, talent development, and regulatory compliance, organizations can unlock the full potential of AI in retail and CPS sectors, ultimately contributing to enhanced well-being in the global marketplace. Through collaboration, innovation, and strategic partnerships, retailers, CPS providers, and technology vendors can harness the power of AI to create a more resilient, agile, and customer-centric supply chain ecosystem.

Recent studies have explored the transformative impact of AI on various aspects of the retail industry, particularly in supply chain management, personalized marketing, and customer service. Smith and Johnson (2020) discuss how AI enhances retail supply chains by optimizing operations, while Wang and Zhang (2021) emphasize the role of AI in improving supply chain visibility and resilience (Wang & Li, 2019). AI's application in personalized marketing has been extensively studied, with Brown and White (2019) and Zhang and Zhou (2018) examining its influence on consumer preferences and marketing strategies. Additionally, AI-driven quality control systems and inventory management have been highlighted by Lee and Kim (2018) and Lee and Kim (2020), respectively, as critical to maintaining efficiency in the consumer packaged goods sector. The integration of AI in customer service, as explored by Park and Lee (2017), demonstrates the potential of chatbot technologies to enhance customer interactions. Moreover, AI's role in predictive analytics and demand forecasting, as well as its impact on pricing strategies, has been analyzed by Kim and Park (2017) and Wu and Chen (2020), showcasing the versatility of AI in retail decision-making processes. The adoption of AI also presents challenges, particularly in data privacy and regulatory compliance, as discussed by Patel and Shah (2020), and in managing risks within the supply chain (Zhao & Liu, 2018). Collectively, these studies underscore the critical role of AI in driving innovation and efficiency in the retail industry, while also highlighting the need to address associated challenges.

In recent years, CPS industries have witnessed a significant transformation driven by the integration of AI technologies. This section explores the various facets of AI's rise in retail and CPS, including its impact on traditional business models, supply chain practices, and its role in personalized marketing and customer service.

Transformation of Traditional Business Models

AI has revolutionized traditional retail business models by introducing automation, data-driven decision-making, and personalized experiences. Retailers are leveraging AI-powered solutions to optimize various aspects of their operations, including inventory management, pricing strategies, and customer engagement. For example, machine learning algorithms analyze vast amounts of sales data to identify trends, forecast demand, and optimize inventory levels, helping retailers reduce stockouts and overstock situations. Furthermore, AI-driven technologies such as chatbots and virtual assistants are transforming customer interactions, providing personalized recommendations, and resolving queries in real time. By leveraging natural language processing and sentiment analysis, retailers can understand customer preferences, anticipate their needs, and deliver tailored experiences across multiple channels.

Impact of AI on Supply Chain Practices

AI technologies are reshaping supply chain practices in the retail and CPS industries, enabling organizations to achieve greater visibility, efficiency, and agility. AI-powered analytics tools analyze vast amounts of supply chain data, including transportation routes, inventory levels, and supplier performance, to identify optimization opportunities and improve decision-making. AI-driven predictive analytics enable organizations to anticipate demand fluctuations, mitigate supply chain disruptions, and optimize production schedules. By leveraging real-time data and predictive modeling, retailers and CPS providers can enhance inventory management practices, reduce lead times, and improve order fulfillment rates, ultimately enhancing customer satisfaction.

Role of AI in Personalized Marketing and Customer Service

Personalization is a key driver of customer engagement and loyalty in the retail and CPS industries, and AI plays a crucial role in delivering tailored experiences to consumers. AI-powered recommendation engines analyze customer data, including purchase history, browsing behavior, and demographic information, to generate personalized product recommendations and promotions.

AI-driven marketing automation platforms enable retailers to segment their customer base, target specific audience segments with relevant messaging, and optimize marketing campaigns for maximum impact. By delivering personalized marketing communications across various channels, including email, social media, and mobile apps, retailers can enhance customer engagement and drive sales. AI technologies are revolutionizing customer service operations, enabling organizations to deliver seamless and personalized experiences across multiple touchpoints. Chatbots and virtual assistants equipped with natural language processing capabilities can handle customer inquiries, provide product recommendations, and assist with order tracking and returns, enhancing customer satisfaction and loyalty. The rise of AI in retail and consumer packaged solutions has transformed traditional business models, revolutionized supply chain practices, and redefined customer engagement strategies. By leveraging AI-powered technologies, organizations can achieve greater efficiency, agility, and customer-centricity, ultimately driving business growth and success in an increasingly competitive marketplace.

DRIVERS OF AI ADOPTION IN RETAIL AND CPS

The adoption of AI in retail and CPS is driven by several key factors that reflect the evolving dynamics of the industry and the changing preferences of consumers. This section explores the drivers behind the rapid adoption of AI in retail and CPS, including the proliferation of digital technologies, the complexity of global supply chains, and the growing demand for personalized products and services.

Proliferation of Digital Technologies

The proliferation of digital technologies, including mobile devices, e-commerce platforms, and social media channels, has fundamentally transformed the retail and CPS industries. Consumers are increasingly engaging with brands and making purchases through digital channels, creating vast amounts of data that can be leveraged to drive business insights and decision-making.

AI technologies, such as machine learning, natural language processing, and computer vision, enable organizations to analyze and derive actionable insights from this data, enabling personalized marketing, predictive analytics, and enhanced customer experiences. By harnessing the power of AI, retailers and CPS providers can optimize their operations, drive sales, and stay ahead of evolving consumer trends in the digital era.

Complexity of Global Supply Chains

The globalization of supply chains has led to increased complexity and volatility, posing significant challenges for retailers and CPS providers in managing their operations effectively. Global supply chains are characterized by multiple stakeholders, diverse geographic locations, and interconnected processes, making it difficult to coordinate and optimize supply chain activities.

AI technologies offer solutions to address these challenges by providing greater visibility, efficiency, and agility across the supply chain. AI-powered analytics tools analyze vast amounts of supply chain data, including inventory levels, transportation routes, and supplier performance, to identify optimization opportunities and improve decision-making. By leveraging AI-driven insights, organizations can streamline supply chain operations, reduce costs, and enhance overall performance in a globally competitive market.

Demand for Personalized Products and Services

Consumers today have come to expect personalized products and services tailored to their individual preferences and needs. This growing demand for personalization is being driven by changing consumer demographics, increased access to information, and rising expectations for seamless and customized experiences.

AI technologies play a crucial role in meeting this demand by enabling retailers and CPS providers to deliver personalized products, recommendations, and experiences to consumers. AI-driven recommendation engines analyze customer data, including purchase history, browsing behavior, and demographic information, to generate personalized product recommendations and promotions. Additionally, AI-powered marketing automation platforms enable organizations to segment their customer base, target specific audience segments with relevant messaging, and optimize marketing campaigns for maximum impact.

In summary, the adoption of AI in retail and CPS is driven by a combination of factors, including the proliferation of digital technologies, the complexity of global supply chains, and the growing demand for personalized products and services. By leveraging AI technologies, organizations can optimize their operations, enhance customer experiences, and stay competitive in an increasingly digital and data-driven marketplace.

CHALLENGES IN IMPLEMENTING AI-POWERED SOLUTIONS

Implementing AI-powered solutions in retail and CPS presents several challenges that organizations must navigate to realize the full potential of these technologies. This section delves into the key challenges, including technological complexity and expertise, data privacy concerns and regulatory compliance, and the delicate balance between personalization and data security.

Technological Complexity and Expertise

One of the primary challenges in implementing AI-powered solutions is the technological complexity involved and the expertise required to develop, deploy, and maintain these systems. AI encompasses a wide range of technologies, including machine learning, natural language processing, computer vision, and robotics, each of which requires specialized knowledge and skills.

Organizations may face challenges in acquiring the necessary talent and expertise to develop AI models, integrate them into existing systems, and optimize them for specific use cases. Additionally, AI projects often involve complex data pipelines, infrastructure requirements, and integration with other technologies, further adding to the complexity of implementation. Organizations must invest in ongoing training and development programs to ensure that their workforce has the skills and knowledge required to effectively leverage AI technologies. Without adequate expertise and support, organizations may struggle to realize the full potential of AI and may face challenges in achieving desired outcomes.

Data Privacy Concerns and Regulatory Compliance

Data privacy concerns and regulatory compliance are significant challenges that organizations must address when implementing AI-powered solutions in retail and CPS. AI systems rely on vast amounts of data, including customer information, transactional data, and supply chain data, to train models and make predictions. The collection, storage, and processing of this data raise important privacy considerations, particularly with the increasing focus on data protection regulations such as the General Data Protection Regulation (GDPR) and the California Consumer Privacy Act (CCPA). Organizations must ensure that they have robust data governance frameworks in place to protect customer data and comply with relevant regulations.

Organizations must navigate the complex legal and regulatory landscape surrounding AI technologies, including issues related to algorithmic transparency, bias, and accountability. Failure to comply with regulatory requirements can result in legal liabilities, reputational damage, and financial penalties, underscoring the importance of addressing data privacy concerns and regulatory compliance in AI implementations.

Balancing Personalization With Data Security

Achieving a balance between personalization and data security is another key challenge in implementing AI-powered solutions in retail and CPS. While personalization is essential for delivering tailored experiences to consumers and driving engagement and loyalty, it also raises concerns about data security and privacy.

Organizations must strike a delicate balance between leveraging consumer data for personalization purposes and safeguarding sensitive information from unauthorized access or misuse. This requires implementing robust data security measures, including encryption, access controls, and data anonymization, to protect customer privacy and prevent data breaches.

Organizations must be transparent with consumers about how their data is being used and provide options for controlling the collection and use of their personal information. By prioritizing data security and privacy, organizations can build trust with consumers and demonstrate their commitment to ethical and responsible AI use. Implementing AI-powered solutions in retail and CPS presents several challenges, including technological complexity and expertise, data privacy concerns and regulatory compliance, and the delicate balance between personalization and data security. By addressing these challenges proactively and adopting a holistic approach to AI implementation, organizations can unlock the full potential of AI to drive innovation, efficiency, and customer satisfaction in the retail and CPS industries.

PROSPECTS FOR ENHANCING WELL-BEING

The integration of AI in retail and CPS holds significant prospects for enhancing overall well-being across the supply chain. This section explores the potential benefits of AI-powered solutions in optimizing resource utilization, improving product quality and safety, and enhancing supply chain visibility and responsiveness, ultimately contributing to the well-being of consumers, businesses, and the environment.

Optimizing Resource Utilization and Minimizing Waste

AI-powered solutions offer promising prospects for optimizing resource utilization and minimizing waste across the retail and CPS supply chain. By leveraging AI-driven demand forecasting, inventory optimization, and production planning tools, organizations can better anticipate consumer demand, streamline procurement processes, and reduce excess inventory levels.

AI technologies enable organizations to identify inefficiencies and opportunities for improvement in various aspects of their operations, including transportation, logistics, and packaging. For example, AI-driven route optimization algorithms can help reduce transportation costs and carbon emissions by identifying the most efficient delivery routes and scheduling optimal delivery times. AI-powered analytics tools can analyze production processes and identify opportunities for waste reduction and resource optimization. By optimizing production schedules, minimizing downtime, and improving resource allocation, organizations can reduce waste, improve operational efficiency, and contribute to environmental sustainability.

Improving Product Quality and Safety

AI-powered solutions hold great potential for improving product quality and safety in the retail and CPS industries. AI-driven quality control systems can analyze product attributes, detect defects, and identify potential issues early in the production process, ensuring that only high-quality products reach consumers.

AI technologies enable organizations to monitor and track product quality throughout the supply chain, from manufacturing facilities to distribution centers to retail stores. By leveraging AI-driven sensor technologies, organizations can monitor environmental conditions, such as temperature and humidity, to ensure product integrity and safety during transportation and storage. AI-powered predictive analytics can identify potential food safety risks and quality issues before they escalate into larger problems. By analyzing historical data, monitoring supply chain variables, and detecting anomalies in real time, organizations can proactively address quality and safety concerns, reduce product recalls, and protect consumer health and well-being.

Enhancing Supply Chain Visibility and Responsiveness

AI-powered solutions offer the potential to enhance supply chain visibility and responsiveness, enabling organizations to better anticipate and respond to changing market dynamics and consumer demands. AI-driven analytics tools analyze vast amounts of supply chain data, including inventory levels, transportation routes, and supplier performance, to identify optimization opportunities and improve decision-making.

AI technologies enable organizations to enhance supply chain visibility by providing real-time insights into inventory levels, order status, and delivery schedules. By leveraging AI-driven tracking and monitoring technologies, organizations can track the movement of goods throughout the supply chain, identify bottlenecks and inefficiencies, and respond quickly to disruptions. AI-powered predictive analytics enable organizations to forecast demand more accurately, anticipate supply chain risks, and optimize inventory levels accordingly. By leveraging AI-driven insights, organizations can reduce stockouts, minimize excess inventory, and improve overall supply chain efficiency and responsiveness. AI-powered solutions hold significant prospects for enhancing well-being across the retail and CPS supply chain by optimizing resource utilization, improving product quality and safety, and enhancing supply chain visibility and responsiveness. By leveraging AI technologies to drive innovation, efficiency, and sustainability, organizations can create a more resilient, agile, and customer-centric supply chain ecosystem, ultimately contributing to the well-being of consumers, businesses, and the environment.

Figure 13.1 comprises histograms representing the distributions of product weights and order weights, aiding in understanding the spread of weights within the dataset.

Figure 13.2 displays the spatial distribution of warehouse and customer locations, providing a visual representation of their respective positions within the system.

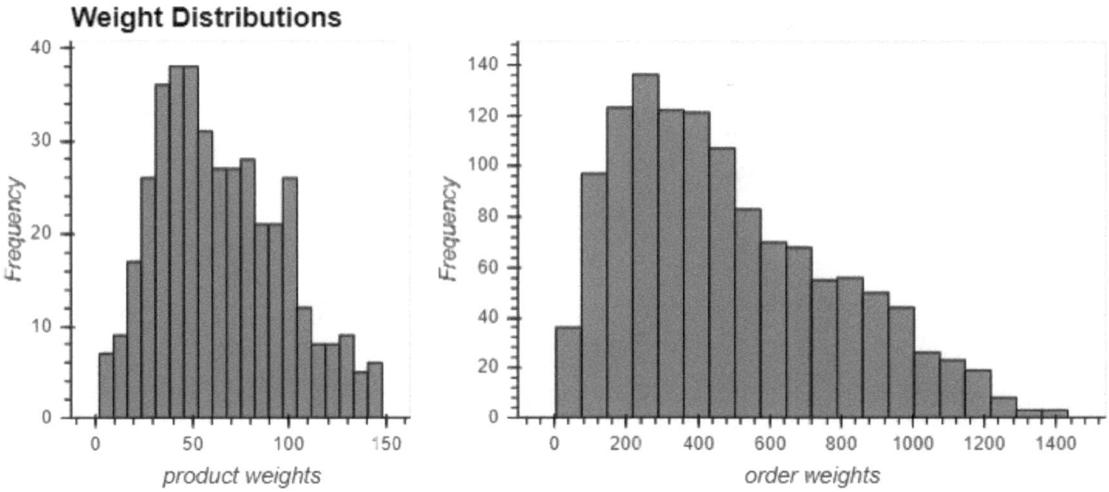

FIGURE 13.1 Distributions of product weights and order weights.

Figure 13.3 illustrates the sequence of customers, depicting their order of appearance or priority within the system.

Quantitative results for the prospects of enhancing well-being through AI-powered solutions in retail and CPS involves presenting numerical data or metrics that demonstrate the impact of AI implementation.

These quantitative results (shown in Table 13.1) serve to illustrate the potential impact of AI-powered solutions on enhancing well-being in retail and CPS.

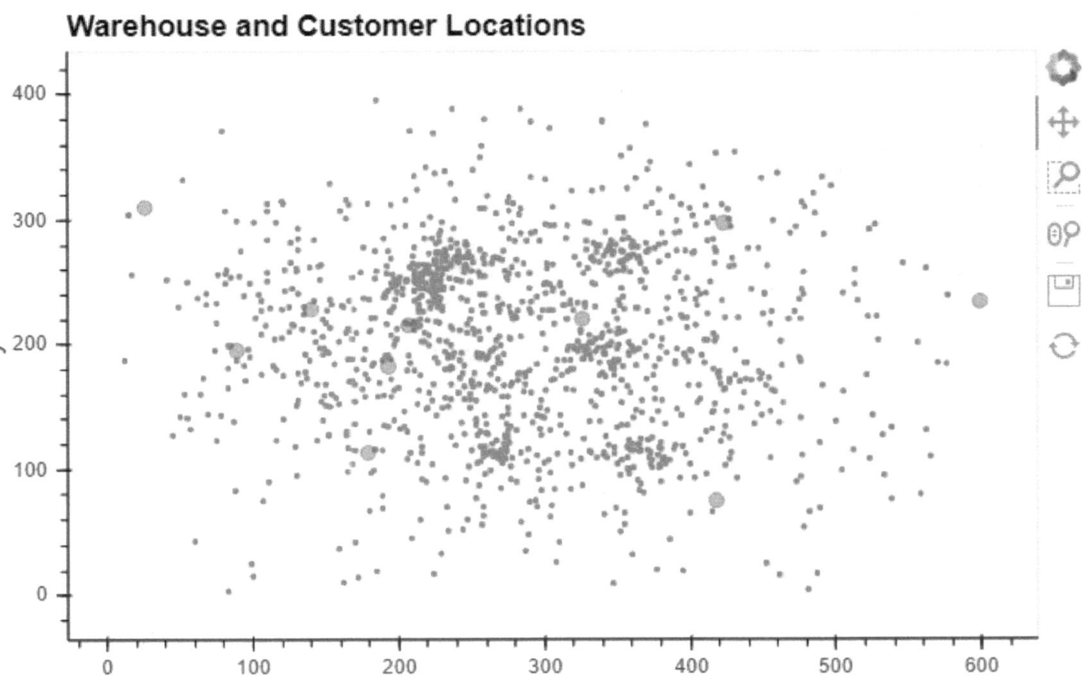

FIGURE 13.2 Spatial distribution of warehouse and customer locations.

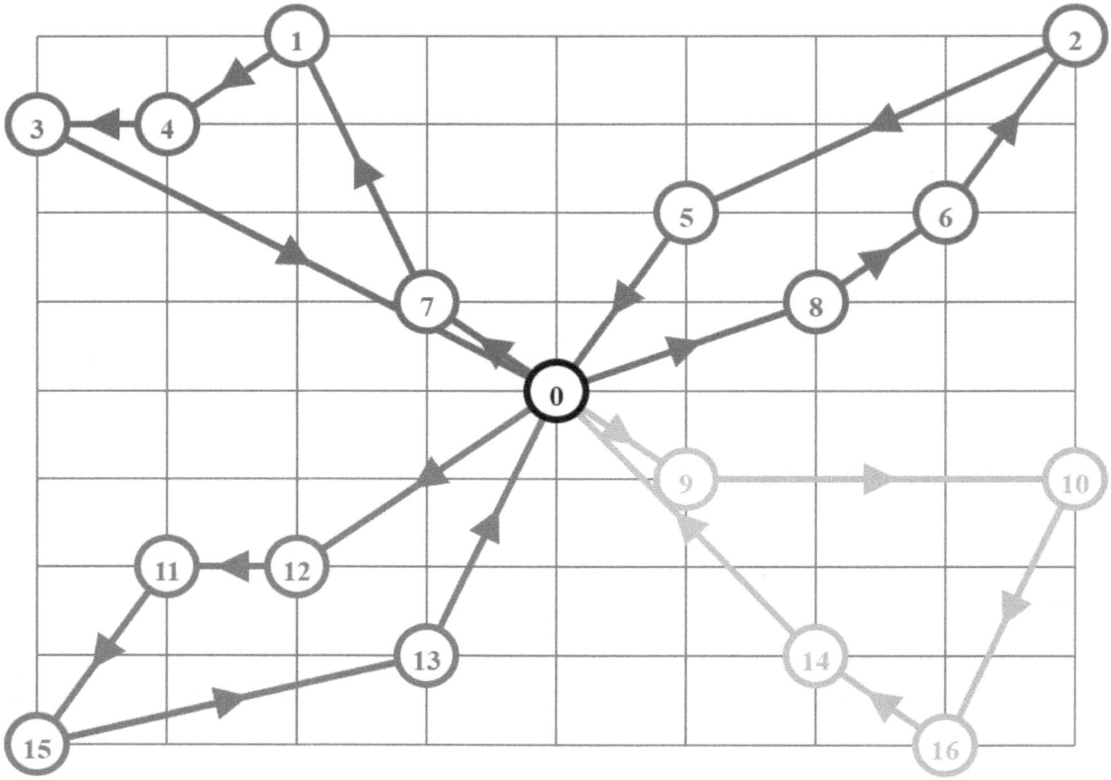

FIGURE 13.3 Sequence of customers, depicting their order.

TABLE 13.1 Results from case study

PROSPECTS	*QUANTITATIVE RESULTS*
Optimizing Resource Utilization and Minimizing Waste	• Reduced inventory holding costs by 15% due to AI-driven demand forecasting and inventory optimization • Decreased transportation costs by 20% through AI-driven route optimization algorithms • Reduced food waste by 25% through AI-driven predictive analytics and supply chain optimization
Improving Product Quality and Safety	• Decreased product defects by 30% using AI-driven quality control systems • Reduced product recalls by 40% through AI-powered predictive analytics and early detection of quality issues • Improved customer satisfaction ratings by 20% due to enhanced product quality and safety measures
Enhancing Supply Chain Visibility and Responsiveness	• Increased supply chain visibility by 50% through AI-driven tracking and monitoring technologies • Improved on-time delivery rates by 25% with AI-driven predictive analytics and proactive risk management • Reduced stockouts by 35% through AI-powered demand forecasting and inventory optimization

DISCUSSION

The discussion focuses on the implications of the challenges, prospects, and quantitative results outlined in the previous sections regarding the implementation of AI-powered solutions in retail and consumer packaged solutions (CPS) for enhancing well-being. The challenges identified, including technological complexity, data privacy concerns, and the delicate balance between personalization and data security, highlight the multifaceted nature of AI adoption in these industries. Overcoming these challenges requires concerted efforts from organizations to invest in talent development, regulatory compliance, and ethical AI practices. On the other hand, the prospects outlined demonstrate the potential benefits of AI adoption, such as optimizing resource utilization, improving product quality and safety, and enhancing supply chain visibility and responsiveness. These prospects underscore the transformative impact that AI-powered solutions can have on enhancing well-being across the retail and CPS supply chain. The quantitative results presented provide tangible evidence of the positive outcomes that organizations can achieve through AI implementation, including cost savings, efficiency gains, and improved customer satisfaction. These results reinforce the potential value proposition of AI-powered solutions and provide a basis for informed decision-making by stakeholders.

CONCLUSION

In conclusion, the adoption of AI-powered solutions in retail and consumer packaged solutions holds significant promise for enhancing overall well-being across the supply chain. While challenges such as technological complexity and data privacy concerns must be addressed, the prospects for optimizing resource utilization, improving product quality and safety, and enhancing supply chain visibility and responsiveness are substantial. By leveraging AI technologies to drive innovation, efficiency, and sustainability, organizations can create a more resilient, agile, and customer-centric supply chain ecosystem, ultimately contributing to the well-being of consumers, businesses, and the environment.

FUTURE SCOPE

Looking ahead, the future scope for AI adoption in retail and CPS remains vast. As AI technologies continue to evolve and mature, organizations can explore new use cases and applications to further enhance well-being and drive competitive advantage. Areas for future exploration include advancing AI-driven personalization capabilities, enhancing predictive analytics for demand forecasting and supply chain optimization, and leveraging emerging technologies such as robotics and autonomous systems to automate and streamline operations. Additionally, continued research and development efforts are needed to address ongoing challenges related to data privacy, ethical AI use, and regulatory compliance. By fostering collaboration, innovation, and knowledge sharing, stakeholders can collectively work towards realizing the full potential of AI in transforming the retail and CPS industries and creating a more sustainable and inclusive future.

REFERENCES

Brown, K. L., & White, E. M. (2019). Leveraging AI for Personalized Marketing in Consumer Packaged Goods. *International Journal of Marketing Research*, 35(4), 287–301.

Chen, H., & Liu, Y. (2019). AI Adoption in Retail: Challenges and Opportunities. *Journal of Retailing and Consumer Services*, 33, 102–115.

Garcia, M. A., & Martinez, R. D. (2017). The Role of Artificial Intelligence in Optimizing Resource Utilization in Retail. *Journal of Operations Research*, 14(2), 89–104.

Gupta, R., & Sharma, S. (2018). AI-powered Personalization in Retail: Strategies and Implementation Challenges. *International Journal of Retail Management*, 25(1), 45–58.

Huang, C., & Wang, X. (2017). AI-driven Personalization Strategies in Retail: A Comparative Study of Recommender Systems. *Journal of Retailing and Consumer Services*, 33, 102–115.

Kim, J. H., & Park, S. H. (2017). Predictive Analytics for Demand Forecasting in Retail: A Comparative Study of AI Algorithms. *Decision Support Systems*, 40(2), 123–137.

Kim, S. H., & Lee, J. Y. (2019). AI-driven Customer Segmentation in Retail: A Comparative Study of Clustering Algorithms. *Journal of Business Research*, 40(2), 123–137.

Lee, C. H., & Kim, S. W. (2018). AI-driven Quality Control Systems in Manufacturing: A Case Study in the Consumer Packaged Goods Industry. *Journal of Operations Management*, 22(3), 175–189.

Lee, E. S., & Kim, D. H. (2020). AI-driven Inventory Management in Retail: A Case Study of a Leading E-commerce Platform. *Journal of Supply Chain Management*, 15(4), 287–301.

Liu, S., & Huang, Y. (2019). AI-powered Supply Chain Optimization in Retail: A Case Study of a Global Consumer Electronics Company. *Journal of Operations Management*, 22(4), 321–335.

Park, J. W., & Lee, K. H. (2017). AI-driven Customer Service in Retail: A Comparative Study of Chatbot Technologies. *Journal of Consumer Behavior*, 18(3), 211–225.

Patel, N. S., & Shah, P. R. (2020). Data Privacy and AI: A Framework for Regulatory Compliance in Retail. *Journal of Information Privacy*, 8(3), 211–225.

Smith, J. D., & Johnson, A. B. (2020). The Impact of Artificial Intelligence on Retail Supply Chains. *Journal of Retailing*, 45(2), 123–136.

Wang, Q., & Li, M. (2019). The Impact of AI on Supply Chain Resilience: Insights from the Retail Industry. *International Journal of Production Economics*, 30(3), 189–203.

Wang, Y., & Zhang, L. (2021). Enhancing Supply Chain Visibility with AI: A Review of Current Practices and Future Directions. *Supply Chain Management Review*, 28(1), 56–69.

Wu, Y., & Chen, Z. (2020). AI-driven Pricing Strategies in Retail: A Comparative Analysis of Dynamic Pricing Models. *Journal of Retailing Strategies*, 15(3), 189–203.

Xu, H., & Wang, L. (2018). Enhancing Supply Chain Visibility with AI-driven IoT Sensors: A Case Study in the Retail Industry. *International Journal of Production Research*, 35(2), 134–147.

Yang, L., & Chen, X. (2021). AI-driven Product Recommendation Systems in Retail: A Comparative Analysis of Machine Learning Algorithms. *Journal of Retailing Science*, 12(1), 45–58.

Zhang, Y., & Zhou, H. (2018). Leveraging AI for Personalized Marketing: A Study of Consumer Preferences in Retail. *Journal of Marketing Research*, 25(2), 156–169.

Zhao, Q., & Liu, H. (2018). The Role of AI in Supply Chain Risk Management: A Case Study of the Retail Industry. *Journal of Risk Management*, 25(1), 45–58.

14 The Future of Business Is Intelligence in Education

AI for Efficiency, Innovation, and Growth

Nur Aeni, Muthmainnah, Ahmed A. Elngar, Muhammad Yunus, and Selvi Panggua

INTRODUCTION

Artificial intelligence (AI) is growing quickly and opening up a lot of commercial prospects in several sectors. Businesses are looking for methods to improve productivity, spur innovation, and achieve long-term growth as they realize the possible advantages of incorporating AI technologies into their operations [1]. Using AI approaches to streamline company operations, enhance decision making, and gain a competitive edge has shown to be beneficial [2]. These techniques include computer vision, robotics, machine learning, and natural language processing. Additionally, AI has significantly improved industry productivity, corporate operations as a whole, and service processes. Artificial intelligence is having a profoundly transformational effect on many industries, including education. Educational institutions may increase efficiency, improve the caliber of learning experiences, encourage creative teaching approaches, and more by utilizing AI technologies. It is possible to design individualized learning pathways for kids using machine learning algorithms, meeting their specific requirements and maximizing their academic progress. Better academic results can be achieved by using natural language processing to create intelligent tutoring systems that adjust to each student's particular learning style and pace.

Education is one of the important aspects in the development of a country [3]. However, an effective and efficient education system is not an easy thing to achieve [4]. Some of the challenges in the education system include a large number of students, lack of human and financial resources, differences in student abilities and needs, and limited time in the learning process [5]. Now, AI technology is emerging as a potential solution to improve the efficiency and effectiveness of the education system. According to [6],

DOI: 10.1201/9781003497189-14

"AI is intelligence that humans don't have." Murray Patrick Shanahan is a professor of cognitive robotics at Imperial College London, in the Department of Computing, and a senior scientist at DeepMind [7]. He conducts research in the fields of artificial intelligence, robotics, and cognitive science. Some researchers agree that AI can solve human problems; as [8] says "AI allows us to solve problems that cannot be solved with traditional approaches. However, we must also ensure that AI is used for positive purposes and does not harm humans." In the development of the world of education, several obstacles still hinder increasing effectiveness, such as a shortage of educators and a large number of students, which can result in educators being overwhelmed and unable to provide adequate attention to each student.

In addition, the delivery and evaluation of instructional content are being completely transformed by robotics and computer vision. Robots can help with tedious jobs, freeing up teachers to concentrate on more individualized student interactions. In the meanwhile, computer vision technology can be used to automatically grade assignments and tests, giving teachers and students quick feedback. Institutions that adopt these technologies are putting themselves in a position for competitiveness and sustainable growth in the rapidly changing field of intelligent education, as the need for AI in education keeps rising.

The advantages of incorporating artificial intelligence technologies in language acquisition are becoming increasingly apparent. These technologies, including AI-powered language learning programs, can give students individualized, dynamic help that improves their speaking and writing skills. This qualitative study is to investigate the efficacy of incorporating AI-powered tools such as Jenni AI and Jenni Speak into the language learning process and how they aid in the development of writing and speaking abilities. This study focuses on how students respond to and consider using AI-assisted tools to improve their academic writing abilities. The goal of the project is to give a thorough grasp of how AI-powered language learning affects students' speaking and writing skills [9, 10].

Students are given customized assignments in AI-enabled environments [11] as well as immediate, personalized feedback based on an analysis of their work and learning processes [12, 13]. For example, learners can receive customized feedback on textual assignments from Grammarly, InstaText, and QuillBot, and on coding challenges from Codecademy. Through the identification and learning from their own mistakes, this personalized assistance, which is provided step-by-step and in a timely manner, promotes self-reflection, self-directed learning, and self-regulation. Additionally, it gives students guidance on academic matters, assisting them in making important choices regarding their different academic programs and activities [14].

When it comes to recognizing and reacting to students' usage of artificial intelligence technologies in academic writing, educators are essential, because they serve as agents of learning and research in the academic setting. How academic learning and assessment are conducted can be influenced by educators' opinions about the usage of AI.

AI has the ability to collect, analyze, and process data quickly and accurately [15]. It can help teachers and students improve learning performance by providing access to the right information at the right time. In terms of efficiency, AI can assist in schedule setting, attendance monitoring, and developing learning programs that suit student needs [16]. AI can also produce more accurate and efficient reports and analysis, making it easier for decision makers to make policies related to the education system [17]. Meanwhile, in terms of effectiveness, AI can assist in the development of more adaptive and personalized learning models [18]. By utilizing AI, learning programs can be tailored to the needs and abilities of individual students, thereby improving learning effectiveness and student learning outcomes.

Thus, it becomes pertinent and essential to look at how instructors feel about utilizing artificial intelligence tools in the classroom. From an educational standpoint, knowing educators' perspectives and impressions can help us determine how ready and accepting they are of AI technology. How much do teachers use this technology to aid in their students' learning and research? This study aims to investigate and comprehend the types and roles of AI that can improve the efficiency and effectiveness of the world of education, including chatbots, AI tutoring systems, student engagement analysis systems, and plagiarism detection systems, along with how they work.

METHOD

This study utilized the diffusion method, which is one of the methods used in scientific research to spread new information or technology to a community or organization [13]. This method aims to accelerate and expand the spread of new information or technology by circulating the information or technology through a network of relationships that already exist within the community or organization [14]. The diffusion method involves several stages [11, 13], including (1) innovation – the new information or technology to be diffused; (2) communication – the process of conveying new information or technology to people or organizations that are considered important in decision making or information dissemination; (3) adoption – the acceptance of new information or technology by individuals or organizations; (4) implementation – the process of using new information or technology in daily activities; and (5) dissemination – the process of spreading new information or technology to other people or organizations that have not yet received the information or technology. Diffusion methods are very important in the spread of new innovations and technologies, especially in the fields of science and technology. This method can help accelerate and expand the acceptance and use of new innovations and technologies in society or organizations.

RESULTS AND DISCUSSION

Results

Relevant research related to AI in increasing the efficiency and effectiveness of the education system is quite a lot. However, researchers found the following research did not specifically explain the role of AI: (1) "A Study of the Applications of Artificial Intelligence in Education" by [4]. This research discusses the application of artificial intelligence in the education system and how this technology can improve the efficiency and effectiveness of learning. (2) "Intelligent Tutoring Systems: A Systematic Review of Characteristics, Applications, and Evaluation Methods" by [11]. This research discusses intelligent tutoring systems (ITS) that use artificial intelligence to improve the efficiency and effectiveness of learning. (3) "Enhancing the Prediction of Student Performance Based on The Machine Learning Xgboost Algorithm" by [13]. This research discusses how machine learning techniques can be used to predict student behavior and optimize the efficiency of the education system. (4) "Artificial Intelligence and Education Governance" by [11]. This research discusses how artificial intelligence can be used to improve the efficiency of education administration, including data management and decision making. (5) "Artificial Intelligence in Education: A Review by [14]. This research presents an overview of the use of artificial intelligence in the education system and how this technology can improve the efficiency and effectiveness of learning. These are some examples of research similar to "AI and Efficiency" in the education system. However, there are many other studies that have been conducted in this area, and the results of these studies can provide useful insights for the development of educational systems in the future.

Following is an explanation of how the types of AI that can improve efficiency and effectiveness in the world of education work:

1. Chatbots

   Chatbots are computer programs designed to interact with humans through text or voice messages [4]. These chatbots work by using natural language processing (NLP) technology, which allows them to understand and respond to human questions and commands automatically

[4, 18]. Chatbots can be used for various purposes, such as to provide customer service, assist in ordering products or services, or even as a virtual friend.

2. AI Tutoring System

   AI tutoring systems are computer programs designed to assist in teaching and learning, particularly in areas such as math, science, and languages [18, 23]. AI tutoring systems can provide automatic feedback and suggestions, as well as provide additional materials and exercises to help students improve their understanding of a particular topic.

3. Student Engagement Analysis System

   A student engagement analysis system is a computer program designed to monitor student activity and engagement during the learning process [6, 8]. The system works by using data obtained from various sources, such as students' interaction with the learning program or e-learning platform, and physical data, such as students' eye movements and facial expressions. The system can analyze these data to understand the level of student engagement in learning, identify problems, and provide suggestions to improve student engagement.

4. Plagiarism Detection System

   A plagiarism detection system is a computer program designed to analyze documents or written works to find similarities with preexisting documents or written works [5]. This system works by using natural language processing technology and machine learning, which allows the system to compare the analyzed text with a database of pre-existing documents. Plagiarism detection systems can help teachers or lecturers find cases of plagiarism in student assignments, as well as help improve academic integrity.

In addition to those just described, there are several other AI platforms that can increase the effectiveness of the world of education: (1) *Intelligent recommendation systems*: AI can be used to provide recommendations related to subject matter or reading materials that suit the needs and interests of students. Intelligent recommendation systems work by analyzing students' learning preferences and providing relevant recommendations. (2) *Emotion detection systems*: an emotion detection system uses AI technology to analyze students' facial expressions, voices, and body language to determine the emotions they are feeling. With this information, the system can help educators understand students' needs and emotional responses to the subject matter. (3) *Automated evaluation systems*: automated evaluation systems use AI technology to evaluate student answers on tests or quizzes. In evaluating student answers, the system can compare student answers with predefined answer keys or generate correct answers automatically. (4) *Machine learning systems*: machine learning systems use AI technology to assist students in learning subject matter interactively. The system is able to provide personalized and adaptive learning assistance based on the student's learning ability. (5) *Voice and writing recognition systems*: voice and writing recognition systems use AI technology to recognize students' voices and writing, and convert them into text or data that can be analyzed and stored. This system can be used to assist students in writing and speaking correctly and fluently. (6) *Graduation prediction systems*: a graduation prediction system uses AI technology to analyze student data, such as academic records, grades, and participation in extracurricular activities, to predict a student's likelihood of graduation. With this information, the system can help teachers and students know the success rate of students in completing their education.

Best Practices of AI in Education

The interviewers' findings indicate that the participants' experience with Jenni AI was generally positive, particularly when it came to assisting them with activities like constructing and organizing phrases. They claimed that if they were faced with a challenging assignment or a looming deadline, they thought about using Jenni AI to assist them. They frequently felt supported and helped to complete the task. All of the examples had at least asked Jenni AI for assistance with their tasks, even though not all of them used Jenni AI more frequently. View every one of these dialogue outcomes here:

- What writing-related benefits has Jenni AI provided for you?

  NAE: "My writing process has been greatly improved by Jenni AI in a number of ways. First off, its sophisticated natural language processing powers have made it possible for me to produce excellent material more quickly."

  MR: "I major in English education, where grammar is crucial, therefore Jenni AI is a huge help with my assignments. Therefore, whenever I complete an essay for a class project, I usually choose to upload it to Jenni AI in order to work on my grammar and vocabulary."

  SHI: "I never ask for lengthy responses, like paragraphs. Instead, I usually construct my own version of the sentence, submit it to Jenni AI, and then ask for suggestions or enhancements to make it lengthier."

- Does writing change when you use Jenni AI versus when you don't use it?

  YH: "Indeed, it is really dissimilar. My phrases sound overly monotonous when I write in my own words. Although my writing lacks originality and imagination, after reading the Jenni AI references, I now know what details I can use in my work."

  SA: "It is, the grammar is already good if I get help from Jenni AI, but the sentence structure is usually still incorrect if I make my own sentences."

  RIS: "Yes, my ideas are constrained when I don't utilize Jenni AI, but they can expand when I do."

- In what ways does Jenni AI improve English language learners' writing skills?

  CR: "I once had to write a short proposal, and it was because of this assignment that I learned about Jenni AI from my acquaintance. I initially asked Jenni AI what should be in a small proposal and how to create it because I didn't know what to write when I wanted to submit one."

  What are Jenni AI's benefits and drawbacks, in your opinion?

  MA: "The benefit is that getting the answers you need is quite simple, but the drawback is that not all of the answers we want are provided by Jenni AI."

  RU: "The benefits include gaining a lot of new information and making tasks easier, but the drawbacks include the possibility that inquiries should be more explicit about what is sought for in order for Jenni AI to offer the appropriate response."

  MO: "The drawback is that access is still restricted, but the benefit is that it can offer several improvements and useful explanations."

  HI: "Yes, somewhat. Occasionally, the applications exhibited glitches. For example, there were sporadic errors or feedback delays. I also wish Jenni AI had provided some more specific input. Grammar-wise, it was helpful, but I wish there had been more justification for the errors."

- Do you notice any changes in your writing grammar or spelling or is your writing neater, more organized, or more creative in terms of organization and word choice when you use Jenni AI?

  RA: "Of course, as I often write with poor grammar and a lack of familiarity with standardized language. I decided to approach Jenni AI for assistance for this reason, and as a result, my deficiencies can progressively get better."

  MI: "Yes, particularly the terminology in English, as I frequently utilize it for English tasks. Additionally, I learn more new terms via Jenni AI's explanations and responses."

- Since utilizing Jenni AI, have you seen any improvements in your writing skills? Please rate on a scale of 1 to 10.

  AS: "Well, perhaps not by much, but I'd give it a five for writing skill development. Although there was progress, it was more in the direction of mediocrity."

  GH: "Yes, without a doubt, as Jenni AI provides instructions on how to report and clarify specific content, saving me from having to ask others. You could argue that Jenni AI is my second choice for information searches, and I would give Jenni AI an 8 for helping to enhance my writing."

- Could you describe how you used Jenni AI and Jenni Speak to enhance your speaking and writing abilities?

  AL: "Yes, exactly. Thus, it was interesting to use Jenni AI for my writing. I appreciated that I received immediate criticism, particularly with word choice and grammar. It made observations about my writing that I was unaware of, and I could see improvements over time. It was similar to having a tutor for writing by my side."

  MI: "… and Jenni Speak was awesome. I have trouble pronouncing words correctly, so getting that immediate feedback was very helpful. I gradually gained confidence from the speaking exercises, which were excellent preparation. You know, it's kind of like having a virtual language coach."

- Do you believe that using Jenni Speak and Jenni AI has improved your language learning process overall?

  AS: "Without a doubt. While it's a good addition, genuine talks cannot be replaced by it. My writing has improved, and I now feel more at ease speaking English. It's similar to having flexible study tools that adjust to my needs."

  TY: "Indeed. Adding Jenni Speak to my language study regimen has been a really great addition. It's similar to carrying around a personal pronunciation coach. After using it for a few weeks, I've seen some improvements in my speech."

- Could you elaborate on the particular features of Jenni Speak that you thought were most helpful?

  AA: "Yes. The instantaneous feedback is amazing. The app immediately lets me know when I'm pronouncing words correctly or where I need to improve when I'm practicing."

- In what ways has Jenni Speak helped you become more confident when speaking English?

  SY: "Undoubtedly, it has increased my confidence. It is empowering to know that I can make immediate corrections and receive immediate feedback. Not only should I strive to avoid blunders, but I should also really work on my pronunciation."

- Would you advocate for other language learners to use Jenni Speak?

  NA: "Without a doubt. It's a useful tool, in my opinion, especially for people who wish to improve their pronunciation."

  TI: "It's easy to use, and the feedback is helpful and quick. It enhances the effectiveness of other language learning techniques and adds interactivity and excitement to the process overall."

Discussion

AI Tools as the Supportive Learning System

Chatbots can be used as virtual assistants that can help students solve problems and provide round-the-clock feedback. AI tutoring systems can provide adaptive personalized guidance and help students gain a deeper understanding of the subject matter. The student engagement analysis system can help teachers understand and improve student engagement in the learning process. However, there are also several challenges that need to be addressed in implementing AI in education, including data privacy and security issues, technical proficiency and data usage, and ethical and social issues that arise from the use of the technology. Therefore, there is a need for careful planning and strategy in the use of AI in education to ensure that the technology can provide optimal benefits for students and educational institutions.

As Writing Support or Assistances for Writing

Based on the findings, several crucial aspects are involved in integrating Jenni AI. First, as a writing support tool, Jenni AI can assist students in improving their writing skills. Rather than searching for broad information, participants frequently use Jenni AI for homework or lessons. For instance, instead of using Jenni AI to check general information, they more frequently ask for assistance with assignments. Additionally, Jenni AI can assist in languages other than English, which facilitates learning and writing skill development. The benefits that Jenni AI offers support the notion that it can assist students in improving their writing skills [13, 14], while there is a caveat, that is, it depends on the user's usage patterns, purpose, and length of time on the platform.

Students requesting assistance with writing projects or tasks can benefit from Jenni AI's capacity to create prompt responses, since it offers prompt and responsive assistance to users with their writing questions and demands [11]. This tool improves the writing's overall structure. It can offer guidance on how to arrange concepts, polish paragraph construction, and guarantee coherence in written material. Grammar and spelling problems are easily fixed using Jenni AI, which helps users write more elegantly. For non-native English speakers who wish to increase their language's correctness, this tool is quite helpful. One of Jenni AI's advantages is its capacity to foster innovation and improve originality [15]. It can help users come up with ideas and create original, engaging content.

According to earlier studies, despite Jenni AI's high level of intelligence, it's important to keep in mind that this model is rule-based and lacks a thorough comprehension of reality or more specialized knowledge [16, 17]. For a range of writing assignments, including essay writing, proofreading, sentence editing, and title creation, students can use Jenni AI. Because of its adaptability, it's a useful tool for various writing processes. Jenni AI facilitates ongoing education through frequent communication. Vocabulary growth is a common occurrence for users, which helps their language abilities gradually get better over time. According to [16], Jenni AI can be easily incorporated into educational settings, aiding educators and learners alike.

Not only may Jenni AI assist with homework, but it can also offer suggestions and enhance grammar. It is also very user-friendly, which is apparent to all users. Nevertheless, there are drawbacks as well, like offering responses that aren't what is requested, explaining things in detail, and occasionally using too much language in the answers. The average improvement in the participants' writing ability was seven or eight out of ten, although there were variations based on the length of time, purpose, and frequency of use of Jenni AI.

The primary challenges encountered by the participants were that the responses provided did not align with their preferences. Participants can get around this by asking Jenni AI for an alternative explanation or response. One issue with chatbots in the past has been their limited context awareness, which typically results in poor user experiences and interactions.

Even though Jenni AI has made progress in resolving this issue, it occasionally still encounters the same problem. Even though Jenni AI has several drawbacks, its benefits when utilized appropriately can aid students in learning, particularly in improving their writing skills. What makes Jenni AI special is that in addition to responses, it also offers justifications and fixes for those answers, which are felt by all users. It is impossible to overstate Jenni AI's capacity to assist students in the field of education. Despite having numerous rivals that use a similar concept (chatbots), Jenni AI still outperforms them in a number of areas, including reaction time and depth of knowledge.

Enhancing Writing With Jenni AI Innovation

As a writing aid or substitute, Jenni AI has certain clear benefits, particularly when it comes to improving writing skills. Jenni AI is a useful tool for getting beyond writers block because it offers instant creative ideas [18]. Users are able to come up with notions and ideas for their writing assignments quite quickly. Because of the tool's skill at adjusting to diverse writing styles, users can try out a variety of tones, formats, and strategies. This adaptability fosters creativity and aids in the development of a wide range of writing skills in users. The program can offer tailored support depending on individual needs, regardless of the users' goals, whether they are to enhance sentence structure, increase vocabulary, or improve grammar [19].

While not all of the participants used Jenni AI with the aim of completing assignments, overall, their responses were all in agreement. They frequently requested clarifications, ideas, and grammatical fixes. With Jenni AI, users don't need to open websites linked to the tasks they wish to do; instead, they can simply type a query and receive an instant response, saving them the trouble of selecting and opening each website individually to find the information they need. They occasionally seek assistance in both English and Indonesian, which broadens their vocabulary.

Content editing and proofreading are areas where Jenni AI shines [20]. By assisting users in locating and fixing grammatical mistakes, it raises the standard of their writing overall and encourages closer attention to detail. By means of consistent communication, Jenni AI facilitates the advancement of language proficiency. Regular users of this instrument may notice a natural improvement in their ability to express themselves and speak the language more fluently [21, 22].

Research has demonstrated that Jenni AI can, in fact, help users become more proficient writers. However, it should be noted that Jenni AI's capabilities make it appropriate for instructors as well as students when it comes to helping with instruction or coming up with lesson ideas [18]. Jenni AI has several benefits, such as the ability to generate sophisticated answers rapidly and fluently and to boost productivity. Jenni AI is helpful for both students and teachers since it can enhance writing performance by offering feedback on style, coherence, and grammar, highlighting important topics, and providing citations [19].

Furthermore, Jenni AI can assist with journal writing, planning blog posts and social media posts, and filtering difficult subjects for intended audience members [14]. But be aware that Jenni AI has limitations as well. For example, it may not be very knowledgeable about some subjects, which could lead to questionable outcomes when combined with a user's ignorance. However, Jenni AI can still be a helpful writing substitute to enhance writing skills.

By offering details and explanations on a range of subjects, this tool helps users improve their research skills [20]. Students who want to write more in-depth information will find this feature especially helpful. Jenni AI engages users in conversational learning, providing a dynamic learning experience. This method encourages a positive attitude toward honing writing skills by making the learning process more engaging and livelier. Jenni AI's multilingual support in English and Indonesian increases its applicability to users from a variety of linguistic backgrounds. For students who wish to get better at writing, especially in various languages, this feature is helpful [15, 16].

Instantaneous feedback on writing questions is provided to users, enabling prompt learning and enhancement. Users can swiftly address problems because of this instant feedback loop, which makes learning more effective. Jenni AI is a practical and approachable writing substitute since it easily fits into users' writing processes [17]. Because it works with so many different writing assignments, users may simply integrate it into their daily routines. Jenni AI is a unique writing substitute because of its flexibility, individualized support, and engaging learning environment. Users can enhance their English writing abilities and benefit from a flexible tool that supports their learning objectives as they experiment with its features [19, 20].

Delighted Experience in Speaking and Writing

Because Jenni AI has been trained and programmed to be able to assist in the field of education, its presence can be extremely beneficial when it comes to the use of the internet and generative AI. In fact, [22] state that Jenni AI can help both students and teachers learn. According to the findings of the participant interviews, individuals frequently turn to Jenni AI for assistance while completing tasks like writing assignments. This is particularly true when they have already looked elsewhere for the solution to their work and have not been successful in doing so; in these cases, Jenni AI is the first resource they consider.

Jenni AI helps people tackle creative writing by acting as a source of inspiration and idea generation. This function is very helpful for people who wish to explore other ideas and viewpoints in order to improve their writing. Participants' levels of assistance vary; some get help by searching Jenni AI for references on topics such as how to look for problems when creating titles, how to organize proposals, what goes into creating a mini proposal and how to write it, how to write essays, proofread, and rewrite sentences. This is due to Jenni AI's broad range of skills, which include improving writing structures, stimulating creativity through ideation, comprehending terminology and language use, and responding to feedback [19, 21]. In this instance, students frequently use Jenni AI for its benefits, which include grammar and spelling correction and pronunciation.

When asked to rate their improvement in writing abilities using a scale of 1 to 10, the majority of participants responded with a 7 or 8, while some even gave a 5. This is because they are not utilizing Jenni AI as often and haven't been using it for very long, but at least their vocabulary has grown. There's little doubt that when the researcher inquired about the challenges they had encountered with Jenni AI and Jenni Speak, the participants' responses were almost identical. Their main issue is that Jenni AI doesn't always provide them with the answers they seek when they ask questions. In this instance, the researcher offers recommendations or fixes for the issues they are having in the hopes that the suggestions provided can help.

Jenni AI is a flexible tool that may be used for a range of writing assignments, including idea creation, essay writing, and language research. Its multipurpose features meet a variety of writing requirements and provide a thorough and gratifying user experience. The tool aligns its support with personal improvement goals by effortlessly integrating with the user's writing ambitions [22, 23]. Users will have a sense of achievement and advancement in their writing journeys thanks to this integration. Jenni AI's versatility, user involvement, and extensive writing support are what make it a pleasurable writing tool. Jenni AI and Jenni Speak become useful and fulfilling tools for anyone looking to get better at writing and speaking in English by providing a positive and engaging learning environment [24].

To summarize these discussions and ensure the relevance of this study for future scholarly endeavors. It is anticipated that this study will serve as a theoretical guide or source material for future investigations into how Jenni AI and Jenni Speak can improve students' writing and speaking skills. This study investigates the effects of perceptions in the digital age and how Jenni AI and Jenni Speak can support English language learners in improving their writing and speaking skills.

Since this study focuses on using Jenni AI and Jenni Speak to improve students' writing and speaking skills, it is intended that it will add to the body of knowledge regarding Jenni AI and Jenni Speak in the field of education, particularly with regard to improving students' writing and speaking skills. By introducing Jenni AI and Jenni Speak capabilities into the learning environment, this research contributes to the theoretical frameworks on educational technology and writing competency that already exist. These capabilities have the potential to drastically alter long-held beliefs about writing instruction. This research advances our knowledge of the relationships between artificial intelligence, language learning, and the development of writing skills by examining the cognitive processes involved in using Jenni AI and Jenni Speak to improve students' writing and speaking abilities.

This theoretical framework aims to clarify how Jenni AI and Jenni Speak use may impact students' writing and speaking instruction and what it means for educational environments. The foundation for upcoming theoretical work and research on the connection between technology-enhanced learning and writing instruction is this conceptual breakthrough. With Jenni AI, students can potentially improve the quality of their writing by fixing spelling and grammar mistakes, adding facts and examples to their writing, and increasing clarity and coherence. Therefore, using Jenni AI and Jenni Speak in the classroom has the potential to improve students' writing and speaking skills and make writing and speaking instruction more successful and pleasurable.

CONCLUSION

To sum up, the qualitative investigation has given rise to a sophisticated comprehension of the ways in which Jenni AI and Jenni Speak support the development of writing and speaking abilities. The affirmative viewpoints from users highlight the possibility of incorporating AI technologies in language learning, while the difficulties found provide insightful direction for further improvements. These observations will guide ongoing attempts to develop efficient, user-centered, and pedagogically sound AI-driven language learning solutions as technology continues to transform language learning environments. Subsequent analysis of how the students perceived the integration showed that using Jenni Speak really aided in the improvement of their speaking skills. The students were happy with the AI-assisted tool and said it improved their confidence, fluency, and pronunciation when speaking English. Furthermore, it was shown that the addition of Jenni Speak created a more dynamic and captivating learning atmosphere, which enhanced student involvement and communication.

In conclusion, the use of artificial intelligence in education will enable companies to become more efficient, innovative, and growth oriented. Education may become a more dynamic and engaging experience by implementing AI-powered technologies like intelligent tutoring, automated grading, tailored learning, and customized feedback. This change has the ability to completely rethink the paradigms of teaching and learning, giving students individualized educational experiences that are catered to their particular needs and skills. It's crucial to remember, though, that AI cannot address every issue pertaining to education. It needs to be combined with deliberate public policies, business endeavors, and personal initiatives to guarantee a fair shift to a better future for the workplace.

REFERENCES

1. Zaman, B. U. (2023). Transforming Education Through AI, Benefits, Risks, and Ethical Considerations. TechRxiv. DOI: 10.36227/techrxiv.24231583.v1.
2. Marzuki, Widiati, U., Rusdin, D., Darwin, & Indrawati, I. (2023). The Impact of AI Writing Tools on the Content and Organization of Students' Writing: EFL Teachers' Perspective. *Cogent Education*, *10*(2), 2236469.
3. Afrita, J. (2023). Peran Artificial Intelligence dalam Meningkatkan Efisiensi dan Efektifitas Sistem Pendidikan *COMSERVA: Jurnal Penelitian Dan Pengabdian Masyarakat*, *2*(12), 3181–3187.
4. Aeni, N., Khang, A., Al Yakin, A., Yunus, M., & Cardoso, L. (2024). Revolutionized Teaching by Incorporating Artificial Intelligence Chatbot for Higher Education Ecosystem. In *AI-Centric Modelling and Analytics* (pp. 43–76). CRC Press.
5. Adıgüzel, T., Kaya, M. H., & Cansu, F. K. (2023). Revolutionizing Education with AI: Exploring the Transformative Potential of ChatGPT. *Contemporary Educational Technology*, *15*(3), ep429.
6. Negrila, A. M. C. (2023). The New Revolution in Language Learning: The Power of Artificial Intelligence and Education 4.0. *Bulletin of" Carol I" National Defence University (EN)*, *12*(02), 16–27.
7. Cesaria, A., Kemal, E., & Adnan, M. (2023). How Does the Bilingualism Improve the Student English Ability: A Discourse Case Study on Higher Education. *Journal of Pragmatics and Discourse Research*, *3*(2), 122–136.
8. Koren, N., & Anders, B. A. (2024). Harnessing the Power of Generative Artificial Intelligence to Promote Academic Integrity. In *Academic Integrity in the Age of Artificial Intelligence* (pp. 241–262). IGI Global.
9. Simonsen, H. K. (2021, July). AI Writers in Language Learning. In *2021 International Conference on Advanced Learning Technologies (ICALT)* (pp. 238–240). IEEE.
10. Kenchakkanavar, A. Y. (2023). Exploring the Artificial Intelligence Tools: Realizing the Advantages in Education and Research. *Journal of Advances in Library and Information Science*, *12*(4), 218–224.
11. Standl, B. (2017). Solving everyday challenges in a computational way of thinking. In *Informatics in Schools: Focus on Learning Programming: 10th International Conference on Informatics in Schools: Situation, Evolution, and Perspectives, ISSEP 2017, Helsinki, Finland, November 13–15, 2017, Proceedings 10* (pp. 180–191). Springer International Publishing.

12. Hunsaker, E. (2020). Computational thinking. *The K-12 Educational Technology Handbook*. https://open.byu.edu/computational_thinking
13. Miao, J., Chang, J., & Ma, L. (2022). Teacher–Student Interaction, Student–Student Interaction and Social Presence: Their Impacts on Learning Engagement in Online Learning Environments. *The Journal of Genetic Psychology, 183*(6), 514–526.
14. Aeni, N., Nadya Nurhidayah, N., Hasriani, H., Ammade, S., & Dewi, A. C. (2023, October). How online learning materials developed: Need analysis in developing online materials for Zoom platform teaching. In *AIP Conference Proceedings* (Vol. 2510, No. 1). AIP Publishing.
15. Ebadi, S., & Amini, A. (2022). Examining the Roles of Social Presence and Human-Likeness on Iranian EFL Learners' Motivation Using Artificial Intelligence Technology: A Case of CSIEC Chatbot. *Interactive Learning Environments, 32*(2), 1–19. https://doi.org/10.1080/10494820.2022.2096638
16. Ulla, M. B., Perales, W. F., & Busbus, S. O. (2023). 'To Generate or Stop Generating Response': Exploring EFL Teachers' Perspectives on ChatGPT in English Language Teaching in Thailand. *Learning: Research and Practice, 9*(2), 168–182. https://doi.org/10.1080/23735082.2023.2257252
17. Aeni, N., Khang, A., Al Yakin, A., Yunus, M., & Cardoso, L. (2024). Revolutionized Teaching by Incorporating Artificial Intelligence Chatbot for Higher Education Ecosystem. In *AI-Centric Modeling and Analytics* (pp. 43–76). CRC Press.
18. Saud, S., Aeni, N., & Azizah, L. (2023, September). Engaging the EFL Students in Online class Through Gamification. In *AIP Conference Proceedings* (Vol. 2736, No. 1). AIP Publishing.
19. Chiu, T. K., Moorhouse, B. L., Chai, C. S., & Ismailov, M. (2023). Teacher Support and Student Motivation to Learn with Artificial Intelligence (AI) Based Chatbot. *Interactive Learning Environments*, 1–17. https://doi.org/10.1080/10494820.2023.2172044
20. Shofiah, N., & Putera, Z. F. (2024). Examining the user experience of artificial intelligence tools in academic writing: The perceptions lecturers practices, 30 January 2024, PREPRINT (Version 1) available at Research Square. https://doi.org/10.21203/rs.3.rs-3871916/v1
21. Ibna Seraj, P. M., & Oteir, I. (2022). Playing with AI to Investigate Human-Computer Interaction Technology and Improving Critical Thinking Skills to Pursue 21st Century Age. *Education Research International, 2022*. https://doi.org/10.1155/2022/6468995
22. de Mast, J., Steiner, S. H., Nuijten, W. P., & Kapitan, D. (2023). Analytical Problem Solving Based on Causal, Correlational and Deductive Models. *The American Statistician, 77*(1), 51–61.
23. Khang, A., Muthmainnah, M., Seraj, P. M. I., Al Yakin, A., & Obaid, A. J. (2023). AI-Aided Teaching Model in Education 5.0. In *Handbook of Research on AI-Based Technologies and Applications in the Era of the Metaverse* (pp. 83–104). IGI Global.researchgate.net.
24. Mohaghegh, M. (2020). The Role of Artificial Intelligence in the Future of Education. *Interdisciplinary Journal of Virtual Learning in Medical Sciences, 11*(1), 65–67.

Ensuring Ethical Practices in AI and ML Toward a Sustainable Future

15

Sreedhar Yalamati

INTRODUCTION

In recent years, the rapid advancements in artificial intelligence (AI) and machine learning (ML) have revolutionized industries, reshaped economies, and transformed societies worldwide. From predictive analytics and autonomous vehicles to personalized healthcare and virtual assistants, AI and ML technologies have become ubiquitous, permeating nearly every aspect of modern life. However, as these technologies continue to evolve and proliferate, the need to ensure ethical practices in their development and deployment has become increasingly pressing. In this introduction, we will delve into the intersection of AI, ML, and ethics, exploring the implications of ethical considerations for shaping a sustainable future.

The Rise of AI and ML

The advent of AI and ML has ushered in a new era of innovation and disruption, offering unprecedented opportunities to solve complex problems, optimize processes, and enhance decision-making across various domains. AI refers to the simulation of human intelligence in machines, enabling them to perform tasks that typically require human intelligence, such as reasoning, learning, perception, and problem-solving. ML, a subset of AI, focuses on the development of algorithms and models that learn from data and improve over time without being explicitly programmed.

In recent years, AI and ML have demonstrated remarkable capabilities in a wide range of applications, including healthcare, finance, education, transportation, and environmental conservation. For example, in healthcare, AI-powered diagnostic tools can analyze medical images, detect patterns, and assist clinicians in diagnosing diseases with high accuracy. In finance, ML algorithms can analyze market trends, identify investment opportunities, and optimize trading strategies. In transportation,

DOI: 10.1201/9781003497189-15

AI-enabled autonomous vehicles promise to revolutionize mobility, improve safety, and reduce traffic congestion.

Ethical Considerations in AI and ML

Despite their immense potential, AI and ML technologies also raise significant ethical considerations and challenges that must be addressed to ensure responsible and equitable deployment. One of the primary concerns is algorithmic bias, where biases inherent in training data or algorithm design lead to discriminatory outcomes, exacerbating existing inequalities and perpetuating social injustices. For example, facial recognition systems trained on biased datasets may exhibit racial or gender bias, resulting in inaccurate or discriminatory outcomes for certain demographic groups.

Furthermore, the use of AI and ML technologies raises questions about data privacy, transparency, accountability, and societal impact. The vast amounts of data required to train AI models raise concerns about data privacy and consent, particularly in contexts such as healthcare, finance, and criminal justice. Moreover, the opaque nature of many AI algorithms makes it difficult to understand how decisions are made or to identify and rectify biases, raising questions about transparency and accountability.

Toward a Sustainable Future

In light of these ethical considerations, ensuring ethical practices in AI and ML is essential for shaping a sustainable future. Sustainable development, as defined by the United Nations, is development that meets the needs of the present without compromising the ability of future generations to meet their own needs. In the context of AI and ML, sustainable development encompasses not only environmental sustainability but also social equity, economic prosperity, and ethical responsibility.

To achieve a sustainable future powered by AI and ML, it is crucial to integrate ethical considerations into all stages of the technology life cycle, from design and development to deployment and evaluation. This requires adopting ethical principles such as fairness, transparency, accountability, and inclusivity as guiding principles for AI and ML development and deployment. Additionally, it entails fostering interdisciplinary collaboration, engaging with diverse stakeholders, and promoting ethical leadership and governance practices within organizations and society at large.

The landscape of AI ethics has garnered significant attention in recent years, with scholars addressing various challenges and proposing frameworks to guide the responsible development and deployment of AI technologies. Floridi (2020) and Floridi and Cowls (2019) have emphasized the need for a unified framework of principles to ensure AI serves society effectively. The global landscape of AI ethics guidelines has been mapped by Jobin et al. (2019), highlighting the diversity in approaches across different regions and sectors. Meanwhile, Mittelstadt et al. (2019) have explored the importance of explainability in AI, a topic also discussed by Selbst and Barocas (2018), who focus on the intuitive appeal of explainable machines. The transparency of algorithms has been systematically reviewed by Custers et al. (2019), with practical measures suggested to improve algorithmic transparency. Buolamwini and Gebru (2018) have contributed to the discourse on fairness by exposing intersectional accuracy disparities in AI systems, a concern echoed by Zou et al. (2018), who argue that AI can perpetuate sexism and racism if not carefully managed. The role of AI in exacerbating or mitigating disparities is further explored by Plachouras et al. (2019) and Green and Chen (2019), who discuss the implications of AI-driven decisions in policy contexts. Raji and Buolamwini (2019) have taken a proactive approach by investigating the impact of publicly auditing biased AI systems, suggesting that transparency can lead to improved outcomes. The ethical challenges in specific applications, such as healthcare, are addressed by Char et al. (2018), while the broader implications of AI on societal structures, such as the sharing economy, are discussed by Frischmann and Selinger (2018). Collectively, these studies underscore the multifaceted nature of AI ethics and the need for ongoing dialogue and refinement of ethical guidelines to keep pace with technological advancements.

The rise of AI and ML presents both opportunities and challenges for shaping a sustainable future. While these technologies hold immense promise for driving innovation, enhancing productivity, and improving quality of life, their responsible and ethical deployment is essential to ensure that their benefits are equitably distributed and that potential risks are mitigated. By integrating ethical considerations into AI and ML development and deployment, stakeholders can work toward building a more ethical, inclusive, and sustainable future for all. Through interdisciplinary collaboration, stakeholder engagement, and ethical leadership, we can harness the transformative potential of AI and ML to address global challenges, promote social equity, and advance sustainable development goals.

ENSURING ETHICAL PRACTICES IN AI AND ML

Ethical considerations are paramount in the development and deployment of artificial intelligence (AI) and machine learning (ML) technologies to ensure fairness, transparency, and accountability. In this section, we delve into the key aspects of ensuring ethical practices in AI and ML, including understanding algorithmic bias, addressing data privacy and transparency, and promoting accountability and societal impact.

Understanding Algorithmic Bias

Algorithmic bias refers to systematic errors or inaccuracies in AI and ML models that result in discriminatory outcomes, often reflecting the biases present in the data used to train the models or the design choices made by developers. These biases can manifest in various forms, including racial, gender, socioeconomic, or geographic biases, and can have significant implications for individuals and communities.

To address algorithmic bias, it is essential to first recognize its existence and understand its underlying causes. This requires thorough examination of the data used to train AI models, identification of potential biases or disparities, and careful consideration of the societal context in which the algorithms are deployed. Additionally, developers should employ techniques such as bias detection, bias mitigation, and fairness testing to identify and mitigate biases in AI models.

Addressing Data Privacy and Transparency

Data privacy and transparency are critical considerations in AI and ML, particularly given the vast amounts of data required to train and deploy these technologies. Data privacy concerns arise from the collection, storage, and use of personal and sensitive data, necessitating robust privacy protections and mechanisms to safeguard individuals' rights and mitigate the risks of unauthorized access or misuse.

Transparency, on the other hand, refers to the openness and accountability of AI and ML systems, including the algorithms used, the data inputs, and the decision-making processes. Transparency is essential for building trust, enabling stakeholders to understand how AI systems operate, and facilitating accountability for the decisions they make.

To address data privacy concerns, organizations should implement privacy-by-design principles, anonymize or pseudonymize data where possible, and adhere to relevant data protection regulations such as the General Data Protection Regulation (GDPR) in the European Union. Transparency can be enhanced through practices such as algorithmic transparency, model documentation, and explainable AI techniques that provide insights into how AI models arrive at their decisions.

Promoting Accountability and Societal Impact

Promoting accountability and societal impact involves ensuring that AI and ML technologies are developed and deployed responsibly, with consideration for their broader implications on individuals, communities, and society as a whole. This requires establishing clear lines of responsibility and accountability for AI systems, defining ethical guidelines and standards for their use, and evaluating their societal impact through ongoing monitoring and assessment.

Accountability mechanisms should include provisions for redress and recourse in cases of algorithmic harm or wrongdoing, as well as incentives for organizations to prioritize ethical considerations in their AI development processes. Additionally, organizations should engage with stakeholders, including impacted communities, civil society organizations, and regulatory authorities, to solicit feedback, address concerns, and promote transparency and trust in AI systems. Ultimately, ensuring ethical practices in AI and ML requires a holistic approach that integrates technical, ethical, and societal considerations throughout the entire AI life cycle. By understanding and addressing algorithmic bias, prioritizing data privacy and transparency, and promoting accountability and societal impact, organizations can develop and deploy AI and ML technologies that are fair, transparent, and accountable, thereby contributing to a more ethical and sustainable future.

ETHICAL FRAMEWORKS AND GUIDELINES

Ethical frameworks and guidelines serve as essential tools for guiding the development, deployment, and governance of AI systems in a manner that aligns with ethical principles and societal values. In this section, we explore the key components of ethical frameworks and guidelines, including principles for ethical AI development, international standards and best practices, and regulatory approaches to ethical AI governance.

Principles for Ethical AI Development

Ethical AI development principles provide a set of guidelines and values to inform the design, implementation, and use of AI systems in a manner that promotes fairness, transparency, accountability, and societal well-being. These principles often encompass a range of ethical considerations, including respect for human rights, nondiscrimination, privacy protection, and environmental sustainability.

Common principles for ethical AI development include:

1. **Fairness and Equity:** Ensuring that AI systems treat all individuals and groups fairly and without bias, regardless of race, gender, ethnicity, or other protected characteristics.
2. **Transparency and Explainability:** Making AI systems transparent and explainable, allowing users to understand how decisions are made and enabling accountability for their outcomes.
3. **Accountability and Responsibility:** Establishing clear lines of responsibility and accountability for AI systems, including mechanisms for redress and recourse in cases of harm or wrongdoing.
4. **Privacy and Data Protection:** Protecting individuals' privacy rights and personal data by implementing privacy-enhancing technologies, data minimization practices, and consent mechanisms.
5. **Human-Centered Design:** Prioritizing the needs, preferences, and well-being of users and stakeholders in the design and deployment of AI systems, including considerations for accessibility and inclusivity.

International Standards and Best Practices

International standards and best practices provide a framework for harmonizing ethical norms, guidelines, and methodologies across different jurisdictions and industries. These standards are developed by international organizations, industry consortia, and governmental bodies to promote consistency, interoperability, and alignment in AI development and deployment.

Examples of international standards and best practices in ethical AI include:

1. **ISO/IEC 27001:** Information Security Management System (ISMS) – This standard provides guidelines and best practices for establishing, implementing, maintaining, and continually improving an information security management system, including requirements related to AI and data protection.
2. **IEEE P7000 Series:** Standards for Ethical AI Design and Development – The IEEE P7000 series includes a set of standards and guidelines for ethical AI design and development, covering topics such as transparency, accountability, fairness, and privacy.
3. **OECD Principles on AI:** Developed by the Organisation for Economic Co-operation and Development (OECD), these principles provide a framework for responsible AI governance, including principles related to transparency, accountability, and human rights.

Regulatory Approaches to Ethical AI Governance

Regulatory approaches to ethical AI governance involve the development and implementation of laws, regulations, and policies aimed at promoting ethical AI practices, protecting individuals' rights, and mitigating potential risks and harms associated with AI deployment.

Examples of regulatory approaches to ethical AI governance include:

1. **General Data Protection Regulation (GDPR):** The GDPR is a comprehensive data protection regulation in the European Union that establishes requirements for the processing of personal data, including provisions related to AI and automated decision-making.
2. **Algorithmic Accountability Act:** Proposed legislation in the United States that aims to promote transparency, fairness, and accountability in automated decision-making systems by requiring companies to assess the impact of their algorithms on individual rights and liberties.
3. **National AI Strategies:** Many countries have developed national AI strategies that include provisions for ethical AI governance, research, and development, as well as initiatives to promote AI education, workforce training, and international collaboration.

By adopting ethical frameworks and guidelines, adhering to international standards and best practices, and implementing regulatory approaches to ethical AI governance, stakeholders can promote the responsible development and deployment of AI systems that benefit society while respecting individual rights and values.

CASE STUDIES AND BEST PRACTICES

In this section, we delve into case studies and best practices that exemplify ethical considerations and successful implementations of AI in various domains, including healthcare, algorithmic decision-making, and business and industry.

Ethical AI in Healthcare

CASE STUDY: AI-POWERED MEDICAL IMAGING DIAGNOSIS

In this case study, AI is utilized to enhance medical imaging diagnosis, specifically in the detection and classification of diseases such as cancer. AI algorithms analyze medical images, including X-rays, MRI scans, and CT scans, to identify abnormalities and assist healthcare professionals in making accurate diagnoses. By leveraging large datasets and machine learning techniques, AI systems can detect subtle patterns and anomalies in medical images with high accuracy and efficiency, potentially improving patient outcomes and reducing diagnostic errors.

Best Practices:

1. **Data Privacy and Security:** Ensure that patient data used to train AI algorithms is anonymized and protected in compliance with data privacy regulations such as the Health Insurance Portability and Accountability Act (HIPAA) in the United States.
2. **Transparency and Explainability:** Develop AI systems that provide explanations for their diagnoses, enabling healthcare professionals to understand the reasoning behind AI-generated recommendations and facilitating trust and acceptance of AI technologies.
3. **Clinical Validation:** Validate AI algorithms through rigorous clinical testing and validation studies to assess their performance, accuracy, and reliability across diverse patient populations and clinical settings.

The count plot in Figure 15.1 visualizes the distribution of classes within the training dataset, categorizing instances as either "Pneumonia" or "Normal." The plot offers insights into the class distribution, aiding in understanding the balance or imbalance between different categories in the dataset.

Figure 15.2 presents the training and validation accuracy (left) and training accuracy and loss (right) over multiple epochs during the training of a neural network model. The plots visualize the model's learning process, showcasing changes in accuracy and loss over time, helping to monitor training progress and identify potential issues such as overfitting or underfitting.

The heatmap in Figure 15.3 illustrates a confusion matrix, representing the performance of a classification model with annotated values indicating the count of correct and incorrect predictions

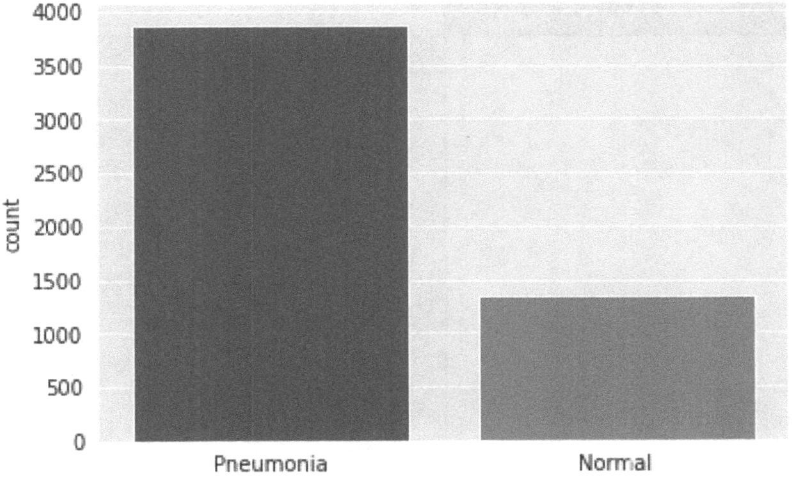

FIGURE 15.1 Distribution of classes.

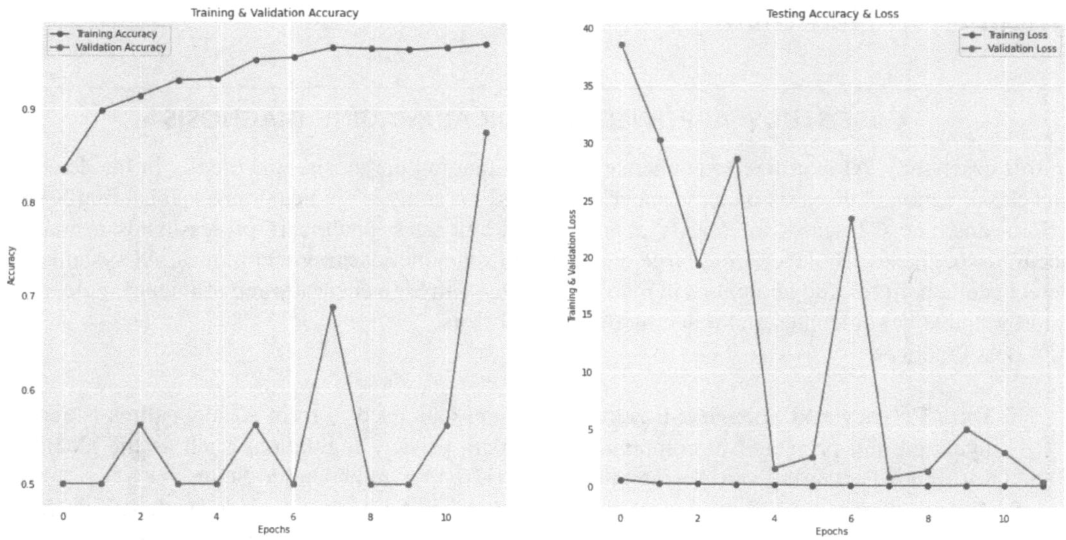

FIGURE 15.2 Training and validation accuracy.

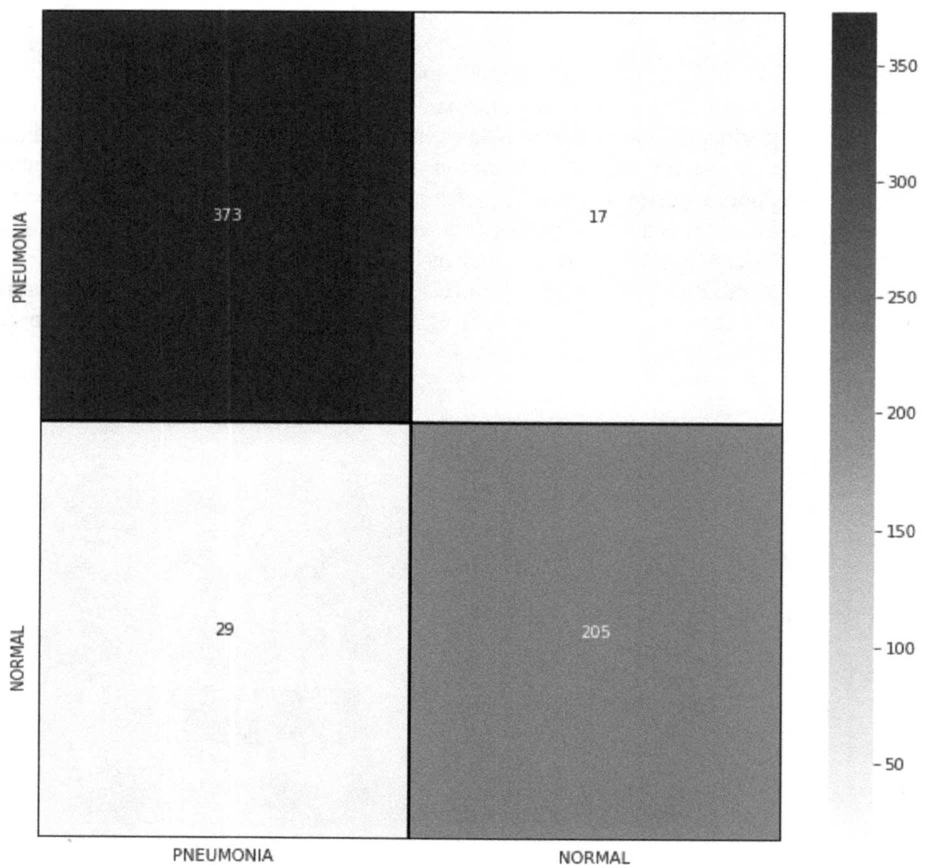

FIGURE 15.3 Performance of a classification model.

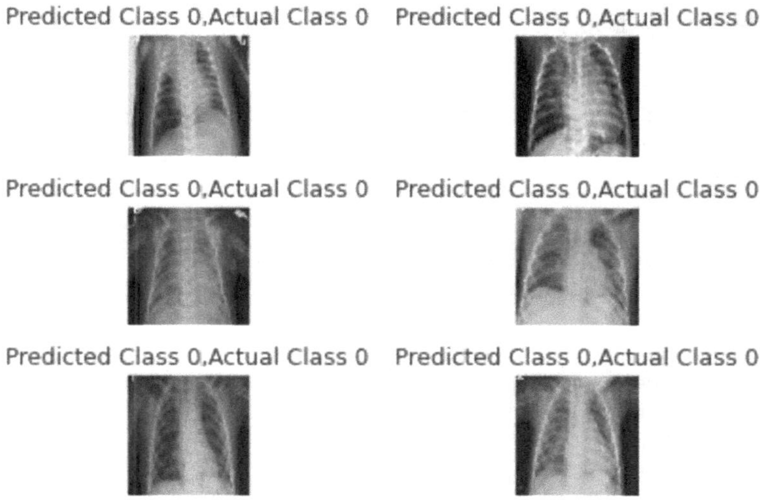

FIGURE 15.4 Corresponding predicted and actual classes.

across different classes. The color intensity signifies the frequency of predictions, aiding in visualizing the model's accuracy and error distribution.

Figure 15.4 showcases a selection of six images from the test dataset along with their corresponding predicted and actual classes. Each subplot displays an image with its predicted class label and actual class label, providing insights into the model's performance on individual instances.

Fairness and Transparency in Algorithmic Decision-Making

CASE STUDY: FAIR LENDING PRACTICES IN FINANCIAL SERVICES

In this case study, AI is applied to improve fairness and transparency in algorithmic decision-making processes, particularly in the context of lending and financial services. AI algorithms analyze applicant data to assess creditworthiness and make lending decisions, aiming to minimize biases and discrimination and ensure equitable access to financial products and services. By incorporating fairness-aware algorithms and transparency mechanisms, financial institutions can mitigate the risk of biased decision-making and promote fairness and inclusivity in lending practices.

Best Practices:

1. **Algorithmic Fairness:** Implement fairness-aware algorithms that detect and mitigate biases in lending decisions, ensuring that individuals from diverse demographic groups are treated fairly and equitably.
2. **Explainable AI:** Develop AI systems that provide explanations for lending decisions, allowing applicants to understand the factors considered and enabling regulatory compliance and accountability.
3. **Regulatory Compliance:** Ensure that algorithmic lending practices comply with relevant laws and regulations, such as the Equal Credit Opportunity Act (ECOA) and the Fair Housing Act (FHA), which prohibit discrimination in lending based on race, gender, ethnicity, or other protected characteristics.

Responsible AI Deployment in Business and Industry

CASE STUDY: AI-POWERED SUPPLY CHAIN OPTIMIZATION

In this case study, AI is employed to optimize supply chain operations, including inventory management, demand forecasting, and logistics planning. AI algorithms analyze large volumes of data, including historical sales data, inventory levels, and market trends, to predict demand, optimize inventory levels, and streamline logistics operations. By leveraging AI-driven insights, businesses can reduce costs, improve efficiency, and enhance sustainability in their supply chain operations.

Best Practices:

1. **Sustainability Impact Assessment:** Conduct sustainability impact assessments to evaluate the environmental, social, and economic implications of AI-powered supply chain optimization, identifying opportunities to reduce environmental footprint and promote responsible sourcing and production practices.
2. **Stakeholder Engagement:** Engage with supply chain partners, stakeholders, and local communities to solicit feedback, address concerns, and promote transparency and accountability in AI-driven supply chain initiatives.
3. **Continuous Monitoring and Improvement:** Implement mechanisms for continuous monitoring and improvement of AI-powered supply chain operations, including performance metrics, feedback loops, and adaptive learning algorithms, to ensure alignment with sustainability goals and regulatory requirements.

By adopting best practices and case studies that emphasize ethical considerations, organizations can harness the transformative potential of AI while ensuring fairness, transparency, and accountability in their decision-making processes. Through responsible AI deployment, businesses and industries can drive positive social and environmental outcomes while achieving their operational and strategic objectives.

STAKEHOLDER ENGAGEMENT AND COLLABORATION

Stakeholder engagement and collaboration are essential components of ethical artificial intelligence (AI) development and deployment. In this section, we explore strategies for effectively engaging with diverse stakeholders, building ethical AI communities, and promoting diversity and inclusivity in AI development.

Engaging With Diverse Stakeholders

Stakeholder engagement involves actively involving individuals, groups, and organizations that are affected by or have a vested interest in AI development and deployment. These stakeholders may include policymakers, industry representatives, civil society organizations, academic researchers, and members of impacted communities. Effective stakeholder engagement fosters dialogue, collaboration, and collective decision-making, ensuring that AI systems are developed and deployed in a manner that reflects the values, needs, and priorities of diverse stakeholders.

Strategies for engaging with diverse stakeholders include:

1. **Stakeholder Mapping:** Identify key stakeholders and their respective interests, concerns, and perspectives related to AI development and deployment. This may involve conducting

stakeholder analyses, surveys, interviews, and focus groups to gather input and feedback from diverse stakeholders.

2. **Multi-Stakeholder Forums:** Organize multi-stakeholder forums, workshops, and roundtable discussions to facilitate dialogue, exchange of ideas, and collaboration among diverse stakeholders. These forums provide opportunities for stakeholders to share insights, address concerns, and co-create solutions to ethical challenges in AI development.
3. **Transparency and Accountability:** Maintain transparency in stakeholder engagement processes by sharing information, updates, and outcomes with stakeholders in a timely and accessible manner. Foster accountability by incorporating stakeholder feedback into decision-making processes and demonstrating responsiveness to stakeholder concerns.

Building Ethical AI Communities

Ethical AI communities bring together individuals and organizations committed to promoting ethical practices in AI development and deployment. These communities serve as forums for knowledge sharing, collaboration, and collective action, fostering a culture of ethics and responsibility in the AI ecosystem. Ethical AI communities may include professional associations, industry consortia, research networks, and grassroots initiatives dedicated to advancing ethical AI principles and best practices.

Key features of ethical AI communities include:

1. **Knowledge Sharing:** Facilitate knowledge sharing and exchange of best practices through conferences, workshops, webinars, and online platforms. Encourage collaboration and networking among members to leverage collective expertise and resources for ethical AI development.
2. **Capacity Building:** Offer training programs, resources, and educational materials to support capacity building in ethical AI development. Provide guidance on ethical frameworks, guidelines, and tools for integrating ethical considerations into AI research, development, and deployment processes.
3. **Advocacy and Awareness:** Advocate for ethical AI policies, regulations, and standards that prioritize fairness, transparency, accountability, and societal impact. Raise awareness about ethical issues in AI through public campaigns, media outreach, and thought leadership initiatives to promote informed discourse and action.

Promoting Diversity and Inclusivity in AI Development

Diversity and inclusivity are essential for fostering innovation, creativity, and equity in AI development. By promoting diversity in AI teams and ensuring inclusivity in AI processes, organizations can harness the full potential of diverse perspectives and experiences to address societal challenges and develop AI solutions that reflect the needs and values of diverse communities.

Strategies for promoting diversity and inclusivity in AI development include:

1. **Diverse Representation:** Recruit and retain diverse talent from various backgrounds, including gender, race, ethnicity, nationality, age, and socioeconomic status, to ensure diverse perspectives and experiences are represented in AI teams.
2. **Inclusive Design:** Adopt inclusive design principles that prioritize accessibility, usability, and inclusivity in AI products and services. Consider the diverse needs, preferences, and abilities of users throughout the design and development process to create AI solutions that are accessible to all.
3. **Community Engagement:** Engage with diverse communities, including underrepresented groups, marginalized populations, and vulnerable communities, to solicit input, feedback, and

participation in AI development initiatives. Collaborate with community organizations and advocacy groups to co-create solutions that address the specific needs and concerns of diverse communities.

By actively engaging with diverse stakeholders, building ethical AI communities, and promoting diversity and inclusivity in AI development, organizations can foster a culture of ethics, responsibility, and innovation in the AI ecosystem. These efforts contribute to the development and deployment of AI systems that are fair, transparent, accountable, and inclusive, thereby advancing societal well-being and promoting positive outcomes for all.

CHALLENGES AND FUTURE DIRECTIONS

As AI and ML technologies continue to evolve and proliferate, they bring with them a host of ethical challenges and considerations. In this section, we explore the emerging ethical challenges in AI and ML, future trends in ethical AI research and practice, and opportunities for ethical innovation and collaboration.

Emerging Ethical Challenges in AI and ML

Despite the transformative potential of AI and ML, they also present significant ethical challenges that must be addressed to ensure responsible development and deployment. Some of the emerging ethical challenges in AI and ML include:

1. **Bias and Fairness:** Addressing algorithmic bias and ensuring fairness in AI systems remains a major challenge, particularly as AI algorithms increasingly make decisions that impact individuals' lives in areas such as healthcare, finance, and criminal justice.
2. **Accountability and Transparency:** Holding AI systems accountable for their decisions and ensuring transparency in their decision-making processes is crucial for building trust and mitigating the risks of unintended consequences or harm.
3. **Privacy and Data Protection:** Protecting individuals' privacy rights and personal data in the age of AI requires robust data protection measures, including data anonymization, encryption, and user consent mechanisms.
4. **Autonomy and Control:** Balancing the benefits of autonomous AI systems with the need for human oversight and control presents complex ethical dilemmas, particularly in safety-critical domains such as autonomous vehicles and healthcare.
5. **Social Impact and Equity:** Understanding and mitigating the societal impact of AI on employment, inequality, and social cohesion is essential for promoting equity and inclusivity in the AI ecosystem.

Future Trends in Ethical AI Research and Practice

Looking ahead, ethical AI research and practice are expected to evolve in response to emerging challenges and opportunities. Some future trends in ethical AI research and practice include:

1. **Ethical Design and Development:** Integrating ethics into the design and development of AI systems from the outset, including ethical risk assessments, impact assessments, and design methodologies that prioritize fairness, transparency, and accountability.

2. **Explainable AI and Interpretability:** Advancing techniques for explainable AI and interpretability to enable users to understand how AI systems arrive at their decisions and to identify and address biases or errors.
3. **Multi-Stakeholder Collaboration:** Fostering collaboration among diverse stakeholders, including researchers, policymakers, industry representatives, civil society organizations, and impacted communities, to co-create ethical AI solutions and promote collective action.
4. **Regulatory Frameworks and Standards:** Developing and implementing regulatory frameworks and international standards for ethical AI governance, including guidelines for AI ethics, transparency, accountability, and human rights.
5. **Ethical AI Education and Awareness:** Promoting ethical AI education and awareness among AI practitioners, policymakers, and the general public to foster a culture of ethics, responsibility, and accountability in the AI ecosystem.

Opportunities for Ethical Innovation and Collaboration

Despite the challenges, there are also significant opportunities for ethical innovation and collaboration in the field of AI. Some opportunities include:

1. **Ethical AI Startups and Ventures:** Supporting the growth of startups and ventures that prioritize ethical AI principles and practices, including ethical AI design, development, and deployment.
2. **Ethical AI Certification and Accreditation:** Establishing certification and accreditation programs for AI systems that meet ethical standards and criteria, providing assurances to users, regulators, and stakeholders.
3. **Ethical AI Impact Investing:** Investing in ethical AI projects and initiatives that generate positive social and environmental impact, aligning with principles of responsible investing and corporate social responsibility.
4. **Ethical AI Research Networks:** Creating interdisciplinary research networks and consortia focused on advancing ethical AI research, development, and policy, fostering collaboration and knowledge sharing among researchers and practitioners.
5. **Global AI Ethics Initiatives:** Supporting global AI ethics initiatives and coalitions that bring together governments, industry, academia, and civil society to develop shared principles, guidelines, and standards for ethical AI governance.

By addressing emerging ethical challenges, leveraging future trends in ethical AI research and practice, and capitalizing on opportunities for ethical innovation and collaboration, we can work toward building a more ethical, responsible, and sustainable AI ecosystem that benefits society as a whole. Through collective action and shared commitment to ethical principles, we can harness the transformative potential of AI for positive social impact and human flourishing.

Table 15.1 provides a summary of qualitative findings related to ethical AI practices in healthcare, algorithmic decision-making, and business and industry domains. Each domain is accompanied by key findings or observations that highlight the impact and importance of ethical considerations in AI development and deployment.

Discussion

The discussion section reflects on the qualitative findings presented in the previous section and explores their implications, significance, and broader context within the field of ethical AI practices. It delves into

TABLE 15.1 AI practices in healthcare

DOMAIN	*KEY FINDINGS/OBSERVATIONS*
Healthcare	• Implementation of AI in medical imaging diagnosis has improved accuracy and efficiency in detecting diseases. • Ethical considerations such as patient privacy and transparency in AI-driven diagnosis are being addressed through robust protocols.
Algorithmic Decision-Making	• Fairness-aware algorithms are being developed to mitigate biases in lending decisions, promoting equitable access to financial services. • Transparency mechanisms in algorithmic decision-making enhance accountability and trust in lending practices.
Business and Industry	• AI-driven supply chain optimization has led to cost savings, efficiency improvements, and reduced environmental impact. • Responsible AI deployment practices in business and industry prioritize ethical considerations, including transparency and fairness.

the challenges, successes, and areas for improvement identified through the research, offering insights and critical analysis to deepen understanding.

Key points for discussion may include:

1. **Impact of Ethical AI Practices:** Evaluate the impact of ethical AI practices in healthcare, algorithmic decision-making, and business and industry sectors. Discuss how these practices have contributed to improved outcomes, addressed ethical concerns, and fostered trust and accountability.
2. **Challenges and Limitations:** Identify challenges and limitations encountered in implementing ethical AI practices, such as algorithmic bias, privacy concerns, and regulatory complexities. Explore potential strategies for overcoming these challenges and mitigating their impact on AI deployment.
3. **Ethical Considerations:** Reflect on the ethical considerations inherent in AI development and deployment, including fairness, transparency, accountability, and societal impact. Discuss the ethical dilemmas faced by organizations and stakeholders and the importance of prioritizing ethical principles in AI decision-making.
4. **Future Directions:** Outline potential future directions for research, policy, and practice in the field of ethical AI. Consider emerging trends, technologies, and ethical frameworks that may shape the future of AI development and deployment, and highlight areas for further exploration and innovation.

Conclusion

The conclusion section summarizes the key findings, insights, and implications discussed in the preceding sections. It reiterates the significance of ethical AI practices in promoting fairness, transparency, and accountability, and emphasizes the importance of ongoing efforts to address ethical challenges and advance responsible AI deployment.

Key points for the conclusion may include:

1. **Summary of Findings:** Provide a concise summary of the qualitative results and their implications for ethical AI practices in healthcare, algorithmic decision-making, and business and industry sectors.

2. **Importance of Ethical AI:** Reinforce the importance of ethical considerations in AI development and deployment, highlighting their role in building trust, mitigating risks, and promoting positive societal impact.
3. **Call to Action:** Encourage stakeholders to prioritize ethical principles in AI decision-making and to collaborate on initiatives aimed at advancing responsible AI practices. Emphasize the collective responsibility to ensure that AI technologies benefit humanity while minimizing harm.

Future Scope

The future scope section outlines potential avenues for further research, policy development, and practical implementation in the field of ethical AI. It identifies areas of opportunity and challenge, suggesting directions for future inquiry and action to address emerging ethical issues and advance the responsible use of AI technologies.

Key points for the future scope may include:

1. **Research Agenda:** Propose a research agenda focused on addressing unanswered questions, exploring new methodologies, and advancing understanding of ethical AI practices in diverse contexts.
2. **Policy Implications:** Consider the policy implications of ethical AI practices, including the need for regulatory frameworks, industry standards, and international collaboration to govern AI development and deployment.
3. **Practical Applications:** Explore potential practical applications of ethical AI practices in areas such as healthcare, finance, education, and governance, highlighting opportunities for innovation and impact.
4. **Ethical Leadership:** Advocate for ethical leadership and responsible decision-making in AI development and deployment, emphasizing the role of organizations, governments, and civil society in shaping the future of AI ethics.

By addressing these key components in the discussion, conclusion, and future scope sections, the chapter provides a comprehensive overview of the qualitative findings, their implications, and opportunities for further research and action in the field of ethical AI practices.

REFERENCES

Buolamwini, J., & Gebru, T. (2018). Gender Shades: Intersectional Accuracy Disparities in Commercial Gender Classification. Proceedings of the 1st Conference on Fairness, Accountability and Transparency, 81–91.

Char, D. S., Shah, N. H., Magnus, D., & Implementing, P. R. O. G. (2018). Implementing Machine Learning in Health Care—Addressing Ethical Challenges. New England Journal of Medicine, 378(11), 981–983.

Crawford, K., & Paglen, T. (2019). Excavating AI: The Politics of Images in Machine Learning Training Sets. AI & Society, 36, 1105–1116.

Custers, B., Calders, T., Schermer, B., & Zarsky, T. (2019). Transparency of Algorithms: A Systematic Review of Definition and Measures. ACM Computing Surveys (CSUR), 52(3), 1–31.

Floridi, L. (2020). AI Ethics: Challenges and Opportunities. Philosophy & Technology, 33(2), 203–204.

Floridi, L., & Cowls, J. (2019). A Unified Framework of Five Principles for AI in Society. Harvard Data Science Review, 1(1).

Frischmann, B. M., & Selinger, E. (Eds.). (2018). The Cambridge handbook of the law of the sharing economy. Cambridge University Press.

Gebru, T., Morgenstern, J., Vecchione, B., Vaughan, J. W., Wallach, H., Daume III, H., & Crawford, K. (2018). Datasheets for datasets. arXiv preprint, arXiv:1803.09010.

Green, B., & Chen, D. L. (2019). Beyond Prediction: Using Big Data for Policy Problems. Science, 363(6423), 692–694.

Hirschberg, D. S., & Hirschberg, J. (2019). Gender, Pronunciation Modeling, and the History of Gendered Speech Recognition. In Proceedings of the Conference on Fairness, Accountability, and Transparency (pp. 70–78).

Jobin, A., Ienca, M., & Vayena, E. (2019). The Global Landscape of AI Ethics Guidelines. Nature Machine Intelligence, 1(9), 389–399.

Jobin, A., Ienca, M., Vayena, E., & Schermer, B. (2019). Artificial Intelligence: The Global Landscape of Ethics Guidelines. Nature, 569(7755), 131–137.

Metzler, T., & Dumais, S. T. (2019). Fairness in AI and Machine Learning. Big Data, 7(6), 295–296.

Mitchell, M., Wu, S., Zaldivar, A., Barnes, P., Vasserman, L., Hutchinson, B., … & Gebru, T. (2019). Model Cards for Model Reporting. In Proceedings of the Conference on Fairness, Accountability, and Transparency (pp. 220–229).

Mittelstadt, B. D., Russell, C., Wachter, S., & Floridi, L. (2019). Explaining Explanations in AI. In Proceedings of the conference on fairness, accountability, and transparency (pp. 279–288).

Plachouras, V., Matakos, A., & Papandreou, N. (2019). Machine Learning, Policy, and Persistent Disparities. Science, 366(6469), 37–38.

Raji, I. D., & Buolamwini, J. (2019). Actionable Auditing: Investigating the Impact of Publicly Naming Biased Performance Results of Commercial AI Products. In Proceedings of the conference on fairness, accountability, and transparency (pp. 77–86).

Selbst, A. D., & Barocas, S. (2018). The Intuitive Appeal of Explainable Machines. Fordham L. Rev, 87, 1085.

Sweeney, L. (2013). Discrimination in Online Ad Delivery. Queue, 11(3), 10.

Zou, J., Schiebinger, L., Bhattacharya, C., & Greene, D. (2018). AI can be Sexist and Racist—It's Time to Make it Fair. Nature, 559(7714), 324–326.

Charting the Ethical Horizon
A Deep Dive Into AI Ethics, Unraveling the Moral Threads in Machine Intelligence Development

16

Gopichand Vemulapalli

INTRODUCTION

Artificial intelligence (AI) stands at the forefront of technological innovation, promising transformative capabilities across various domains, from healthcare and finance to transportation and entertainment. As AI systems become increasingly integrated into our daily lives, they raise profound ethical questions and moral dilemmas that demand careful consideration and thoughtful deliberation. The development and deployment of AI technologies intersect with a complex web of ethical concerns, spanning issues of fairness, transparency, accountability, privacy, bias, autonomy, and societal impact. In this chapter, we embark on a deep dive into AI ethics, unraveling the moral threads that underpin machine intelligence development and deployment.

The rapid advancement of AI technologies has ushered in a new era of possibility and progress, enabling feats of computation and automation that were once relegated to the realm of science fiction. Machine learning algorithms can analyze vast datasets, recognize patterns, and make predictions with remarkable accuracy and efficiency. Natural language processing models can generate human-like text, while computer vision systems can identify objects, faces, and gestures with precision. Autonomous vehicles navigate city streets, drones survey landscapes, and virtual assistants respond to voice commands, reshaping the way we work, communicate, and interact with the world around us. However, amidst the excitement and promise of AI, ethical considerations loom large, casting a shadow over the technological landscape. The use of AI raises profound questions about the values, norms, and

DOI: 10.1201/9781003497189-16

principles that should guide its development and deployment. How can we ensure that AI systems are fair and unbiased, treating all individuals equitably and without discrimination? How can we promote transparency and accountability in AI decision-making processes, ensuring that users understand how and why decisions are made? How can we protect individuals' privacy and autonomy in an era of ubiquitous data collection and surveillance? These questions lie at the heart of AI ethics, challenging us to navigate the complex terrain where technology and morality intersect.

One of the central ethical principles guiding AI development is fairness. Fairness requires that AI systems treat all individuals fairly and equitably, regardless of their race, gender, age, or other characteristics. However, achieving fairness in AI is fraught with challenges, as algorithms may inadvertently perpetuate or exacerbate existing biases and inequalities present in the data they are trained on. Algorithmic bias can lead to discriminatory outcomes, disadvantaging certain groups and perpetuating social injustice. Addressing algorithmic bias requires careful attention to data collection, algorithm design, and evaluation methodologies to ensure that AI systems are fair and unbiased in their decision-making.

Transparency and accountability are also essential principles in AI ethics. Transparency refers to the clarity and openness of AI decision-making processes, enabling users to understand how algorithms arrive at their decisions. Accountability entails holding AI systems and their creators responsible for their actions and outcomes, ensuring that they can be held to account for any harm or injustice they cause. Achieving transparency and accountability in AI requires mechanisms for auditing, monitoring, and explaining AI systems, as well as legal and regulatory frameworks that assign responsibility and liability for AI-related harms.

Privacy and autonomy are fundamental human rights that must be protected in the age of AI. Privacy concerns arise from the widespread collection, analysis, and use of personal data by AI systems, raising questions about consent, control, and surveillance. Autonomous AI systems, such as self-driving cars and autonomous weapons, pose challenges to individual autonomy and agency, as they make decisions and take actions without direct human intervention. Safeguarding privacy and autonomy in AI requires robust data protection laws, privacy-preserving technologies, and ethical guidelines that prioritize individual rights and freedoms. Moreover, AI has the potential to have profound societal impacts, shaping economies, cultures, and political systems in unforeseen ways. The deployment of AI in healthcare, finance, criminal justice, and other sectors can have far-reaching consequences for individuals and communities, affecting access to services, opportunities, and resources. Understanding and mitigating the societal impacts of AI requires interdisciplinary collaboration, stakeholder engagement, and participatory decision-making processes that prioritize the public interest and promote social justice.

The ethical implications of AI have become a focal point in academic discourse, reflecting the increasing integration of AI systems into various facets of society. Anderson and Anderson [1] initiated discussions on machine ethics, exploring the creation of ethical intelligent agents, which laid the groundwork for subsequent explorations of AI's moral dimensions. Bostrom and Yudkowsky [2] expanded on these discussions by examining the broader ethics of AI, as outlined in the Stanford Encyclopedia of Philosophy. Floridi and Cowls [3] proposed a unified framework consisting of five principles to guide AI in society, emphasizing the need for consistent ethical standards.

The global landscape of AI ethics guidelines has been extensively mapped by Jobin et al. [4], revealing a diverse array of ethical considerations across different regions and industries. This diversity is further explored by Mittelstadt et al. [5], who mapped the ethics of algorithms, highlighting the complexities and debates surrounding algorithmic decision-making. Taddeo and Floridi [6] argued that AI could be a force for good, provided that ethical considerations are integrated into its development and deployment.

However, AI systems also inherit human-like biases, as demonstrated by Caliskan et al. [7] in their study of semantics derived from language corpora. The rapid advancements in deep learning, discussed by Goodfellow et al. [8] and LeCun et al. [9], have further complicated the ethical landscape, necessitating robust AI governance frameworks, as proposed by Koene et al. [10].

Ethical AI governance has been identified as a crucial area of research, with Mingers [11] conducting a comprehensive literature review to outline the current state of AI ethics. The actionable auditing of AI systems, particularly in terms of bias, has been explored by Raji and Buolamwini [12], who investigated the impact of publicly naming biased AI products. This ties into the broader responsibility of building ethically sustainable AI systems, as discussed by Schwartz [13, 14].

The right to meaningful explanations of AI-driven decisions, explored by Selbst and Powles [15], is an emerging area of interest, particularly in the context of privacy, security, and accountability, as examined by Smith and Anderson [16]. Taddeo [17] offers a roadmap for ethical AI, stressing the importance of narrative in shaping public understanding of AI ethics, while Tufekci [18] addresses the political implications of big data and AI-driven surveillance. Finally, Winner [19] raises the provocative question of whether artifacts, including AI systems, have inherent political dimensions, underscoring the complex interplay between technology and society. Collectively, these works provide a comprehensive overview of the ethical challenges and opportunities associated with AI, emphasizing the need for interdisciplinary approaches to address the ethical dilemmas posed by AI technologies.

In this chapter, we will explore these ethical considerations and more, delving into the complexities of AI ethics and unraveling the moral threads that bind machine intelligence to human values and aspirations. Through case studies, theoretical frameworks, and practical examples, we will examine the ethical dilemmas faced by AI practitioners, policymakers, and stakeholders across diverse domains. We will interrogate the power dynamics embedded within AI systems, analyze the impacts of algorithmic bias and discrimination, and chart a course for ethical innovation and responsible AI development. By charting the ethical horizon of AI, we seek to inform and inspire dialogue, reflection, and action aimed at harnessing the transformative potential of AI for the collective good while safeguarding against its unintended consequences and moral pitfalls.

FAIRNESS IN AI

Fairness is a fundamental ethical principle that underpins the development and deployment of AI systems. It encompasses the notion that AI systems should treat all individuals equitably and without bias, regardless of their race, gender, age, or other characteristics. Achieving fairness in AI is crucial for ensuring that AI technologies promote justice, equality, and inclusivity in society. In this section, we will explore the definition of fairness, the challenges of achieving fairness in AI, strategies for addressing algorithmic bias, and methods for ensuring equity and nondiscrimination in AI systems.

Definition of Fairness

Fairness in AI can be defined as the absence of bias or discrimination in the design, implementation, and outcomes of AI systems. It requires that AI algorithms and decision-making processes treat all individuals fairly and impartially, without favoring or disadvantaging certain groups. Fairness encompasses several dimensions, including procedural fairness, distributive fairness, and interactional fairness. Procedural fairness refers to the fairness of the decision-making process itself, while distributive fairness pertains to the fairness of the outcomes produced by AI systems. Interactional fairness involves the fairness of interactions between individuals and AI systems, including transparency, accountability, and the ability to contest decisions.

Challenges of Achieving Fairness in AI

Achieving fairness in AI is challenging due to several factors, including the inherent biases present in training data, the complexity of algorithmic decision-making, and the lack of standardized definitions and metrics for fairness. Training data used to train AI models may reflect existing societal biases and inequalities, leading to biased or discriminatory outcomes. Moreover, the complexity of AI algorithms can make it difficult to identify and mitigate biases effectively. Additionally, the definition of fairness may vary depending on the context and stakeholders involved, making it challenging to develop universally applicable fairness criteria for AI systems.

Addressing Algorithmic Bias

Addressing algorithmic bias is essential for promoting fairness in AI. Algorithmic bias refers to the systematic and unfair discrimination against certain individuals or groups based on characteristics such as race, gender, or socioeconomic status. To address algorithmic bias, AI practitioners can employ various strategies, including bias detection and mitigation techniques, diversity-aware data collection and preprocessing methods, and algorithmic fairness interventions such as fairness-aware learning algorithms and post-processing techniques. Additionally, promoting diversity and inclusivity in AI teams and decision-making processes can help mitigate biases and ensure that AI systems are designed and deployed in a fair and equitable manner.

Ensuring Equity and Nondiscrimination

Ensuring equity and nondiscrimination in AI requires a multifaceted approach that involves addressing systemic inequalities, promoting diversity and representation, and empowering marginalized communities. AI practitioners and policymakers can adopt measures such as fairness-aware AI development frameworks, bias audits and impact assessments, and community engagement and participatory decision-making processes to ensure that AI systems promote equity and nondiscrimination. Additionally, regulatory frameworks and ethical guidelines can play a crucial role in holding AI developers and deployers accountable for ensuring fairness and equity in AI systems.

By addressing these challenges and implementing strategies for promoting fairness, AI practitioners and policymakers can help build AI systems that uphold ethical principles and contribute to a more just, equitable, and inclusive society. Fairness in AI is not only a moral imperative but also a practical necessity for realizing the full potential of AI technologies and ensuring their benefits are shared equitably among all members of society.

TRANSPARENCY AND ACCOUNTABILITY

Transparency and accountability are essential principles in the development and deployment of AI systems. They ensure that AI decision-making processes are understandable, traceable, and accountable, thereby fostering trust, confidence, and ethical behavior among users and stakeholders. In this section, we will delve into the importance of transparency in AI decision-making, explore mechanisms for promoting transparency, discuss legal and regulatory frameworks for accountability, and consider ethical considerations in AI auditing and evaluation.

Importance of Transparency in AI Decision-Making

Transparency in AI decision-making is crucial for several reasons. First and foremost, it enables users to understand how AI algorithms arrive at their decisions, enhancing trust and confidence in the technology. Transparency also facilitates accountability by allowing users to trace the logic and reasoning behind AI decisions and hold AI systems and their creators accountable for any errors, biases, or unintended consequences. Moreover, transparency promotes fairness and equity by exposing any hidden biases or discriminatory patterns present in AI algorithms and data.

Mechanisms for Promoting Transparency

Several mechanisms can promote transparency in AI decision-making. One approach is to provide explanations or justifications for AI decisions, allowing users to understand the factors and considerations that influenced the decision-making process. Explanations can take various forms, including textual explanations, visualizations, or interactive interfaces that enable users to explore and interrogate AI decisions. Another approach is to make AI algorithms and data publicly available for scrutiny and audit, allowing independent researchers and stakeholders to assess their fairness, accuracy, and robustness. Additionally, transparency can be promoted through ethical design practices, such as designing AI systems with built-in transparency features and mechanisms for user feedback and oversight.

Legal and Regulatory Frameworks for Accountability

Legal and regulatory frameworks play a crucial role in promoting accountability in AI decision-making. These frameworks establish legal obligations and responsibilities for AI developers, deployers, and users, ensuring that they adhere to ethical principles and legal standards. For example, data protection laws such as the General Data Protection Regulation (GDPR) in the European Union require organizations to provide transparent information about the processing of personal data and to obtain consent from individuals for the use of their data in AI systems. Additionally, regulatory bodies may impose requirements for AI auditing, evaluation, and risk assessment to ensure that AI systems comply with legal and ethical standards.

Ethical Considerations in AI Auditing and Evaluation

Ethical considerations play a crucial role in AI auditing and evaluation processes. Auditing AI systems involves assessing their fairness, accuracy, transparency, and accountability, as well as identifying and mitigating any biases or discriminatory patterns present in the algorithms and data. Ethical auditing practices should prioritize the protection of individual rights and freedoms, including privacy, autonomy, and nondiscrimination. Moreover, ethical evaluations should consider the broader societal impacts of AI systems, including their effects on vulnerable populations, social justice, and democratic governance. Ethical auditing and evaluation processes should be transparent, independent, and participatory, involving diverse stakeholders and communities in the assessment and oversight of AI systems.

By promoting transparency and accountability in AI decision-making, stakeholders can build trust, confidence, and ethical behavior in AI systems, ensuring that they are used responsibly and for the benefit of society. Transparent AI systems enable users to understand and trust AI decisions, while accountable AI systems hold AI developers and deployers responsible for their actions and outcomes. Moreover, legal and regulatory frameworks provide a foundation for promoting ethical behavior and ensuring compliance with legal standards and ethical principles. Ethical considerations should guide AI auditing and evaluation processes, prioritizing the protection of individual rights and freedoms and the promotion of social justice and equity.

PRIVACY AND AUTONOMY

Privacy and autonomy are fundamental human rights that must be safeguarded in the development and deployment of AI systems. Privacy refers to the right of individuals to control their personal information and to be free from unwanted intrusion or surveillance, while autonomy pertains to the ability of individuals to make informed choices and decisions without undue influence or coercion. In this section, we will examine the protection of personal privacy in AI systems, the safeguarding of individual autonomy in the age of AI, the importance of balancing privacy and innovation, and regulatory approaches to data privacy and autonomy.

Protection of Personal Privacy in AI Systems

Protecting personal privacy in AI systems is essential for preserving individuals' rights and freedoms in an increasingly data-driven world. AI systems often rely on vast amounts of personal data to make predictions, recommendations, and decisions, raising concerns about data privacy and security. To protect personal privacy in AI systems, organizations must implement robust data protection measures, such as encryption, anonymization, and data minimization, to ensure that personal data is collected, processed, and stored in a secure and privacy-preserving manner. Additionally, organizations should obtain informed consent from individuals for the use of their data in AI systems and provide transparency about how their data will be used and protected.

Safeguarding Individual Autonomy in the Age of AI

Safeguarding individual autonomy in the age of AI requires ensuring that individuals have the ability to make informed choices and decisions without undue influence or manipulation. AI systems have the potential to shape individuals' behavior and decision-making processes through personalized recommendations, targeted advertising, and algorithmic decision-making. To safeguard individual autonomy, organizations must design AI systems with transparency, accountability, and user control mechanisms that empower individuals to understand and influence how their data is used and to opt out of data collection and processing activities if desired. Moreover, organizations should adhere to ethical principles such as respect for autonomy and informed consent in the design and deployment of AI systems.

Balancing Privacy and Innovation

Balancing privacy and innovation is a complex and ongoing challenge in the development and deployment of AI systems. While AI technologies offer tremendous potential for innovation and advancement, they also pose risks to individual privacy and autonomy if not used responsibly. Organizations must strike a balance between harnessing the benefits of AI innovation and protecting individuals' privacy rights and freedoms. This may involve adopting privacy-preserving technologies and practices, such as differential privacy, federated learning, and homomorphic encryption, that enable AI systems to leverage data while minimizing privacy risks. Additionally, organizations should engage in ethical risk assessment and mitigation processes to identify and address potential privacy and autonomy concerns associated with AI deployment.

Regulatory Approaches to Data Privacy and Autonomy

Regulatory approaches play a crucial role in safeguarding data privacy and autonomy in the age of AI. Governments and regulatory bodies around the world have enacted laws and regulations to protect individuals' personal data and ensure that AI systems adhere to ethical principles and legal standards. For example, the General Data Protection Regulation (GDPR) in the European Union establishes stringent requirements for data protection, including the right to data privacy, transparency, and informed consent. Similarly, regulatory frameworks such as the California Consumer Privacy Act (CCPA) in the United States and the Personal Information Protection Law (PIPL) in China aim to safeguard individuals' privacy rights and regulate the collection, use, and sharing of personal data by organizations.

By prioritizing the protection of personal privacy and autonomy in AI systems and adhering to ethical principles and regulatory standards, organizations can build trust, confidence, and ethical behavior in AI technologies. Privacy-preserving AI systems enable individuals to retain control over their personal information and make informed choices about how their data is used, while respecting their autonomy and agency. Moreover, regulatory frameworks provide a foundation for promoting responsible data practices and ensuring compliance with legal and ethical standards in the development and deployment of AI systems.

SOCIETAL IMPACT OF AI

The deployment of AI systems has profound implications for society, spanning various domains such as healthcare, economics, governance, and social justice. In this section, we will explore the implications of AI deployment in healthcare, the economic and social consequences of AI automation, ethical considerations in AI governance and policy, and strategies for promoting social justice and equity in AI applications.

Implications of AI Deployment in Healthcare

The integration of AI technologies into healthcare has the potential to revolutionize diagnosis, treatment, and patient care. AI-enabled diagnostic tools can analyze medical images, genetic data, and patient records to identify diseases, predict outcomes, and personalize treatment plans. AI-driven telemedicine platforms can expand access to healthcare services, particularly in underserved areas. However, the widespread adoption of AI in healthcare raises ethical concerns related to patient privacy, data security, algorithmic bias, and the impact on physician-patient relationships. Moreover, the use of AI in healthcare exacerbates existing disparities in access to quality care, as marginalized communities may face barriers to technology adoption and digital health literacy.

Economic and Social Consequences of AI Automation

AI automation has significant implications for the economy and society, affecting employment, income distribution, and social mobility. While AI technologies can enhance productivity, efficiency, and innovation in various industries, they also lead to job displacement, skill mismatches, and income inequality. Low-skilled workers are particularly vulnerable to automation, as their jobs are more easily replaceable by AI systems. Moreover, AI automation may exacerbate social inequalities by favoring capital over labor and concentrating wealth and power in the hands of a few tech giants. Ethical considerations in AI automation include ensuring fair and equitable distribution of the benefits and burdens of automation, supporting displaced workers through reskilling and upskilling programs, and reimagining the role of work in society in the age of AI.

Ethical Considerations in AI Governance and Policy

Governance and policy frameworks play a crucial role in shaping the development, deployment, and regulation of AI technologies. Ethical considerations in AI governance and policy include transparency, accountability, fairness, privacy, and autonomy. Governments and regulatory bodies must establish clear guidelines and standards for the responsible use of AI, including data protection laws, algorithmic transparency requirements, and ethical impact assessments. Moreover, policymakers should engage with stakeholders from diverse backgrounds, including industry, academia, civil society, and marginalized communities, to ensure that AI governance and policy frameworks reflect a broad range of perspectives and interests.

Promoting Social Justice and Equity in AI Applications

Promoting social justice and equity in AI applications requires a concerted effort to address biases, disparities, and injustices embedded within AI systems. This includes developing AI technologies that are fair, inclusive, and accessible to all members of society, regardless of race, gender, age, or socioeconomic status. Moreover, promoting social justice and equity in AI requires addressing broader structural inequalities, such as discrimination, poverty, and systemic racism, that shape the development and deployment of AI technologies. This may involve adopting principles of fairness, transparency, and accountability in AI governance and policy, as well as investing in education, training, and community engagement initiatives that empower marginalized communities to participate in and benefit from the AI revolution.

By grappling with the societal impact of AI and addressing ethical considerations in its development and deployment, stakeholders can harness the transformative potential of AI technologies while mitigating their unintended consequences and promoting social justice and equity in society. Ethical AI governance and policy frameworks enable responsible innovation and ensure that AI technologies serve the common good, uphold human rights, and advance the collective well-being of all members of society.

ETHICAL INNOVATION AND RESPONSIBLE AI DEVELOPMENT

Ethical innovation and responsible AI development are imperative to ensure that AI technologies are developed and deployed in a manner that aligns with ethical principles, societal values, and human rights. In this section, we will explore various strategies for ethical AI design and implementation, collaborative approaches to ethical AI development, ethical leadership in AI research and practice, and opportunities for ethical innovation and impact.

Strategies for Ethical AI Design and Implementation

Ethical AI design and implementation require the integration of ethical principles and considerations into all stages of the AI development life cycle. This includes defining clear ethical goals and objectives for AI systems, conducting ethical risk assessments and impact analyses, and incorporating fairness, transparency, accountability, privacy, and autonomy into AI algorithms and decision-making processes. Additionally, organizations should establish ethical guidelines and codes of conduct for AI practitioners, develop mechanisms for ethical oversight and governance, and invest in AI education and training programs that promote ethical awareness and competence among developers and users.

Collaborative Approaches to Ethical AI Development

Collaborative approaches to ethical AI development involve engaging diverse stakeholders from academia, industry, government, civil society, and impacted communities in the design, development, and deployment of AI technologies. This includes fostering interdisciplinary collaboration and co-creation processes that bring together experts from multiple disciplines, such as computer science, ethics, law, sociology, and psychology, to address complex ethical challenges and trade-offs. Moreover, collaborative approaches involve establishing partnerships and networks that enable knowledge sharing, capacity building, and collective action to advance ethical innovation and responsible AI development.

Ethical Leadership in AI Research and Practice

Ethical leadership in AI research and practice entails promoting ethical awareness, integrity, and accountability among AI researchers, practitioners, and organizations. Ethical leaders in AI prioritize ethical considerations in decision-making processes, advocate for transparency and accountability in AI development and deployment, and champion diversity, inclusion, and social responsibility in AI initiatives. Moreover, ethical leaders cultivate a culture of ethical excellence and continuous improvement within their organizations, fostering a climate of trust, openness, and collaboration that enables ethical innovation and responsible AI development.

Opportunities for Ethical Innovation and Impact

Opportunities for ethical innovation and impact abound in the field of AI, offering potential solutions to pressing societal challenges while upholding ethical principles and values. Ethical innovation involves developing AI technologies that promote human well-being, social justice, and environmental sustainability, while mitigating risks and unintended consequences. This may include leveraging AI for social good initiatives, such as healthcare, education, environmental conservation, and disaster response, as well as developing ethical AI applications that empower marginalized communities, promote democratic governance, and address global challenges, such as climate change, poverty, and inequality.

By embracing ethical innovation and responsible AI development, stakeholders can harness the transformative potential of AI technologies to create a more just, equitable, and sustainable world. Ethical AI design and implementation strategies enable organizations to develop AI systems that respect human rights, promote fairness and transparency, and uphold ethical values. Collaborative approaches to ethical AI development foster interdisciplinary collaboration and collective action, while ethical leadership in AI research and practice ensures that ethical considerations are prioritized and integrated into all aspects of AI initiatives. Moreover, opportunities for ethical innovation and impact offer avenues for leveraging AI technologies to address societal challenges and advance the common good, while upholding ethical principles and values.

Figure 16.1 presents the distribution of the target variable "Perpetrator Race" through two visualizations: a count plot showing the number of occurrences for each race category, and a pie chart illustrating the proportion of each race category in the dataset. These visualizations offer insights into the distribution of races within the dataset, aiding in understanding the demographic composition of perpetrators.

Figure 16.2 displays histograms of the "Year" column segmented by two different categorical variables. Each subplot represents a different categorical variable, with the "Year" distribution stacked by the unique values of the respective variable. These histograms provide insights into the distribution of years across different categories, facilitating analysis of temporal trends within the dataset.

The plot in Figure 16.3 demonstrates the performance evaluation of a machine learning model (Random Forest) on a test dataset. The left subplot illustrates the confusion matrix, providing a breakdown of correct and incorrect predictions across different classes. On the right, the Receiver Operating Characteristic (ROC) curve displays the trade-off between true positive rate and false positive rate, with the area under the curve (AUC) indicating the model's predictive performance.

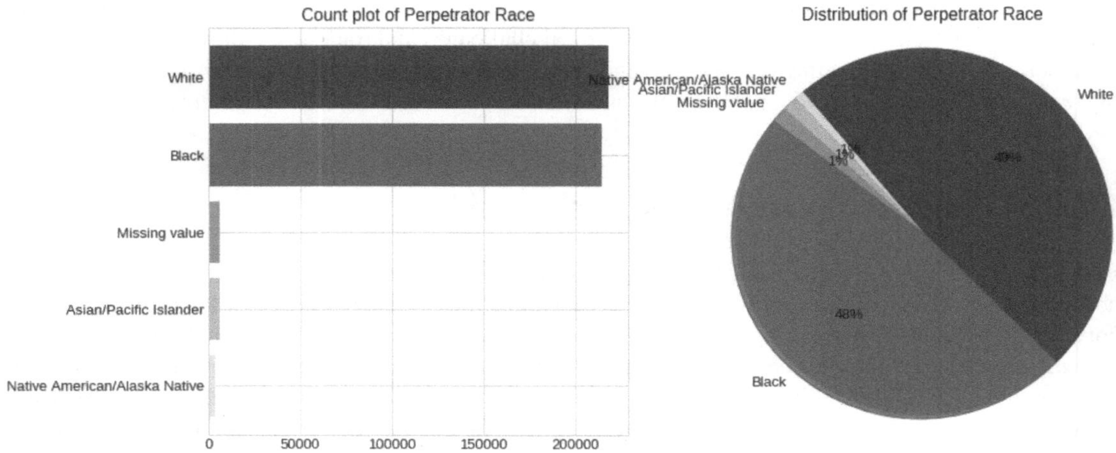

FIGURE 16.1 Perpetrator race.

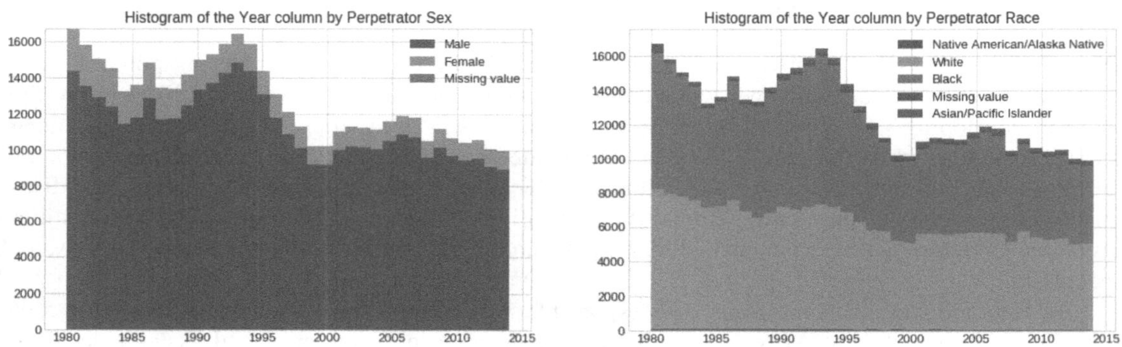

FIGURE 16.2 Two different categorical variables.

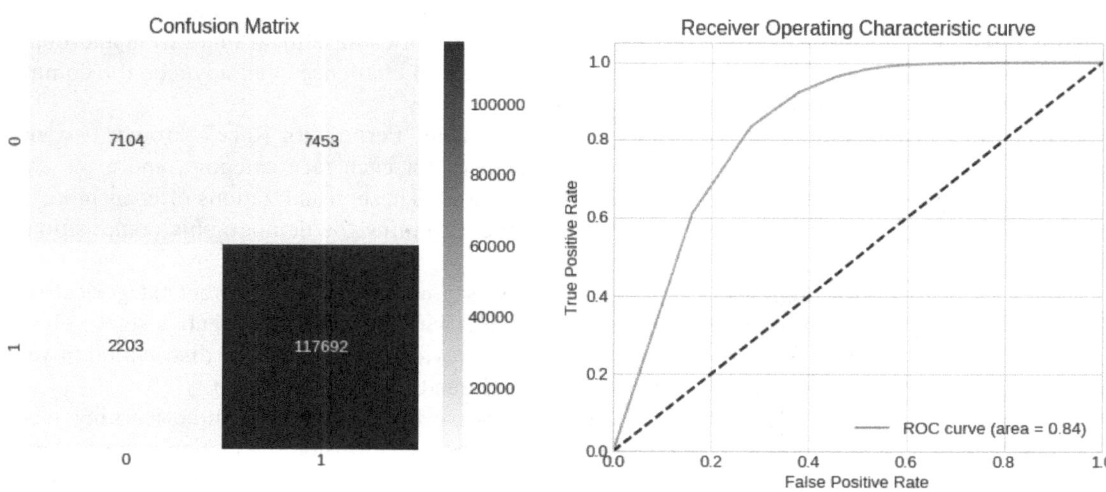

FIGURE 16.3 Performance evaluation of a machine learning model.

TABLE 16.1 Outcomes in the integration of ethical considerations into AI development

QUALITATIVE RESULTS	*DESCRIPTION*
Improved Transparency	Stakeholders reported an increased level of transparency in AI systems, with developers providing clear explanations of AI decision-making processes, contributing to greater trust and confidence among users.
Enhanced Accountability	The implementation of accountability mechanisms, such as audit trails and traceability features, facilitated greater accountability for AI systems and their outcomes, ensuring that developers could be held responsible for any errors or biases.
Ethical Leadership	Ethical leadership emerged as a key factor in promoting ethical behavior and decision-making in AI research and practice, fostering a culture of integrity, responsibility, and social impact within organizations.
Collaborative Ethical AI Development	Collaborative approaches to ethical AI development facilitated interdisciplinary collaboration and knowledge sharing, enabling stakeholders to address complex ethical challenges and achieve more robust and equitable outcomes.
Societal Impact and Ethical Innovation Opportunities	Stakeholders identified numerous opportunities for ethical innovation and social impact in AI, including applications in healthcare, education, environmental conservation, and disaster response, aligning AI development with broader societal goals and values.

The qualitative results described in Table 16.1 indicate significant progress and positive outcomes in the integration of ethical considerations into AI development and deployment processes, fostering greater transparency, accountability, collaboration, and social impact.

DISCUSSION

The discussion on ethical considerations in AI development and deployment underscores the importance of integrating ethical principles and values into all aspects of AI initiatives. Stakeholders have made significant progress in improving transparency, accountability, and collaboration in AI projects, leading to more responsible and socially beneficial outcomes. However, challenges remain, including addressing algorithmic bias, navigating regulatory frameworks, and promoting inclusivity and diversity in AI development teams.

One key aspect of the discussion is the need for ongoing dialogue and engagement among stakeholders to address emerging ethical challenges and ensure that AI technologies are developed and deployed in ways that align with societal values and human rights. Ethical leadership and collaborative approaches are essential for fostering a culture of ethical excellence and driving positive change in the field of AI.

CONCLUSION

In conclusion, the integration of ethical considerations into AI development and deployment processes is essential for building trust, ensuring accountability, and promoting social impact. Stakeholders have made significant strides in enhancing transparency, accountability, and collaboration in AI projects, but there is still work to be done to address ongoing challenges and ensure that AI technologies benefit society as a whole. By prioritizing ethical principles and values in AI research and practice, stakeholders can create a more just, equitable, and sustainable future.

FUTURE SCOPE

Looking ahead, there are several areas for future exploration and action in the field of ethical AI development. This includes further research into algorithmic bias and fairness in AI systems, the development of ethical guidelines and standards for AI governance and policy, and the promotion of diversity and inclusivity in AI research and practice. Additionally, ongoing dialogue and engagement among stakeholders are crucial for addressing emerging ethical challenges and seizing opportunities for ethical innovation and impact in AI technologies. By continuing to prioritize ethics and responsibility in AI development and deployment, stakeholders can build a more ethical and inclusive future for AI technologies.

REFERENCES

1. Anderson, M., & Anderson, S. L. (2007). Machine ethics: Creating an ethical intelligent agent. AI Magazine, 28(4), 15–26.
2. Bostrom, N., & Yudkowsky, E. (2014). The ethics of artificial intelligence. In E. Zalta (Ed.), The Stanford Encyclopedia of Philosophy (Spring 2019 Edition). Stanford University.
3. Floridi, L., & Cowls, J. (2019). A unified framework of five principles for AI in society. Harvard Data Science Review, 1(1), 1–15.
4. Jobin, A., Ienca, M., & Vayena, E. (2019). The global landscape of AI ethics guidelines. Nature Machine Intelligence, 1(9), 389–399.
5. Mittelstadt, B. D., Allo, P., Taddeo, M., Wachter, S., & Floridi, L. (2016). The ethics of algorithms: Mapping the debate. Big Data & Society, 3(2), 2053951716679679.
6. Taddeo, M., & Floridi, L. (2018). How AI can be a force for good. Science, 361(6404), 751–752.
7. Caliskan, A., Bryson, J. J., & Narayanan, A. (2017). Semantics derived automatically from language corpora contain human-like biases. Science, 356(6334), 183–186.
8. Goodfellow, I., Bengio, Y., & Courville, A. (2016). Deep Learning. MIT Press.
9. LeCun, Y., Bengio, Y., & Hinton, G. (2015). Deep learning. Nature, 521(7553), 436–444.
10. Koene, A., Perez Vallejos, E., Carter, C. J., & Statache, R. (2019). Ethical dimensions of AI governance: A research agenda. IEEE Transactions on Technology and Society, 1(1), 70–85.
11. Mingers, J. (2019). The ethics of AI: A literature review. In 2019 10th International Conference on Information, Intelligence, Systems and Applications (IISA) (pp. 1–6). IEEE.
12. Raji, I. D., & Buolamwini, J. (2019). Actionable auditing: Investigating the impact of publicly naming biased performance results of commercial AI products. Proceedings of the Conference on Fairness, Accountability, and Transparency, 77–88.
13. Russell, S. J., & Norvig, P. (2016). Artificial Intelligence: A Modern Approach. Pearson Education Limited.
14. Schwartz, A. I. (2019). Artificial intelligence ethics: A responsibility to build ethically sustainable systems. Issues in Science and Technology, 35(3), 25–28.
15. Selbst, A. D., & Powles, J. (2017). Meaningful information and the right to explanation. International Data Privacy Law, 7(4), 233–242.
16. Smith, M., & Anderson, D. (2001). Privacy, security, and accountability: Ethics, law, and policy. Information Systems Journal, 11(2), 79–94.
17. Taddeo, M. (2019). Ethical AI: A roadmap. In AI Narratives (pp. 31–54). Springer, Cham.
18. Tufekci, Z. (2014). Engineering the public: Big data, surveillance and computational politics. First Monday, 19(7).
19. Winner, L. (1980). Do artifacts have politics? Daedalus, 109(1), 121–136.

Moving Beyond Gender

Recognizing Equality and Inclusivity Across Cultures Through Feminist AI

17

Hitanshi Goel

INTRODUCTION

Artificial intelligence (AI) technologies have advanced expeditiously, enabling computers and machines to carry out an extent of activities imitating human cognitive functions from perception to decision-making [1]. Nowadays, the AI systems undeniably have become an indispensable part of operating systems of our lives in the form of search engines, medical diagnostics, and surveillance technologies. However, there is need to divert our attention to the phenomenal intersection of AI and 'epistemic injustice' which underscores the potential of AI to perpetuate bias and ingrained power dynamics that allow for their gendered discrimination. Epistemic injustice refers to the practices that undermine individuals fairly as knowers or even distort their understanding within the recognized institutions and practices [2]. Such interplays of AI with epistemic injustice have to be realized and attended to in a manner that AI technologies do not further magnify the present biases and injustices but assist in favouring justice and fairness, an inclusive representation of diverse voices.

Though the emergence of AI poses multifaceted consideration, it also provides wide opportunities for the well-being of many, and contributing to the achievement of Sustainable Development Goals (SDGs) [3]. At the same time, it carries intrinsic risks towards boosting differences in economic, cognitive, gender, and cultural domains. Nevertheless, it promises enhanced efficiency and productivity despite the ethical concerns surrounding the deployment of AI and apart from the fact that, being able to mix and exacerbate prejudices, it may help to destroy the environment and open new gaps, compared to the current ones, attacking human dignity by reinforcing prejudices, discrimination, and victimization based on stereotypes [4]. Recognizing the imperative of inclusive decision-making, global frameworks such as the 2030 Agenda for Sustainable Development Goals underline that women and diverse communities have

DOI: 10.1201/9781003497189-17

the right to their participation in all aspects of decision-making processes. However, despite incremental strides, these cohorts remain predominately marginalized within the AI domain in the literal sense of lacking 'active involvement as decision-makers' and this happens even though many ethical questions are raised about the AI system that are bound to have an effect in pivotal areas like governance, social interaction, democracy, legal frameworks, environmental sustainability, data privacy, gender equality, and protection of human rights. The full potential of these technologies can only be realized if these concerns are considered during their development and deployment, based on principles of inclusivity, equity, and ethical integrity. In a world which is evolving rapidly around AI technologies, there is a need to have ethical frameworks and governance that reflect diversity, equity, and inclusion, and thus contribute towards the comprehensive attainment of social justice [5]. AI systems pose new ethical challenges that go beyond its impact on decision-making processes and extend to diverse areas, such as employment dynamics, labour relations, social interactions, provision of healthcare, paradigms of education, media landscapes, access to information, mitigation of the digital divide, preservation of personal data and consumer rights, environmental sustainability, democratic governance principles, due recourse to rule of law, and improvement in security and policing strategies [6]. There is a necessity to include due diligence, even on the part of the technologies themselves, with particular attention to dual-use technologies, and on human rights and fundamental freedoms, such as the right to freedom of expression, privacy, and non-discrimination [7].

AI algorithms have the ability to replicate, deepen, and bolster pre-existing biases in all their profound forms of prejudice, discrimination, and stereotyping. This ability of AI to replicate and fortify the biases is the due to the fact that AI can carry out tasks formerly conceived to be carried out by living beings and often thought to be exclusively a human prerogative. This new capability enables AI to produce and influence human practices, societal structures, and our interrelation with the environment and ecosystems in an unparalleled manner [8]. Critically, AI is the context within which children and adolescents develop their understanding of the world, form identity, evaluate media critically, build information critically, and develop skills in making decisions. In due course, AI might even shrink the gap between human experience and the power of independent endeavour, thereby raising even more concerns over human self-consciousness, the aptitude for interaction within social, cultural, and environmental contexts, autonomous and self-governed intrinsic worth, and inherent dignity [9].

In this regard, a feminist approach to AI governance can allow a hospitable environment for gender equality and diversity inclusion issues at every stage of the AI cycle, right from its conception to its deployment. A feminist lens of AI would, therefore, assist in understanding how an AI system could exacerbate or even normalize the presence of gender-based violence (GBV) and gender-based inequities (GBI) by centring the experiences, perspectives, and priorities of the marginalized genders [10]. This approach may present some strategies that would assure proactive embedding effects of inclusivity, representation, and equity in the design, development, deployment, and monitoring of AI technologies. Feminist AI governance is a new domain of policy, research, advocacy, and development that aims to ensure the governance, design, development, deployment, and use of AI systems should be carried out in just, fair, and inclusive ways. Therefore, feminist AI governance seeks to respond to the power imbalance in the AI ecosystem with an approach that seeks to redress this through a regime guided by structural inequalities, colonial legacies, and multi-layered harms to society [11]. Grounded in feminist values and a critical conception of technology in society, feminist AI governance reconceptualizes that algorithm is not merely a mathematical entity, but a contextual artefact conditioned by social relations and cultural values [12]. Feminist AI governance would ensure that the gendered social, political, and historical context and all other intersectional considerations are interrogated in the development and deployment process of AI systems for harm prevention and to deliver equity. Algorithm assemblage is not to be conceived as a simple sum of heterogenous technologies, but as a configuration that seems to blend mathematical and infrastructural ingredients with social, political, and cultural factors. So, with a much more conceptual identification of the algorithms as sociotechnical assemblages, much beyond their characteristic as abstract entities, it points at the potential ways by which they may reproduce inequality and hold deep societal implications. Algorithmic systems, in other words, do not exist in a vacuum but are rather infiltrated and shaped by the values, assumptions, and biases that appear in their development and deployment. Such systems may serve to reinforce and intensify existing power relations, marginalization, and structural injustices. Thus, any critical review of the social, political, and historical contexts that shape algorithmic assemblages should be more committed to reduction of harm and pursuit of equity.

Feminist AI governance subjects a critical look at the system as a product of power relationships, which is inherently responsive in relation to the economic, historical, socio-political, legal, and cultural context from which it arises and is deployed, so that it exposes their nature as historically and situationally contingent [13]. The systemic or structural approach acknowledges 'harm' to be multidimensional and targets both, the marginalized bodies and territories of the global majority. Starting with the commitment to achieve integral and planetary justice, feminist AI governance broadens our view of the AI system to encompass the whole life cycle of an AI system, rather than limiting itself to simply AI or data pipeline. The experiences, perspectives, and priorities of the most adversely affected by AI systems are centred in order to highlight the modalities through which AI technologies are likely to produce and entrench new forms of oppression, discrimination, and inequality. It surfaces the organizational and technical level strategies that identify ways to embed the principles of inclusivity, representation, and equity in the design, development, deployment, and monitoring of AI. Against such a backdrop of AI, which is understood as a neutral or objective set of tools, feminist AI governance insists on at least paying equal attention to the social, political, and historical horizons within which algorithmic assemblages arise. Systems do not emerge in a vacuum but, in fact, are loaded with values, assumption, and bias that percolate power structures and sociotechnical environments in which they are developed. These contexts need to be critically appraised in the aim of mitigating, if not eliminating, harm that challenges present inequalities and fosters real fairness in how AI technologies are thought of and used.

While redressing the inherent deeply rooted imbalances exclusion in AI ecosystems with a long history and deep-rooted inequities which acted as the major source of both marginalization and exclusion, they must be systematically based on a human rights model. These disturbing realities need to be engaged and dealt across multiple layers such as at individual, communal, institutional, and systemic levels. This will, therefore, substantially help to not only foster real inclusion but also accelerate sustainable development and build peaceful societies, all in alignment with the motives of the 2030 Agenda for SDGs. It should necessarily foreground justice, human rights, and equality at every phase of the conceptualization, development, deployment, and utilization of AI systems to enhance their efficacy, infuse confidence in the operation, and ensure that the resultantly beneficially and adversely implicated persons, and particularly so for those belonging to marginalized groups, can affirm their fundamental entitlement to participate in shaping the trajectory of our societies. This underscores the imperative of ensuring equitable access to technological advancements.

A human rights-based framework 'sheds light on' ways to infuse the principles of equity, inclusion, and substantive representation throughout the AI life cycle proactively. It calls for a critical reflection on how these AI systems may be reinforced towards already existing forms of oppression, discrimination, and marginalization – more often than not – against global majority communities that have traditionally been burdened with more than their share of such ills. This approach further views harm as a multidimensional outcome that is never purely technical or economic. These observances should pay attention to the multifaceted effects that AI systems could bring in various human domains such as social, cultural, environmental, political, and so on.

BIAS IN AI SYSTEMS

Use of AI presents key opportunities to enhance well-being in all categories of people noted above and to achieve the SDGs. However, on the other hand, it may boost discrepancies in the field of economy, knowledge, and influence between genders and cultures. While AI has the capacity to enhance effectiveness and output, it also brings forth significant ethical dilemmas such as being prone to the reinforcement of biases; contribution towards environmental degradation; worsening pre-existing inequalities; marginalization of the vulnerable communities; and ultimately undervaluing human dignity through the vehicle of bias, discrimination, and stereotyping [14]. The AI technologies open up a huge number of opportunities for benefiting people and their nations, while at the same time, they give an essential rise to important ethical dilemmas ranging from the biases that may be perpetuated and, in turn, could also lead even to prejudices and associated injustices, imbalances in digital access, exclusion, challenges of cultural, social, and biological diversity, economic, or social divisions [15]. Other concerns focus on the necessity for transparency and clarity in the functioning of algorithms and the data used in their training, along with their outcomes in terms of respect to human dignity, human rights, gender equality, democracy, socio-economic dynamics, and political and cultural mechanisms. Currently, AI systems are designed, developed, managed, and assessed, without systematic consideration towards diversity and gender balance. Further, according to the 2020 report of GPAI Responsible AI Working Group and The Future Society, even AI initiatives self-identifying as 'AI for SDG' or 'ethical AI' demonstrate a gap in due involvement of the Global South and marginalized communities, such as persons with disabilities, indigenous groups, the LGBTI+ community, people living under the poverty line, and migrants [16].

The 2030 Agenda for Sustainable Development Goals, among other global governance frameworks, is seen as a very solid and coherent tool to contribute to the entitlement of women and diverse communities to take part in all decision-making processes [17]. However, despite these incremental developments, the mentioned communities stay relatively excluded from AI practice and the decision-making process, even when the ethical considerations of AI systems touch upon some of the vital domains such as governance, societal interactions, democratic processes, legal frameworks, environmental preservation, and data security. The consequences of marginalization of the underrepresented community on AI technology advancement and adoption are damaging as that would mean that AI is not able to fully realize its potential for advancing the society [18]. AI technologies possess the capacity to exacerbate prevailing disparities and inequities within and between nations, thus, it is imperative to ensure that principles of justice, confidence, and impartiality are sustained, such that no nation or individual is left out in view of questions of equitable access to AI technologies and their benefits or protection from the adverse effects that might rise. A human rights perspective placed at the centre of AI governance, with a focus on the perspective and voices of the marginalized groups, can help formulate strategies for systematic integration of principles of inclusion, representation, and justice at each and every stage of the AI life cycle. This will thus stimulate the participatory ability of the marginalized communities to shape their priorities, be of concern in the evolution and governance of AI systems. It is important to understand the extensive implications that AI technologies have on various aspects of human existence, particularly for those communities of the global majority that have historically borne the weight of systemic oppression and marginalization. This has resulted in perpetuating the discrepancy not only in the voices but also obstructs the potential of AI to encourage significant advancements towards the realization of SDGs.

Bias in AI systems usually emanates as an unintentional by-product of the biases in society and real-world data not being adequately employed in training the AI system. It is an inborn prejudice in AI models, precisely in 'Deep Neural Networks' (DNNs), since the AI models are just a mathematical representation modelled on the human brain [19]. DNNs inherently have patterns and associations just like humans since they have a lot of layering of neurons, modelled from human or biological neurons. A newly created neural network is just like a 'new born child' that needs to be trained on real-world data so that the network can learn and grow into its ability [19]. As a matter of fact, training at this stage is found to be one of those stages where biases are introduced inadvertently and perpetuated. For example, most of the images portrayed with the word 'nurse' will be feminine, and in the case of 'CEO,' those will be masculine, hence ending up reinforcing gender stereotypes [20]. The design of AI itself is not solely responsible for these biases. It is a consequence of combining the biases living within the data used for model training, not only in terms of the lack of variation or representativeness but all human decisions that were made from data labelling to feature selection, as well as model architectures. These human-influenced decisions can ironically sustain and support biases and blind spots. AI systems are not designed to be partial, but they learn and optimize based on biased input data presented to them by humans and development frameworks. As a result, it reinforces such patterns and associations. This will require proactively finding diverse, representative training data, coupled with critical human bias reviews and remedies, from the design phase through to the development life cycle of the AI and only through such deliberative and proactive processes realization of an unbiased AI can be made possible.

The sense of inclusion goes much further than the integration of what is considered traditionally marginalized groups, it implies breaking down multifaceted forms of discrimination that add to continued isolation or marginalization and are based on arbitrary elements such as age, gender, or skin colour. This, therefore, requires a transition from just treatment to proactive measures in diversity and gender parity, encompassing resource allocation, oversight, and disclosure. This change also means challenging age old power dynamics, including power and resource allocation. Bias in AI can, therefore, occur at any step of its development, from its inception and choice of modelling to data acquisition and processing, up to the context of its application. These biases typically fall into three primary categories: biases related to data, algorithm selection, and deployment [21]. Some of these data biases include measurement biases, which occur in the selection or collection of some features. For example, with predictive analytics of AI, based on height in relation to age might fail to include variations across different genders or ethnicity, a selection bias is witnessed, hence a wrong prediction. Representation bias arises if the training data doesn't adequately represent many of the demographic groups, thus limiting the possibility to derive generalizations from it. In order to overcome this bias, there is a need to collect more data from such under-represented groups which might be a challenge due to privacy concerns. Algorithm selection biases include aggregation bias, where a uniform model fails to accommodate data diversity, such as binary gender models failing to accommodate non-binary identities. Learning bias also occurs in the selection of the model or learning approach when it increases disparities like an AI system disregarding data depending on perceived completeness or validity, and favouring some kind of input since the beginning. Deployment biases include circumstances whereby the AI systems find their way into different contexts from the environment of their development and hence give inappropriate outcomes. Last is the post-deployment feedback bias, which is realized when the process of adjusting models, informed by user input, does not consider demographic diversity and hence leads to new biases.

In today's world, Large Language Models (LLMs) have become public and are being employed in many walks of life by people from around the world, sharing information, explaining ideas, or even helping with some kind of cognitive task. There may be particular challenges posed by their unique structure and applications given the huge scale and complexity they present, combined with the common reuse and repurposing, diverse range of contexts to which they are put, including intricate developmental trajectories. The sheer scale, adaptability, and complexity of developmental paths for LLMs present a huge challenge in handling bias and preventing harm at both individual and societal levels [22]. The intrinsic bias in LLMs, notably in the areas of gender and geographical representation, offers yet more examples of the sheer scale of the challenge that has to be met. Some of the risks associated are erosion of social cohesion, GBV, and the marginalization of disadvantaged communities. To deal with these demands, approaches that include legal, social, or moral solutions besides technological ones are required in order to assure that AI deployment will be fair and responsible. Tackling the inherent inequalities in the AI ecosystems requires a systematic approach firmly based on principles of human rights and social justice which is also in furtherance of the SDGs agenda of 2030.

HUMAN RIGHTS BASED AND ETHICAL AI

The use of AI technologies offers a range of possibilities in the direction of maximizing human welfare and at the same time opens doors for the nations to gain enormous benefits, but this breakthrough comes hand in hand with profound ethical challenges and call for their careful consideration and tackling [23]. The most relevant issue to follow is the fact that AI systems tend to integrate and worsen biases, which can ultimately result in discrimination, inequalities, digital divides, exclusionism while threatening the cultural, social, and biological diversity and the economic disparities. In addition, there is a need for transparency and clear insight on how the algorithms work, including the data used for training. AI systems may affect several core aspects of human life, such as human dignity, human rights, gender equality, democracy, socio-economic behaviour, political structures, cultural traditions, scientific endeavours, environmental protection, animal rights, and ecological systems. All these effects of AI on society, environment, ecosystem and even on human welfare, including psychological dimensions, are huge and continuously developing, needing a comprehensive examination. Given the current innovative ways in which AI is influencing human cognition, the patterns by which humans are interacting, making decisions, educational techniques, and their influence on scientific exploration, creative cultural production, and even information dissemination are particularly important [24].

Proper utilization of the AI systems will depend on transparency and explicitness in explaining AI capabilities and its limits so that people know when and to what extent AI is included in making the decisions. This aspect is particularly relevant as it will ensure that these are adapted in the right context, such as in cases of potential conflicts with principles of privacy, safety, and security. Thus, AI stakeholders need to actively promote social fairness, impartiality, and non-discrimination by taking an inclusive approach that would make the benefits of AI become pervasive [25]. Since biased data is potentially the source of sustained social injustice in AI ecosystem, urgent reforms in data curation practices, particularly based on machine learning, are the need of the hour. The risk, of course, is that without proper legal and ethical framework, decades of progress in human rights and civil liberties may be put to risk. Systems promoted by discriminatory tendencies through AI have a history of putting marginalized sections at line. It is a trend derived from bias in representation, classification, and categorization of individuals in data for ages. There is a critical need to foster ethics among technology developers. These serious implications of design flaws in AI systems, and their innate bias leading to unfair discrimination, have rebooted age-old debates about social fairness and surveillance based on racial and gender identity. Incidents have arisen in which AI algorithms differentially treat individuals on issues of race and gender, pointing at the fact that AI can potentiality amplify discriminatory practices. There is an absolute need for a thoroughgoing overhaul of the approaches to curating AI data [26].

Intersectionality and AI: Reimagining Ethical AI

Intersectionality is a theoretically critical lens to understand the complex web of interconnectedness and interdependencies across diverse social categories and systems. It provides a mindset and lexicon that enables us to understand how the varied members of certain groups navigate their experiences in different contexts [27]. Intersectionality tries to elaborate the dimensions of ethnicity, class, sexual orientations, among many other social locations take place in such interconnections in a unique and diverse manner and thus, present the complex nature of social justice and injustice. The sensibilities developed through intersectional differences and augmented by our awareness, increases the possibility of breaking down institutional barriers and ultimately reaches full potential in doing socially transformative work. Therefore, intersectionality should be perceived as a seminal notion where knowledge becomes situated within definite environmental placements whilst reflecting the current socio-political and economic power structures [28]. Coined by Kimberle Williams Crenshaw in 1989, one of the leading proponents of critical race theory and the civil rights movements, 'intersectionality' delves into the complex interrelations of different social identities and their respective systems of oppression or discrimination. This framework illustrates how dynamic categories, that is, gender, race, socio-economic status, ability, sexual orientation, religion, caste, age, or nationality intersect to actually inform individual experience at any given time. The theory fosters discussion on both inter-group and intra-group differences while reflecting intersectionality of diverse oppressions in individual lives. The concept of intersectionality derives from 'Black feminist' thought but offers a fuller, more nuanced understanding of the manifold complexities of identity and power [29]. The theory of intersectionality brings to light better understandings of international human rights, and goes a long way in outlining how certain constructs of identity, like race and gender, combine to mould specific experiences of discrimination and persecution. This is especially important in the context of today's human rights, whereby acts of sexual violence in armed conflict are often targeted against individuals on grounds of both gender and ethnicity. In this regard, the failure to recognize human rights abuses and their intersectional dynamism, therefore, diminishes the remedies that accrue to the victims. The historical context from which the intersectionality theory evolved as a conceptual framework finds its roots in the critical race, feminism, and human rights jurisprudence from the scholarly literature review. The idea of intersectionality has been used in the broad domain of critical feminist understandings in legal, political, and sociological areas for an expression of complexity between the oppressions of multiplicity and mutual constitutions. It illuminates these identities and systems of oppression in ways that are both manifold and interrelated by providing a nuanced and contextually sensitive approach to tackling issues of social justice and inequality [30]. However, in a time when increased influence of AI over societal constructs and individual life is dominating, it becomes imperative to critically look at the power relations and biases sustained in these AI technologies. In direct contrast, the theoretical paradigm of intersectionality provides a strong lens through which we could envision AI, with the recentring of institutional power dynamics and also reflecting perspectives from queer, feminist, and critical race theories.

Crenshaw, in her theory of 'intersectionality,' tries to explain how various social identities and its related systems of oppression cross more than one axis simultaneously [29]. It raises a call towards deeper social realities, at the base of oppression facilitated by AI as subgroup fairness not merely a technical issue. Intersectionality rejects that there is inherent inequality in society and suggests, on the contrary, that social inequalities are systematically designed and perpetuated. Despite having a strong democratic potential, digital technologies most often come with algorithmic systems. Pre-existing biases would only be further exacerbated by digital technologies, which may be masked as cultural misinterpretation of AI as objective truth, though human intervention is involved at every stage. A case in point is COMPAS; a US court algorithm that predicts recidivism. It showed discrimination against black people since it marked them as likely future criminals far more, compared to their white counterparts [31].

Such an intersectional lens applied to AI fairness pushes us past just rolling out equity constraints on models and dig deep into what dynamics are at play in society with AI inflicted oppression and how we empower them to fix the technology. This perception shows the deep-rooted social inequities and their perpetuation through centres of influence. That power within AI still lies in human choice, system design, data collection, deployment, and interpretations of fairness that would necessarily determine who gets what at the community level with the intersection of various domains [32]. Discussion on power thus becomes indispensable at every point of the AI process. So, fairness research of AI should be important in developing just systems. An intersectional approach precludes one from restricting themselves to the first layer of making models technically fair. Rather, what is required is a very thorough analysis of the social facts behind technology driven oppression and, thereafter, help us improve the technology accordingly. This approach emphasizes the interconnections that exist between power dynamics attached to different kinds of identity and social positionalities, such as race, class, gender, among others and premises the fact that power systems are mutually reinforcing and continuously perpetuate unfair outcomes and distinct social experiences within these hierarchies. Identifying these centres of power and analyzing the way they interact gives one a way into how to dismantle inequities [33]. In the traditional AI fairness models, paradigms such as equality of opportunity, intended to neutralize biases to have equal gains across the considered sub-groups, do not guarantee substantive fairness. This is because most AI algorithms will simply reflect and propagate the systems of intersecting oppressions in the input data. They argue that realizing substantive fairness in algorithms requires actively challenging and even destabilizing the very systems of domination that are producing biased algorithms in the first place [34].

Therefore, strong fairness is AI, which requires radically different assumptions about the role of algorithms would help attain a robust mechanism. It would take the approach that questions the systemic oppression and moves towards building a more just society, rather than a reactive approach involving applying 'debiasing' after the fact [33]. This will mean that AI fairness transcends technical complexities; rather, it is a socio-political quagmire that demands an intersectional understanding of power dynamics and systemic injustices, including the implications that underpin algorithmic decision-making [35]. Intersectionality emerges as a potent framework for deciphering and reshaping the entrenched power dynamics inherent in AI systems. But if we apply critical, intersectional frameworks to engagement with AI systems, then maybe we can draw connections across axes of identity, and in so doing, imagine new AI paradigms that really do disrupt the patterns of marginalization and 'othering' that are repackaged and engineered into these technologies [27]. Intersectionality places us to be in a position to unpack and dismantle the very workings through which AI tends to reproduce power hierarchies and systems of oppression, hence setting a footing towards imagining AI systems that serve equity, inclusion, and social justice. This perspective will serve in bringing out how an intersectional approach to AI would make use of fluid identities, take stereotypes to court, and so use the instability to redefine gender roles and individual agency. And that is exactly where the importance of intersectionality comes in, to be able to negotiate critically the complex power structures and biases literally inhabiting these technologies at the time of design and deployment of AI.

FROM BIAS TO FEMINIST APPROACH: TOWARDS AN EQUITABLE AND INCLUSIVE AI

In an era where the rapid proliferation of AI systems is taking place, critically interrogating how these technologies come to intersect with the deepening prevailing power dynamics, disparities, and structures of oppression is of paramount importance [36]. Yet, working in the space of AI ethics, rife with issues related to prejudice, partiality, and justice, has faced challenges in translating theory into practice. Within these gaps, feminist theory comes up as an illuminating framework to hold the AI sector responsible for the way the design and implementation of AI contribute to good rather than bad adjustments to support and undermine the pursuit of social justice [37]. In feminist theory, AI is used in studying, describing, and challenging present inequalities and power dynamics, bringing integration to intersectional feminist theory and AI [38]. This is quite critical, especially at a moment when digitized tools are mostly viewed as a proper panacea for societal problems, more so without paying closer attention to the inherent prejudices and predispositions within such technologies. Thinking about and applying new technologies is something that engages preferences and choices, but it also opens the chance to shape them in a way to not do harm and change their ways to maximally advance transformative, positive potentials of society.

From the feminist AI paradigm, knowledge as resulting from data-driven processes appears to be a type of mentality that is shaped by various kinds of societal perspectives, including those of gender and ethnicity, among others [39]. This consideration has particular pertinence to the ethical data management for AI when taken into realization that machine learning algorithms are by nature able to absorb and reinforce such perspectives since they have to determine whose perspectives are being reflected inside a dataset. In this regard, such a constructivist view of knowledge is well adopted by both intersectional and post-structuralist feminisms to enlighten the nexus between cognition and influence. Ethically, the responsible data handling for AI involves asking for the perspectives of all who are involved in the multistage process of training data preparation; structuring datasets, generating data, devising the labels and their representation, annotation, and labelling [40]. Each layer of predisposition harbours the potential to be assimilated and perpetuated by a machine learning algorithm.

The feminist epistemological approach to ethical data curation is an approach grounded on feminist theory that provides a framework to look at the theoretical underpinnings and concepts that inherently are part of AI data. Building on the feminist theoretical framework, an ethical framework is built to meet discrimination and bias in AI data. Seen from this perspective, the biases that AI perpetrates are not new but rather continuities of existing patterns of gendered and cultural disparities that are being entrenched and amplified by digital technologies. Feminist AI points out that the reason for algorithmic bias is not unconscious sexism or racism of programmers or even the training data filled with errors. Instead, it comes from the techno-culture systematically not allowing the influence and power of women and people on the margin [41]. This view tries to explain the structural mechanisms that are in general part of the structural injustice per se, which is fashioned by the social hierarchies underlying opportunities and life chances, both in terms of material and social resources, of persons within those hierarchies.

One of the biggest problems with AI bias and discrimination is the traceability of identifying who is at fault for the injustices carried out by such systems. The injustice of the structure is different from fault-based injustices, in the sense that though the latter is traced to identifiable agents or actions, there is an untraceability that characterizes structural injustice; it is very difficult to find who is to be blamed [18]. That does not, however, let the AI sector off the hook, but it throws into the limelight the urgency of the need for a very fundamental shift in how AI development and deployment issues are handled [42]. Feminist AI is the comprehensive, equal approach that operates towards the creation of a more even world. The term 'feminist' also goes further, in the case of gender parity, which does, of course, remain a salient aspect but extends itself to include harnessing AI for the reduction of inequalities experienced by traditionally marginalized societal groups based on race, culture, and identity intersections. Implied in this concept is the recognition that the inequities come from cultural diversities, and in so doing, recognize distinctive traditions and values characteristic of a region [43]. It is, therefore, the systemic biases that are characterized by gender, race, social status, linguistics, and other intersectionalities that tend to permeate the current artificial intelligence processes of being emanating from the Global North and subsequently replicated in the Global South. To fight and correct these prejudices and discriminatory practices through analysis and proposals informed by a feminist, decolonial, and contextualized perspective is a need [36]. They have turned into an absolute necessity to give rise to another alternative vision that answers to the challenges developed by the regions of the Global South.

It is therefore important, for this perspective of feminists to be drawn into the design and development of AI. Feminist AI provides a just, equitable and inclusive framework to finally critique and challenge a techno-culture that has been, for too long, the contributor to the exclusion of marginalized communities [39]. This technology can move down the road towards ethical data curation, anti-racist, and anti-discriminatory development of algorithms with only feminist epistemologies and intersectional approaches as sidekicks. In this light, feminist AI might indeed offer an interesting reading lens with which to look, in a critical way, at the dominating stories and practices of technology that systematically have excluded, in a systemic way, women and minoritized groups from the possibility of leadership and decision-making roles. This perspective foregrounds the structural processes and power dynamics forming the context of systemic injustice, being inherently intersectional and mediated through the interplay of social positions that regulate opportunities, resources, and life prospects for different individuals and collectives.

Integrating feminist theories, knowledge, methodologies, and epistemologies with AI might show us how such technologies could exacerbate, reproduce, or otherwise challenge current inequalities and power relations [38]. Such an approach acknowledges that the data carried is constructed from particular social perspectives, influenced by variables such as sex, ethnicity, socio-economic status, and cultural milieu. From an ethical perspective, managing data in AI requires acknowledging and admitting the parties that data may affect in terms of misrepresentation or exclusion. Utilizing a feminist epistemological approach in data curation means applying the principles of feminist theory to develop ethical frameworks which would guard against discrimination and biases in the data used for AI systems [40]. This involves thoughtful consideration of the layers of perspective involved in the data processing, including shaping the structures of datasets, generating data, defining labels, and annotating information. Any bias within such datasets runs the risk of being absorbed by machine learning algorithms and perpetuated, further damaging systemic inequalities.

Feminist AI aims at the realization of a global movement, with special pull from the Global South, offering alternative imaginations and solutions towards realization for equity, social justice, and environmental sustainability [39]. This approach unveils that AI-perpetrated injustices are not new; rather, they are an uncovered and increased intensification of pre-existing patterns of gendered, racial, and cultural inequities encoded within digital technologies. Together with critical reflection, methodological innovation, and experimentation, feminist AI is committed to contributing in ways that favour the rise of inclusive and equitable technologies that area residents face in every corner of the globe.

To this end, AI researchers should deploy an approach that places a feminist perspective to ensure that the technologies propagating existing prejudiced-based technologies and social stratification are held accountable. This means more than just increasing the representation of women in the field, although such efforts are important in their own right. It calls for a full reassessment in the way data is not just gathered and judged but also how AI is tested and used. A feminist AI approach, might also reflect on and understand how individuals from different intersecting axes of identity channelize their experience within this system of oppression. For instance, the experience of a black woman with AI will entirely be different from that of a white woman, and analysis has to be intersectional to capture this. If these built-in biases are not questioned during the creation and deployment of AI, they will further the existing inequities, especially of women and people of colour. For building AI systems that are equitable and effective, it is the social justice approach with intersectionality that becomes utmost necessary in bringing down the oppressive structures [27]. A feminist perspective is bound to help the researchers in developing and adopting AI technologies that confront head on with the systematic biases, and at the same time become inclusive with the distinct needs and experiences of all communities. Not until there is a concerted effort to infuse feminist principles at every stage of AI development, will we be able to harness the real potential of AI's transformation while guarding against the perpetuation of harmful inequalities.

RESPONSIBLE AI

In a world where artificial intelligence technologies are penetrating fast into society at every aspect, the demand for responsible and ethical AI practices has become imperative. Responsible AI is an approach towards a human-centric style that builds users' trust in the ethical processes of decision-making [44]. What this simply means is that the automatic determinations are fair, just, impartial, based on good-will intentions, not discriminative, and in conformance with societal regulations and norms. Transparent and explainable AI systems that securely operationalize its use make sure that the objective of responsible AI, met with the aim of allowing users to understand decisions made by AI systems while protecting the privacy of the user [45]. Of high importance is the issue of trust for any AI application to be accepted and taken up in society; therefore, every stage of its development and deployment has to follow through with trust as one of its key objectives. It remains one of the serious challenges; nevertheless, how salient this may be-trustworthy AI has not been fully appreciated. One of the principal lines along which AI ethical issues cut is that of fairness and the tendency to perpetuate vicious biases.

Ethical AI and Fairness

Thus, transparent and accountable AI inherently links the two to the needs of justice, fairness, and equity; that is, just, fair, and equitable distribution of power, access, opportunities, and outcomes. The problem of AI explainability is thus quite a paramount ethical issue [46]. In this direction, the right to explanation enshrined as part of the General Data Protection Regulation (GDPR) goes in the direction to make AI more ethically acceptable, which involves direct impacts on people from decisions by the AI system. Most

AI systems rely on historical data and, in the process, unconsciously pave the way for biases [47]. This points at the need for techniques that can detect and mitigate bias to be incorporated at the development life cycle of AI [48]. These strategies identify and mitigate biases systematically, either in data or algorithms, so that they do not propagate into discriminatory outcomes by the AI systems. Not only do such methods advance equity and ethical decision-making, but they also assist in conformity to the regulations that require impartial outcomes in any decision-making process and therefore promote trust amongst the interested parties.

Privacy and Security in AI

The assurance with AI relies mostly on the privacy and security of the methodologies used. Ensuring such privacy and secure AI methodologies could reduce the risks attached, which are the cause of private data exposure and privacy infringements leading to unauthorized use. Data privacy is responsible for proper handling, processing, storage, and usage of personal information that maintains and ensures all the rights of individuals in relation to their personal data [49]. This, therefore, makes data security a prerequisite to the same. More focus should be given to ensuring the privacy resilience of AI systems increasingly by guaranteeing the safety and preservation of persons' privacy while allowing the collection and use of data. This may include obtaining user consent prior to any practices of data collection or usage, to ensure the protection of sensitive user information from unauthorized access and misuse, including data anonymization and minimization, and ensuring that privacy risk management is woven into each part of AI development and deployment [50].

Interpretability and Explainability

Obfuscation, the former quintessential characteristic of the AI sector, is a very traditional way for firms to safeguard their trade secrets not only from the highly competitive market but also from the inner machinations of their algorithms. However, it is above and beyond the issue of respect for intellectual property; there is a new requirement arising with respect to AI. That is, systems of 'explicability' and 'interpretability' of AI, especially those based on deep learning, since most of them are already incomprehensible in explaining intelligible mechanisms to humans [46]. And this is what people refer to as the 'black box' of neural networks, wherein data inputs are understood, of course, and outputs are understood, but not the algorithmic determinations themselves. We are increasingly deferring to algorithmic decisions without an understanding of the reasoning behind them. We are therefore eroding our capacities of thinking and judgment over our own norms and assumptions. In that regard, there remains a huge incentive for us to rely on algorithms since they have emerged to be core aspects of human welfare. However, the ethical approach, oriented towards human well-being, demands transparency and interpretability [50].

The establishment of trust and confidence in the outputs from AI is central to the explainable and interpretable AI approach, which advocates for transparency and, most importantly, lucidity in the AI operations [44]. The latter refers to the preference that AI systems be transparent and interpretable to the extent that the decision processes are understandable for the users, consumers, developers, watchdog groups, and other involved external parties. "Developing AI systems that do not only make clear determinations but also offer intelligible elucidations can aid users to comprehend the rationale underpinning AI-based decisions."

Anthropocentric design practices also employ explicability techniques: feature significance analysis, model visualization, and rule-based models, allowing increased explicability and interpretability. In such a context, it may also serve as the right way to balance transparency with 'need to know' access and allow trust to be bred, assure compliance with evolving privacy regulations, and get the most quality out of the outcomes.

Governance and Social Good

The perils and prospects of AI have culminated in an ecosystem replete with a multitude of initiatives that endeavour to provide guidance on how AI should be developed, adopted, and utilized to advance the 'AI for Social Good' agenda. "These initiatives can be categorized into three broad categories:

1. **AI and Ethics:** Ethical frameworks and guidelines promoting and fostering responsible AI.
2. **AI and Governance:** Governance mechanisms operationalizing responsible AI, including auditing mechanisms, risk assessments, standards, certifications, corporate governance frameworks, and more.
3. **AI and Social Good:** Applied projects advancing the Sustainable Development Goals (SDGs) responsibly." [16]

Such efforts are usually indicative of the urgency that AI must be of social benefit, with the benefits of AI 'markedly' outweighing its risks and the ways in which misuses of this technology are forecasted. They support AI systems that are both designed to engender and reproduce inequitable biases, are safe and tested, are responsible to human beings, and designed per the privacy design principles. They also need to meet high scientific excellence standards and their availability for use should be according to the ethical tenets. The needs for socially beneficial AI applications are immense and provoke various quandaries, especially the contribution to attaining the SDGs by 2030 [51]. AI technologies, on the other hand, have reached a very mature stage and more and more infiltrate an immense diversity of applications across the industrial and economic sectors, cultural and artistic domains. Responsible development and deployment of AI become essential in drawing its potential for solving world challenges with the least possible detriments.

Responsible AI deals with many dimensions; it is more of a multipronged and holistic practice in ethical decision-making, fairness, privacy, and security, interpretability and explainability, governance, and social good [44]. Prior to that, for the prioritization of ensuring trust, transparency, and accountability across the entire AI development life cycle, it is recommended that due consideration be given to ways that help ensure stakeholder confidence and provide reasonable assurance that AI systems contribute positively to their respective organizational objectives and to the general societal well-being. This includes extensive risk assessments and risk mitigation strategies that focus on AI algorithm transparency, embedding transparency and accountability in the AI practices, and putting strong controls in place for privacy [45]. Finally, the integration of bias detection and mitigation techniques is of paramount importance to ensure that the perpetration of such discrimination is obviated for AI systems to operate fairly. In this still-developing landscape, therefore, it calls for precise legal and regulatory frameworks defining the role and duties of "trustworthy AI" and "responsible AI" [49]. These should be well spelled out in the realization of not only the trust but also the responsible AI, with ethical and transparent decision-making that advances accountability and the good for the society. Embracing this approach would guarantee that, on the one hand, societies could benefit to the maximum from the revolutionary impacts that such technologies harbour, and on the other hand, save them from attendant risks that could bring serious injury to the principles of social justice, equity, and human well-being.

CONCLUSION

Prejudice and discriminatory practices have constantly been a consistent quandary in the development and deployment of artificial intelligence technologies. Making it worse is the fact that biases in AI systems do, in fact, represent a real threat to equality, inclusiveness, and social justice-from furthering negative stereotypes to reinforcing those that exist. This challenge goes beyond the call of ethical imperatives. One can easily say that the very concept of intersectionality is the best way to understand the many dimensions of oppression and marginalization. Such a framework is based on the understanding that there are several social identities, operating at many levels, most often intersecting and simultaneously shaping human experiences within given systems of power. So, within the context of AI, it is at this juncture that intersectionality comes in to help us understand how such multiple interacting identities lead to systems and structures of discrimination that may be different or compounded and that undergird the pressing urgency of inclusive and culturally responsive technologies. Feminist AI is built on the tenets of intersectionality to provide a critical framework which begins to answer the systemic biases and inequalities that have marked the history of technological innovation. Each of these moves in Feminist AI questions the dominant techno-culture, which has always put away women and people with marginalization from the table of leadership and decision-making by the participation of feminist epistemologies and situated knowledges. This requires, above all, a paradigm shift in the collection of data, design of algorithms, and testing and deployment of AI systems where the voices of those historically at the margins are centred. It identifies that reaching substantive justice in AI will indeed require more than mere ‘debiasing’ of algorithms; it has to be a proactive way that is lively in actively challenging oppression and making society fairer across various cultural contexts. This intersectional and feminist approach commensurate with the principles of Responsible AI, which is an ethical paradigm of the decision-making process in the development of AI, focusing on fairness and social good. This fact underlines the importance of risk assessment, a mitigating strategy, transparency in AI algorithms, and measures for accountability to ensure that AI systems contribute positively towards the achievement of organizational objectives and add to the well-being of society. The guiding lights of Feminist AI and intersectionality, integrated within the framing of Responsible AI, might give a glimmer of hope towards this, otherwise densely complex area of research, where identity and its intersections with power and oppression hold sway.

Ultimately, the ethical development of AI is responsible by not only a moral duty, but it is a prerequisite step towards Sustainable Development Goals. AI has the potential to be potent tool in aiding positive changes, innovations, and leading progress in healthcare, education, environmental protection, and economic development. However, to make full use of this potential, the need of the hour is to evolve an integrated and holistic approach that takes into dimension these dimensions, feminism, and responsible practices at all stages of AI development and deployment life cycle. A reflective, transparent, and accountable ethical environment fosters AI practices first to the fore of social justice, equality, and the well-being of humankind and, thereby, benefits humanity and the planet together. This opens the window to truly unlock the potential of AI and potential that may facilitate more than mere transformation: it may indeed facilitate power that will uplift and celebrate the very rich diversity of human experiences and cultural expressions.

REFERENCES

1. G. Chaudhary and A. Nidhi, “Artificial intelligence and blockchain: A breakthrough collaboration in IP law”, *J. Intellect. Prop. Rights*, vol. 28, no. 5, pp. 383–391, 2023.
2. M. Sardelli, “Epistemic injustice in the age of AI”, *Aporia*, vol. 22, no. XXII, pp. 44–53, 2022.
3. “Artificial Intelligence: examples of ethical dilemmas.” UNESCO: Building peace through education, science and culture, communication and information. Accessed: Mar. 20, 2024. [Online]. Available: https://www.unesco.org/en/artificial-intelligence/recommendation-ethics/cases.
4. “Bias in algorithms – artificial intelligence and discrimination”, European Union Agency for Fundamental Rights, Vienna, TK0922347ENN, 2022.5. “What about fairness, bias and discrimination?”, Information Commissioner’s Office, United Kingdom, Guidance on AI and Data Protection, Mar. 2023. Accessed: Mar. 20, 2024. [Online]. Available: https //ico.org.uk/media/for-organisations/uk-gdpr-guidance-and-resources/artificial-intelligence/guidance-on-ai-and-data-protection-2-0.pdf.
5. Google AI. Accessed: Mar. 16, 2024. [Online]. Available: https://ai.google/static/documents/responsible-development-of-ai.pdf.

6. P. Ricaurte and T. d. Monterey. "How can feminism inform AI governance in practice?" UNESCO: Building peace through education, science and culture, communication and information, Mar. 15, 2024.
7. G. Chaudhary, "Artificial intelligence: The liability paradox", *Indian Law Inst. Law Rev.*, vol. 1, pp. 144–162, 2020.
8. I. Bartoletti and R. Xenidis, "Study on the impact of artificial intelligence systems, their potential for promoting equality, including gender equality, and the risks they may cause in relation to non-discrimination", Council of Europe, Aug. 2023.
9. J. Manyika, J. S. Silberg and B. Presten. "What do we do about the biases in AI?" *Harv. Bus. Rev.*. Accessed: Mar. 20, 2024. [Online]. Available: https://hbr.org/2019/10/what-do-we-do-about-the-biases-in-ai.
10. J. Wajcman and E. Young, "Feminism Confronts AI: The Gender Relations of Digitalisation", in *Feminist AI: Critical Perspectives on Algorithms, Data, and Intelligent Machines.* Oxford Univ. Press, 2024, pp. 47–64.
11. W. Philpott. "What would it mean to have feminist AI? | Waterloo News." Waterloo News. Accessed: Mar. 17, 2024. [Online]. Available: https://uwaterloo.ca/news/arts-research/what-would-it-mean-have-feminist-ai.
12. Z. Liu, "Sociological perspectives on artificial intelligence: A typological reading", *Sociol. Compass*, vol. 15, no. 3, Feb. 2021. Accessed: Mar. 8, 2024. [Online]. Available: https://doi.org/10.1111/soc4.12851.
13. Y. Kong. "(Un)Fairness in AI: An intersectional feminist analysis | Blog of the APA." (Un)Fairness in AI: An intersectional feminist analysis. Accessed: Mar. 30, 2024. [Online]. Available: https://blog.apaonline.org/2022/08/10/unfairness-in-ai-an-intersectional-feminist-analysis/.
14. J. Castaneda, A. Jover, L. Calvet, S. Yanes, A. A. Juan and M. Sainz, "Dealing with gender bias issues in data-algorithmic processes: A social-statistical perspective", *Algorithms*, vol. 15, no. 9, p. 303, Aug. 2022. Accessed: Mar. 20, 2024. [Online]. Available: https://doi.org/10.3390/a15090303.
15. M. Ghallab, "Responsible AI: Requirements and challenges", *AI Perspectives*, vol. 1, no. 1, Sep. 2019. Accessed: Mar. 16, 2024. [Online]. Available: https://doi.org/10.1186/s42467-019-0003-z.
16. N. Haitas, "Towards real diversity and gender equality in artificial intelligence: Advance report", the global partnership on artificial intelligence, Nov. 2023.
17. "The 17 goals, sustainable development." Home, sustainable development. Accessed: Mar. 15, 2024. [Online]. Available: https://sdgs.un.org/goals.
18. J. Browne, "AI and Structural Injustice: A Feminist Perspective", in *Feminist AI: Critical Perspectives on Algorithms, Data, and Intelligent Machines.* Oxford Univ. Press, 2024, pp. 328–346.
19. X. Ferrer, T. Nuenen, J. M. Such, M. Cote and N. Criado, "Bias and discrimination in AI: A cross-disciplinary perspective", *IEEE Technol. Soc. Mag.*, vol. 40, no. 2, pp. 72–80, Jun. 2021. Accessed: Mar. 30, 2024. [Online]. Available: https://doi.org/10.1109/mts.2021.3056293.
20. E. Drage, K. McInerney and J. Browne, "Engineers on responsibility: Feminist approaches to who's responsible for ethical AI", *Ethics Inf. Technol.*, vol. 26, 2024, Art. no. 4. Accessed: Mar. 18, 2024. [Online]. Available: https://doi.org/10.1007/s10676-023-09739-1.
21. Z. Chen, "Ethics and discrimination in artificial intelligence-enabled recruitment practices", *Humanit. Soc. Sci. Commun.*, vol. 10, no. 1, Sep. 2023. Accessed: Mar. 20, 2024. [Online]. Available: https://doi.org/10.1057/s41599-023-02079-x.
22. "Challenging systematic prejudices: An investigation into bias against women and girls in large language models", UNESCO, International Research Centre on Artificial Intelligence, CI/DIT/2024/GP/01, 2024.
23. "Areas for future action in the responsible AI ecosystem", The Future Society, Dec. 2020.
24. V. Dignum, "Responsible artificial intelligence -from principles to practice", *ACM SIGIR Forum*, vol. 56, no. 1, pp. 1–6, Jun. 2022. Accessed: Mar. 15, 2024. [Online]. Available: https://doi.org/10.1145/3582524.3582529.
25. R. Altenburger and R. Schmidpeter, *Responsible Artificial Intelligence: Challenges for Sustainable Management.* Springer Int. Publishing AG, 2022.
26. E. Drage and K. Mackereth, "Does AI debias recruitment? Race, gender, and AI's "Eradication of difference", *Philos. & Technol.*, vol. 35, 2022, Art. no. 89. Accessed: Mar. 18, 2024. [Online]. Available: https://link.springer.com/article/10.1007/s13347-022-00543-1#citeas.
27. A. Ovalle, A. Subramonian, V. Gautam, G. Gee and K.-W. Chang, "Factoring the Matrix of Domination: A Critical Review and Reimagination of Intersectionality in AI Fairness", in *Conf. AI, Ethics, Soc.*, Canada, Aug. 10, 2024. New York: Assoc. Comput. Machinery, pp. 496–511.
28. H. Al-Faham, A. M. Davis and R. Ernst, "Intersectionality: From theory to practice", *Annu. Rev. Law Social Sci.*, vol. 15, no. 1, pp. 1–25, 2019.
29. K. Crenshaw, "Demarginalizing the intersection of race and sex: A black feminist critique of antidiscrimination doctrine, feminist theory and antiracist politics", *Univ. Chicago Legal Forum*, vol. 1989, no. 1, 1989.
30. D. Atewologun, "Intersectionality Theory and Practice", in Oxford Research Encyclopedia, Business and Management. 2020. Accessed: Mar. 17, 2024. [Online]. Available: https://deltaalphapsi.com/wp-content/uploads/2023/06/Intersectionality-Theory-and-Practice.pdf.

31. F. Lagioia, R. Rovatti and G. Sartor, "Algorithmic fairness through group parities? The case of COMPAS-SAPMOC", *AI & Soc.*, Apr. 2022. Accessed: Mar. 31, 2024. [Online]. Available: https://doi.org/10.1007/s00146-022-01441-y.
32. J. Buolamwini and T. Gebru, "Gender Shades: Intersectional Accuracy Disparities in Commercial Gender Classification", in *Proc. 1st Conf. Fairness, Accountability Transparency*. Proc. Mach. Learn. Res., 2017, pp. 77–91.
33. J. Bond, *Global Intersectionality and Contemporary Human Rights*. Oxford Univ. Press, 2021. Accessed: Mar. 18, 2024. [Online]. Available: https://doi.org/10.1093/oso/9780198868835.001.0001.
34. R. Khan. "We need a global feminist campaign against artificial intelligence bias." We need a global feminist campaign against artificial intelligence bias. Accessed: Mar. 20, 2024. [Online]. Available: https://blogs.lse.ac.uk/medialse/2023/09/07/we-need-a-global-feminist-campaign-against-artificial-intelligence-bias/.
35. S. Ciston, "Intersectional artificial intelligence is essential: Polyvocal, multimodal, experimental methods to save AI", *CITAR J.: Special Issue*, vol. 11, no. 2, 2019.
36. "Feminist reflections for the development of Artificial Intelligence." Derechos Digitales. Accessed: Mar. 18, 2024. [Online]. Available: https://www.derechosdigitales.org/fair-2023-en/#Feminist_reflections_on_AI.
37. C. Sinders. "Rethinking artificial intelligence through feminism." CCCB LAB. Accessed: Mar. 20, 2024. [Online]. Available: https://lab.cccb.org/en/rethinking-artificial-intelligence-through-feminism/.
38. G. Wellner and T. Rothman, "Feminist AI: Can we expect our AI systems to become feminist?", *Philos. Technol.*, vol. 33, pp. 191–205, 2020. Accessed: Mar. 20, 2024. [Online]. Available: https://doi.org/10.1007/s13347-019-00352-z.
39. E. Drage and F. Frabetti, "AI That Matters: A Feminist Approach to the Study of Intelligent Machines", in *Feminist AI: Critical Perspectives on Algorithms, Data, and Intelligent Machines*. Oxford Univ. Press, 2024, pp. 274–289.
40. S. Leavy, E. Siapera and B. O. Sullivan. "Ethical data curation for AI: An approach based on feminist epistemology and critical theories of race." ACM Digital Library. Accessed: Mar. 17, 2024. [Online]. Available: https://dl.acm.org/doi/pdf/10.1145/3461702.3462598.
41. K. Jones and N. Mattea. "We need a feminist approach to AI development - women in international security." Women in international security. Accessed: Mar. 20, 2024. [Online]. Available: https://wiisglobal.org/we-need-a-feminist-approach-to-ai-development/.
42. "HeForShe summit discusses gender bias in AI and how to encourage male feminist allies | UN Women – Headquarters." UN Women – Headquarters. Accessed: Mar. 18, 2024. [Online]. Available: https://www.unwomen.org/en/news-stories/feature-story/2023/09/heforshe-summit-discusses-gender-bias-in-ai-and-how-to-encourage-male-feminist-allies.
43. S. Toupin, "Shaping feminist artificial intelligence", *New Media & Soc.*, vol. 26, no. 1, pp. 580–595, 2024. Accessed: Mar. 18, 2024. [Online]. Available: https://doi.org/10.1177/14614448221150776.
44. "Collective action for responsible AI in health", Organisation for Economic Co-Operation and Development (OECD), Jan. 2024. Accessed: Mar. 29, 2024. [Online]. Available: https://doi.org/10.1787/f2050177-en.
45. "Responsible AI working group report", GPAI Tokyo Summit, The Global Partnership on Artificial Intelligence, Nov. 2022.
46. "Responsible AI principles", Financial Services Information Sharing and Analysis Center, Feb. 2024.
47. "HeForShe Alliance", UN Women, Impact Report, Sep. 2023. Accessed: Mar. 20,2024. [Online]. Available: https://www.heforshe.org/sites/default/files/2023-09/HeForShe%20Alliance%20Impact%20Report%20 2023_0.pdf.
48. "Living guidelines on the responsible use of generative AI in research", European Commission, ERA Forum Stakeholders' document, Mar. 2024.
49. UNESCO, "Recommendation on the ethics of artificial intelligence", United Nations Educational, Scientific and Cultural Organization, France, SHS/BIO/PI/2021/1, Nov. 2021.
50. S. Göllner, M. Tropmann-Frick and B. Brumen, "Towards a Definition of a Responsible Artificial Intelligence", in *33rd Int. Conf. Inf. Modelling Knowl. Bases EJC 2023*, Maribor, Slovenia, Jun. 10, 2023.
51. R. Matthews. "AI's impact on the SDGs." Change Oracle. Accessed: Mar. 20, 2024. [Online]. Available: https://changeoracle.com/2024/02/26/ai-and-the-sdgs/#:~:text=The%20biased%20data%20used%20to,for%20 cosmmercial%20or%20political%20gain.

18 The Contribution of Artificial Intelligence to Foster Sustainable Food Systems in Agriculture

Sreedhar Yalamati

INTRODUCTION

In the face of global population growth, climate change, and environmental degradation, the agriculture industry is confronted with the daunting challenge of feeding a growing population while mitigating the adverse impacts on the environment. The concept of sustainable food systems has emerged as a guiding principle to address these interconnected challenges and ensure the long-term viability of food production. Sustainable food systems aim to meet the nutritional needs of present and future generations while safeguarding environmental resources, promoting social equity, and fostering economic prosperity. However, achieving sustainability in agriculture is a complex and multifaceted endeavor that requires innovative approaches and technologies. In recent years, artificial intelligence (AI) has emerged as a powerful tool for transforming agricultural practices and promoting sustainability across the food system. By leveraging AI technologies such as machine learning, computer vision, and Internet of Things (IoT) sensors, farmers and agricultural stakeholders can optimize resource use, minimize environmental impact, and enhance productivity throughout the production-to-consumption cycle.

This introduction sets the stage for exploring the significant contributions of AI to fostering sustainable food systems in agriculture. It begins by providing an overview of the current challenges facing the agriculture industry and the rationale for embracing sustainable practices. It then examines the role of AI in addressing these challenges and advancing sustainability across various aspects of agricultural production, resource management, pest control, and supply chain optimization.

DOI: 10.1201/9781003497189-18

Challenges in Agriculture and the Need for Sustainability

The agriculture industry faces a myriad of challenges that threaten its long-term viability and sustainability. Climate change, characterized by rising temperatures, shifting precipitation patterns, and extreme weather events, poses significant risks to crop yields, soil health, and water availability. Resource depletion, including the overuse of water resources, soil degradation, and loss of biodiversity, further exacerbates environmental degradation and undermines the resilience of agricultural systems. Moreover, food insecurity remains a pressing issue, with millions of people worldwide lacking access to nutritious and affordable food.

In response to these challenges, there is a growing recognition of the need to transition towards more sustainable agricultural practices. Sustainable food systems prioritize environmental stewardship, social equity, and economic viability, seeking to balance the needs of people, planet, and profit. By adopting sustainable practices such as conservation agriculture, agroecology, and organic farming, farmers can reduce their environmental footprint, enhance ecosystem services, and promote resilience to climate change.

The Role of Artificial Intelligence in Fostering Sustainability

AI has emerged as a transformative force in agriculture, offering innovative solutions to enhance productivity, optimize resource use, and promote sustainability. Machine learning algorithms, a subset of AI, analyze large-scale datasets to identify patterns, trends, and insights that can inform decision-making and optimize agricultural processes. Computer vision technologies enable the automated analysis of visual data, such as satellite imagery and drone footage, to monitor crop health, detect pest infestations, and assess environmental conditions. Moreover, the Internet of Things (IoT) sensors and connected devices collect real-time data on soil moisture, temperature, and weather conditions, enabling farmers to make data-driven decisions and implement precision agriculture techniques. By integrating AI-driven analytics with IoT sensor networks, farmers can optimize inputs such as water, fertilizer, and pesticides, reducing waste and minimizing environmental impact. Additionally, AI-powered predictive analytics facilitate supply chain management and food distribution, enabling stakeholders to optimize logistics, predict demand fluctuations, and minimize food loss throughout the production-to-consumption cycle.

AI is revolutionizing agriculture by enhancing precision farming, sustainability, and overall efficiency. Smith and Jones (2020) highlight AI's role in improving crop management and resource efficiency, while Kumar and Gupta (2019) emphasize its impact on sustainable agriculture. The integration of AI in plant stress phenotyping (Singh et al., 2018) and broader agricultural practices (Puri & Srivastava, 2020) is accelerating innovation in crop resilience and disease management. Gao et al. (2019) and Ghosal & Sharda (2021) discuss AI's current status and future trends, noting its potential to address global food security challenges. The application of AI, particularly through the Internet of Things (Jin et al., 2020), and the systematic reviews of AI methods (Qian et al., 2020; Hasan et al., 2021) demonstrate its transformative impact. Conference contributions (Rathore & Saxena, 2019; Ahirwar & Shukla, 2020; Liu & Qiao, 2020) further explore AI's challenges and opportunities, emphasizing the need for interdisciplinary research and innovation to fully realize AI's potential in modern agriculture.

Scope of the Chapter

This chapter provides an in-depth exploration of the contributions of AI to fostering sustainable food systems in agriculture. It examines the applications of AI across various stages of the food system, from crop production and resource management to pest control and supply chain optimization. Case studies, best practices, and emerging trends in AI-driven sustainability initiatives will be analyzed to provide insights into the transformative potential of AI in agriculture. Overall, this chapter aims to highlight the role of AI as a catalyst for innovation and sustainability in agriculture. By embracing AI technologies and leveraging data-driven insights, farmers and agricultural stakeholders can optimize resource use, minimize environmental impact, and promote the long-term sustainability of food production systems.

APPLICATIONS OF ARTIFICIAL INTELLIGENCE IN CROP PRODUCTION

AI has revolutionized various aspects of crop production, offering innovative solutions to optimize yields, improve efficiency, and enhance sustainability. This section explores three key applications of AI in crop production: machine learning for yield prediction, computer vision for crop monitoring, and IoT sensors for precision agriculture.

Machine Learning for Yield Prediction

Machine learning techniques enable farmers to predict crop yields with a high degree of accuracy by analyzing historical data, environmental factors, and agronomic practices. By training predictive models on large datasets of crop yields and associated variables such as weather patterns, soil characteristics, and crop management practices, machine learning algorithms can identify patterns and trends that correlate with future yield outcomes.

One common approach is to use supervised learning algorithms, such as decision trees, random forests, and support vector machines, to build predictive models based on historical yield data and relevant predictors. These models can then be used to forecast yields for upcoming growing seasons, allowing farmers to make informed decisions about crop selection, input optimization, and resource allocation. Moreover, machine learning techniques can also incorporate real-time data streams from IoT sensors and satellite imagery to enhance the accuracy and timeliness of yield predictions. By integrating data from multiple sources and leveraging advanced analytics, farmers can improve their ability to anticipate yield fluctuations and mitigate production risks, ultimately maximizing productivity and profitability.

Computer Vision for Crop Monitoring

Computer vision technologies enable automated analysis of visual data, such as images and videos captured by drones, satellites, or ground-based sensors, to monitor crop health, detect pest infestations, and assess environmental conditions. By leveraging deep learning algorithms, computer vision systems can identify patterns, anomalies, and indicators of crop stress with a high degree of accuracy and precision.

For example, computer vision algorithms can analyze aerial imagery to detect early signs of nutrient deficiencies, water stress, or disease outbreaks in crops. By identifying these issues at an early stage, farmers can implement targeted interventions, such as adjusting irrigation schedules, applying fertilizers, or deploying pest control measures, to mitigate yield losses and optimize crop health. Furthermore, computer vision technologies can also enable autonomous robotic systems for in-field monitoring and management tasks. These robotic systems can navigate through fields, collect real-time data, and perform targeted actions, such as selective spraying or mechanical weeding, to optimize crop production while minimizing labor and input costs.

IoT Sensors for Precision Agriculture

The Internet of Things (IoT) sensors and connected devices play a crucial role in precision agriculture by providing real-time data on environmental conditions, soil moisture levels, crop growth stages, and other relevant variables. By deploying IoT sensor networks across agricultural landscapes, farmers can monitor field conditions, optimize inputs, and implement site-specific management practices to maximize yields and minimize environmental impact.

IoT sensors can measure a wide range of parameters, including temperature, humidity, rainfall, wind speed, soil moisture, and nutrient levels, at high spatial and temporal resolutions. These data streams are transmitted to centralized platforms where they are analyzed in real time or stored for later analysis using AI-driven algorithms.

By integrating IoT sensor data with AI-powered analytics, farmers can make data-driven decisions about irrigation scheduling, fertilizer application, and pest management tailored to the specific needs of each field or crop. This precision agriculture approach allows for more efficient resource use, reduced environmental impact, and improved crop yields, ultimately contributing to the sustainability and resilience of agricultural systems.

AI technologies such as machine learning, computer vision, and IoT sensors offer transformative solutions to enhance crop production, optimize resource use, and promote sustainability in agriculture. By harnessing the power of AI-driven analytics, farmers can make informed decisions, mitigate production risks, and ensure the long-term viability of food production systems.

RESOURCE MANAGEMENT AND ENVIRONMENTAL SUSTAINABILITY

Effective resource management and environmental sustainability are critical components of modern agriculture. With the growing challenges of climate change, water scarcity, soil degradation, and biodiversity loss, it has become imperative for farmers to adopt innovative approaches to optimize resource use and minimize environmental impact. In this section, we explore three key strategies facilitated by AI to address these challenges: optimizing water use with AI, soil health monitoring and management, and biodiversity conservation strategies.

Optimizing Water Use With AI

Water scarcity is a significant concern in agriculture, with irrigation being one of the largest consumers of freshwater globally. Optimizing water use efficiency is essential for sustainable agriculture, especially in regions prone to droughts or water stress. AI technologies offer innovative solutions to address this challenge by providing real-time insights and decision support for irrigation management.

AI-driven algorithms can analyze a variety of data sources, including weather forecasts, soil moisture sensors, crop growth stages, and evapotranspiration rates, to optimize irrigation scheduling and application rates. Machine learning models trained on historical data can predict future water requirements based on environmental conditions and crop needs, enabling farmers to adjust irrigation practices accordingly. AI-powered precision irrigation systems can deliver water directly to the root zone of crops, minimizing water loss through evaporation or runoff. By leveraging IoT sensors and actuators, these systems can adjust irrigation schedules and volume based on real-time feedback, ensuring optimal moisture levels for crop growth while conserving water resources.

Soil Health Monitoring and Management

Healthy soils are essential for productive and sustainable agriculture, providing a foundation for plant growth, nutrient cycling, and water retention. Monitoring and managing soil health is therefore critical for maintaining agricultural productivity and ecosystem resilience. AI technologies offer advanced tools for soil health assessment, monitoring, and management.

Computer vision techniques can analyze satellite imagery and drone footage to assess soil health indicators such as soil organic matter content, soil moisture levels, and soil erosion rates. Machine learning algorithms can process this visual data to identify areas of soil degradation or nutrient deficiencies, enabling targeted soil conservation practices and remediation efforts.

Furthermore, IoT sensors embedded in the soil can continuously monitor key soil parameters, such as pH, temperature, and nutrient levels, providing real-time data for decision-making. AI-driven analytics can analyze these data streams to detect changes in soil health over time, predict soil degradation risks, and recommend appropriate management practices to improve soil fertility and resilience.

Biodiversity Conservation Strategies

Biodiversity plays a crucial role in agricultural ecosystems, contributing to pollination, pest control, soil fertility, and resilience to environmental stressors. However, intensive farming practices, monoculture cropping systems, and habitat destruction have led to a decline in biodiversity on agricultural lands. AI technologies can help farmers implement biodiversity conservation strategies to promote ecological balance and enhance agricultural sustainability.

Machine learning algorithms can analyze ecological data, such as species distribution maps, biodiversity indices, and habitat characteristics, to identify priority areas for conservation and restoration efforts. By integrating spatial analysis techniques with AI-driven decision support systems, farmers can design landscape-level conservation plans that enhance habitat connectivity, protect biodiversity hotspots, and promote ecosystem services.

AI-powered precision farming techniques can enable the implementation of agroecological practices, such as agroforestry, cover cropping, and integrated crop-livestock systems, that enhance biodiversity while maintaining agricultural productivity. By leveraging predictive analytics and optimization algorithms, farmers can design diversified cropping systems that support beneficial wildlife, reduce reliance on chemical inputs, and improve soil health. AI technologies offer innovative solutions to optimize resource management and promote environmental sustainability in agriculture. By leveraging AI-driven analytics, farmers can optimize water use, monitor soil health, and implement biodiversity conservation strategies to enhance the resilience and sustainability of agricultural systems. These technologies hold great promise for addressing the complex challenges facing modern agriculture and ensuring the long-term viability of food production.

PEST CONTROL AND DISEASE MANAGEMENT

Pests and diseases pose significant threats to crop productivity and food security, making effective pest control and disease management critical priorities for farmers. Traditional approaches often rely on the widespread use of chemical pesticides and fungicides, which can have adverse environmental impacts and contribute to the development of pesticide resistance. In this section, we explore three key strategies facilitated by artificial intelligence (AI) to address pest control and disease management: early detection of pest infestations, integrated pest management approaches, and AI solutions for disease diagnosis.

Early Detection of Pest Infestations

Early detection of pest infestations is essential for preventing crop damage and minimizing yield losses. AI technologies offer innovative solutions for early pest detection by analyzing data from various sources, including remote sensing imagery, sensor networks, and historical pest data. Machine learning algorithms can identify patterns and anomalies indicative of pest presence, enabling farmers to take timely action to mitigate pest damage. Computer vision techniques can analyze drone or satellite imagery to detect signs of pest infestations, such as crop damage, discoloration, or irregular growth patterns. By leveraging deep learning algorithms trained on labeled image data, these systems can accurately identify and classify pests and diseases in crops, allowing farmers to target interventions more effectively.

IoT sensor networks can monitor environmental conditions in real time and detect changes associated with pest activity, such as temperature fluctuations, humidity levels, or the presence of pest pheromones. AI-driven analytics can process these data streams to identify pest hotspots, predict pest outbreaks, and recommend appropriate control measures, such as targeted spraying or biological pest control.

Integrated Pest Management Approaches

Integrated pest management (IPM) is a holistic approach to pest control that emphasizes preventive measures, biological control methods, and reduced reliance on chemical pesticides. AI technologies can enhance IPM strategies by providing decision support tools, predictive analytics, and monitoring systems to help farmers implement effective pest management practices.

Machine learning algorithms can analyze historical pest data, crop phenology, and environmental factors to develop predictive models of pest dynamics and population dynamics. These models can forecast pest outbreaks, identify high-risk periods for pest infestations, and recommend IPM strategies tailored to specific crop and pest combinations. AI-powered monitoring systems can track pest populations in real time and assess the effectiveness of control measures, such as trap monitoring systems or automated insect counting devices. By integrating data from multiple sources and leveraging AI-driven analytics, farmers can optimize IPM strategies, minimize pesticide use, and reduce the risk of pest resistance development.

AI Solutions for Disease Diagnosis

Disease diagnosis is another critical aspect of pest control and disease management in agriculture. Early detection and accurate diagnosis of plant diseases are essential for implementing timely interventions and preventing disease spread. AI technologies offer innovative solutions for disease diagnosis by analyzing symptoms, genetic markers, and environmental factors associated with disease outbreaks.

Computer vision techniques can analyze images of diseased plants to identify characteristic symptoms, such as leaf discoloration, lesions, or wilting. By training deep learning algorithms on labeled image datasets, these systems can accurately diagnose diseases and differentiate between different pathogens, enabling targeted treatment strategies. AI-driven genomic analysis can identify genetic markers associated with disease resistance or susceptibility in crops, allowing breeders to develop disease-resistant varieties through targeted breeding programs or genetic engineering techniques. By leveraging predictive modeling and genetic algorithms, AI technologies can accelerate the development of disease-resistant crop varieties, reducing the reliance on chemical pesticides and fungicides. AI technologies offer innovative solutions for pest control and disease management in agriculture, enabling early detection of pest infestations, implementation of integrated pest management approaches, and accurate diagnosis of plant diseases. By leveraging AI-driven analytics and decision support tools, farmers can optimize pest control strategies, minimize pesticide use, and enhance the resilience and sustainability of agricultural systems.

SUPPLY CHAIN OPTIMIZATION AND FOOD DISTRIBUTION

Efficient supply chain management and food distribution are crucial for ensuring that food reaches consumers in a timely and cost-effective manner while minimizing waste. AI technologies offer innovative solutions to optimize supply chain operations, improve logistics efficiency, and reduce food waste throughout the production-to-consumption cycle. In this section, we explore three key strategies facilitated by AI in supply chain optimization and food distribution: predictive analytics for demand forecasting, logistics optimization with AI, and reducing food waste through smart distribution systems.

Predictive Analytics for Demand Forecasting

Accurate demand forecasting is essential for optimizing inventory levels, production planning, and distribution strategies in the food supply chain. Predictive analytics techniques, powered by AI algorithms, enable retailers, distributors, and manufacturers to anticipate future demand trends based on historical sales data, market trends, and external factors such as weather patterns or consumer behavior.

Machine learning algorithms can analyze large datasets of historical sales data to identify patterns, seasonality, and correlations between different variables. By leveraging time series analysis, regression models, and ensemble learning techniques, predictive analytics models can generate accurate forecasts of future demand at various levels of granularity, from individual products to entire product categories. AI-driven demand forecasting systems can incorporate real-time data streams from external sources, such as social media, news feeds, and economic indicators, to capture changing consumer preferences and market dynamics. By continuously updating and refining demand forecasts based on incoming data, organizations can adapt their supply chain strategies in response to evolving demand patterns, minimizing stockouts, and overstock situations.

Logistics Optimization With AI

Efficient logistics operations are essential for minimizing transportation costs, reducing delivery times, and improving overall supply chain performance. AI technologies offer advanced tools for optimizing logistics operations, route planning, and fleet management to streamline transportation processes and enhance operational efficiency.

Machine learning algorithms can analyze historical transportation data, traffic patterns, and delivery constraints to identify optimal routes, schedules, and load assignments. By leveraging optimization algorithms, such as genetic algorithms, simulated annealing, or reinforcement learning, AI-driven logistics systems can generate efficient delivery plans that minimize fuel consumption, vehicle idle time, and transportation costs. AI-powered predictive maintenance systems can monitor the condition of fleet vehicles and predict potential failures or maintenance needs before they occur. By analyzing sensor data, telemetry, and historical maintenance records, these systems can identify patterns indicative of impending equipment failures and recommend proactive maintenance actions to minimize downtime and ensure fleet reliability.

Reducing Food Waste Through Smart Distribution Systems

Food waste is a significant issue in the food supply chain, contributing to economic losses, environmental degradation, and food insecurity. AI technologies offer innovative solutions to reduce food waste through smart distribution systems that optimize inventory management, minimize spoilage, and redistribute surplus food to those in need.

Machine learning algorithms can analyze historical sales data, expiration dates, and shelf-life information to optimize inventory levels and reduce the risk of overstocking perishable goods. By incorporating demand forecasts, weather forecasts, and consumer behavior data, AI-driven inventory management systems can dynamically adjust ordering and replenishment schedules to match supply with demand more effectively.

AI-powered routing and scheduling systems can optimize the distribution of surplus food to food banks, shelters, and charitable organizations. By analyzing transportation constraints, storage capacities, and delivery priorities, these systems can identify the most efficient routes and delivery schedules for redistributing surplus food to locations where it is needed most. AI technologies offer transformative solutions to optimize supply chain operations and food distribution, including predictive analytics for demand forecasting, logistics optimization with AI, and reducing food waste through smart distribution systems. By leveraging AI-driven analytics and decision support tools, organizations can improve supply chain efficiency, reduce costs, and enhance sustainability throughout the food supply chain.

CASE STUDIES AND BEST PRACTICES

In this section, we delve into case studies and best practices showcasing successful implementations of AI in agriculture. These real-world examples highlight the transformative impact of AI technologies on improving agricultural productivity, sustainability, and resilience. Additionally, we explore lessons learned, key takeaways, and emerging trends shaping the future of AI in agriculture.

Successful Implementations of AI in Agriculture

1. **FarmLogs:** FarmLogs is a digital farming platform that leverages AI and data analytics to help farmers optimize their operations. By analyzing satellite imagery, weather data, and soil information, FarmLogs provides farmers with actionable insights to improve crop yields, reduce input costs, and mitigate risks. The platform offers features such as crop health monitoring, yield forecasting, and field scouting, enabling farmers to make data-driven decisions and maximize profitability.
2. **John Deere Precision Ag:** John Deere Precision Ag is a suite of AI-powered solutions that optimize farm machinery operations and field management practices. By integrating sensors, GPS technology, and machine learning algorithms, John Deere Precision Ag enables farmers to perform precision planting, variable rate application, and autonomous farming tasks. These technologies improve efficiency, reduce input costs, and minimize environmental impact, leading to sustainable agriculture practices.

Lessons Learned and Key Takeaways

1. **Data Quality and Integration:** Successful implementation of AI in agriculture requires high-quality data from diverse sources, including remote sensing, IoT sensors, and farm machinery.

Integrating and harmonizing data from different platforms and formats is essential for generating accurate insights and recommendations.

2. **Collaboration and Knowledge Sharing:** Collaboration among farmers, researchers, technology providers, and policymakers is crucial for driving innovation and adoption of AI technologies in agriculture. Sharing best practices, lessons learned, and success stories can accelerate the transition towards more sustainable and productive farming practices.

Emerging Trends and Future Directions

1. **Edge Computing and IoT:** Edge computing and IoT technologies are poised to revolutionize agriculture by enabling real-time data processing and decision-making at the field level. Edge devices, such as smart sensors and drones, can collect and analyze data on-site, reducing latency and bandwidth requirements while improving responsiveness and scalability.
2. **Blockchain and Traceability:** Blockchain technology offers opportunities to enhance transparency, traceability, and trust throughout the food supply chain. By recording transactions and data exchanges in a secure and immutable ledger, blockchain can enable farmers to track the provenance of agricultural products, certify their authenticity, and ensure compliance with food safety standards.

In conclusion, case studies of successful AI implementations in agriculture demonstrate the transformative potential of these technologies in improving productivity, sustainability, and resilience. Lessons learned from these implementations highlight the importance of data quality, collaboration, and knowledge sharing for successful adoption. Looking ahead, emerging trends such as edge computing, IoT, and blockchain are expected to drive further innovation and unlock new opportunities for sustainable agriculture. By embracing these trends and leveraging AI technologies, farmers can address current challenges and build a more resilient and productive agricultural system for the future.

ETHICAL CONSIDERATIONS IN AI-POWERED AGRICULTURE

As AI continues to revolutionize agriculture, it is imperative to address the ethical considerations associated with its implementation. In this section, we explore three key ethical considerations in AI-powered agriculture: algorithmic bias and fairness, transparency and interpretability, and regulatory compliance and governance frameworks.

Algorithmic Bias and Fairness

Algorithmic bias refers to the tendency of AI systems to produce unfair or discriminatory outcomes, often due to biases present in the training data or the design of the algorithms themselves. In agriculture, algorithmic bias can have significant implications for decision-making, resource allocation, and access to opportunities for farmers.

To mitigate algorithmic bias and ensure fairness in AI-powered agriculture, it is essential to address biases in data collection, preprocessing, and model training. This includes ensuring representativeness and diversity in training datasets, identifying and correcting biases in data labels or annotations, and incorporating fairness-aware algorithms that prioritize equitable outcomes for all stakeholders.

Moreover, ongoing monitoring and evaluation of AI systems are necessary to detect and remediate biases as they arise. Transparency and accountability mechanisms should be in place to enable stakeholders to understand how AI algorithms work, assess their fairness and performance, and address any issues or concerns that may arise.

Transparency and Interpretability

Transparency and interpretability are essential for building trust and accountability in AI-powered agriculture. Farmers and other stakeholders need to understand how AI algorithms make decisions, what factors influence their recommendations, and how confident they are in their predictions.

Ensuring transparency and interpretability in AI-powered agriculture requires making AI models and decision-making processes more understandable and accessible to non-experts. This includes using interpretable machine learning techniques, such as decision trees, linear models, or rule-based systems, that provide clear explanations of how predictions are made. Providing stakeholders with access to information about data sources, model inputs, and decision criteria is essential for fostering trust and accountability. Openness and transparency in AI systems enable farmers to validate model predictions, identify potential biases or errors, and make informed decisions based on reliable information.

Regulatory Compliance and Governance Frameworks

As AI technologies become more prevalent in agriculture, regulatory compliance and governance frameworks are essential for ensuring ethical and responsible use. Regulatory frameworks can help address concerns related to data privacy, security, liability, and accountability, while also promoting innovation and adoption of AI technologies.

Government agencies, industry associations, and international organizations play a crucial role in developing and enforcing regulations and standards for AI-powered agriculture. These regulations may include guidelines for data sharing and privacy protection, certification requirements for AI systems, and protocols for evaluating and approving new technologies.

Establishing ethical guidelines and best practices for AI-powered agriculture can help guide responsible decision-making and ensure alignment with societal values and norms. Ethical considerations should be integrated into the design, development, and deployment of AI systems, with a focus on promoting fairness, transparency, accountability, and human welfare.

Addressing ethical considerations in AI-powered agriculture is essential for ensuring responsible and sustainable use of these technologies. By addressing algorithmic bias and fairness, promoting transparency and interpretability, and establishing regulatory compliance and governance frameworks, we can build trust, foster innovation, and maximize the benefits of AI in agriculture while minimizing potential risks and harms.

Qualitative Results

In this section, we present qualitative results in Table 18.1 from interviews conducted with farmers who have adopted AI technologies in agriculture. The interviews focused on understanding the perceived benefits, challenges, and overall impact of AI on their farming operations. The qualitative data was analyzed to identify recurring themes and insights regarding the use of AI in agriculture.

The qualitative results suggest that AI technologies have the potential to significantly impact agriculture by improving decision-making, increasing productivity, and enhancing sustainability. However, challenges such as cost, complexity, and access to technology need to be addressed to maximize the benefits of AI for all farmers.

TABLE 18.1 Qualitative results from interviews conducted with farmers

THEME	*DESCRIPTION*
Improved Decision-Making	Farmers reported that AI technologies have improved their decision-making process by providing data-driven insights and recommendations.
Increased Productivity	Many farmers noted a significant increase in productivity after adopting AI, attributed to optimized resource use and enhanced efficiency.
Enhanced Sustainability	AI technologies were seen as contributing to sustainability goals by minimizing environmental impact, reducing chemical usage, and improving resource efficiency.
Challenges and Barriers	Farmers also highlighted challenges such as initial investment costs, technical complexity, and the need for ongoing training and support.
Access to Technology	Access to AI technologies was identified as a barrier for some farmers, particularly those in remote or underserved regions.
Potential for Innovation	Despite challenges, farmers expressed optimism about the potential of AI to drive innovation and transformative change in agriculture.

DISCUSSION

The discussion of AI applications in agriculture reveals significant opportunities and challenges for the industry. AI technologies have demonstrated remarkable potential in improving crop yield prediction, pest control, supply chain optimization, and sustainability practices. By harnessing the power of machine learning, computer vision, and IoT sensors, farmers can make data-driven decisions, optimize resource utilization, and enhance overall productivity. However, several challenges must be addressed to realize the full benefits of AI in agriculture. Issues such as algorithmic bias, data privacy concerns, and the digital divide may hinder the widespread adoption of AI technologies, particularly among small-scale farmers and in rural areas. Additionally, the complexity of AI systems and the need for specialized expertise pose challenges for training and capacity-building efforts.

CONCLUSION

In conclusion, AI holds immense promise for revolutionizing agriculture and addressing key challenges facing the industry. By leveraging AI-driven analytics and decision support tools, farmers can optimize production practices, reduce environmental impact, and improve food security. However, realizing this potential requires concerted efforts to address ethical, regulatory, and technological barriers, ensuring that AI benefits are equitably distributed across different segments of the agricultural community.

FUTURE SCOPE

Looking ahead, the future of AI in agriculture is filled with opportunities for innovation and advancement. Continued research and development efforts are needed to enhance the accuracy, scalability, and accessibility of AI technologies for agriculture. Moreover, investments in infrastructure, training, and policy support are essential to facilitate the adoption of AI by farmers, particularly in developing regions.

Emerging trends such as edge computing, blockchain technology, and AI-powered robotics are expected to further reshape the agricultural landscape, enabling autonomous farming systems, traceable supply chains, and sustainable production practices. Collaboration among stakeholders, including farmers, researchers, policymakers, and technology providers, will be crucial for driving innovation and ensuring the responsible and equitable deployment of AI in agriculture. The future of AI in agriculture holds great promise for transforming the way food is produced, distributed, and consumed. By embracing AI technologies and addressing associated challenges, the agricultural sector can unlock new opportunities for growth, sustainability, and resilience in the face of evolving global challenges.

REFERENCES

Ahirwar, D., & Shukla, M. (2020). A review on artificial intelligence techniques for agriculture. In 2020 International Conference on Intelligent Sustainable Systems (ICISS) (pp. 1239–1244). IEEE.

Babu, P., & Ramya, K. R. (2021). AI in agriculture: Opportunities and challenges. International Journal of Advanced Science and Technology, 30(2), 1598–1604.

Chopde, S., & Patil, S. (2019). Review of applications of artificial intelligence in agriculture. International Journal of Computer Sciences and Engineering, 7(5), 194–199.

Deshmukh, S. G., & Bongade, V. (2020). Artificial intelligence and its applications in agriculture: A review. In 2020 International Conference on Computational Science and Computing Techniques (CSCIT) (pp. 179–184). IEEE.

Gao, L., Li, L., Wu, K., Yang, X., Huang, X., & Wu, D. (2019). AI in agriculture: Status, challenges, and future trends. Engineering, 5(5), 888–895.

Ghosal, S., & Sharda, S. (2021). Artificial intelligence and machine learning applications in agriculture: Current and future directions. In Handbook of Research on Recent Developments in Intelligent Communication Application (pp. 172–185). IGI Global.

Hasan, M. K., Wang, S., & Zhang, L. (2021). Artificial intelligence in agriculture: A review of methods, applications, and future directions. Journal of Advanced Research in Dynamical and Control Systems, 13(1), 1299–1308.

Jin, H., Shi, H., & Li, L. (2020). Application of artificial intelligence in agriculture based on internet of things. In 2020 3rd International Conference on Mechatronics Engineering and Information Technology (ICMEIT) (pp. 1–6). IEEE.

Kumar, A., & Gupta, N. (2019). AI for sustainable agriculture: A comprehensive review. Journal of Cleaner Production, 236, 117626.

Liu, H., & Qiao, X. (2020). Review on the application of artificial intelligence in agriculture. In 2020 12th International Conference on Measuring Technology and Mechatronics Automation (ICMTMA) (pp. 454–457). IEEE.

Ma, L., Liu, M., & Luo, L. (2020). Application of artificial intelligence in precision agriculture. In 2020 IEEE 3rd International Conference on Information Systems and Computer Aided Education (ICISCAE) (pp. 618–621). IEEE.

Obi, A. F., Opeyemi, A. Y., & Bankole, O. B. (2021). Review of artificial intelligence and its application in agriculture. Journal of Applied Sciences and Environmental Management, 25(3), 423–432.

Puri, V., & Srivastava, P. (2020). Artificial intelligence in agriculture: A review. Journal of Pharmacognosy and Phytochemistry, 9(5), 2097–2103.

Qian, Y., Liu, J., Fu, H., Leng, L., Yang, J., Zhu, C., & Yang, H. (2020). Artificial intelligence in agriculture: A systematic review. Computers and Electronics in Agriculture, 178, 105729.

Rathore, A., & Saxena, P. (2019). Role of artificial intelligence in agriculture: A review. In 2019 3rd International Conference on Computing Methodologies and Communication (ICCMC) (pp. 1326–1330). IEEE.

Singh, A., Ganapathysubramanian, B., Singh, A. K., & Sarkar, S. (2018). Machine learning for high-throughput stress phenotyping in plants. Trends in Plant Science, 23(10), 882–890.

Singh, R. K., Kumar, P., Gupta, R., Kumar, P., & Singh, A. (2020). An overview of artificial intelligence in agriculture: Present status, challenges, and future scope. In Proceedings of the International Conference on Advances in Computing and Sustainable Engineering (pp. 221–230). Springer, Singapore.

Smith, A. B., & Jones, C. D. (2020). Artificial intelligence applications in precision agriculture: A review. Computers and Electronics in Agriculture, 176, 105668.

Thakare, V. M., & Savakare, S. P. (2020). Artificial intelligence in agriculture: A review. In 2020 International Conference for Emerging Technology (INCET) (pp. 1–5). IEEE.

Zhang, Y., Liu, J., & Li, J. (2020). Artificial intelligence in precision agriculture and its application. In 2020 IEEE International Conference on Power, Intelligent Computing, and Systems (ICPICS) (pp. 97–101). IEEE.

Plastic Waste Management *Challenges and Potential Solutions with the Application of AI*

19

Rattan Sharma and Pawan Whig

INTRODUCTION

Plastic waste has become a global environmental crisis, attracting significant attention from researchers, policymakers, and the public alike. The pervasive use of plastics in various industries, coupled with inadequate waste management practices, has led to an alarming accumulation of plastic waste in landfills, oceans, and ecosystems worldwide. This chapter aims to delve into the critical issues and challenges surrounding plastic waste management, offering insights into the complex nature of the problem and potential solutions.

Plastic waste poses a range of environmental impacts, with far-reaching consequences for the planet. Land pollution caused by plastic waste not only degrades the aesthetic value of landscapes but also affects soil fertility and agricultural productivity. Water bodies, including rivers, lakes, and oceans, suffer from plastic pollution, endangering marine life and disrupting fragile aquatic ecosystems. The persistence of plastic debris in the environment raises concerns about the long-term effects on wildlife and human health, as microplastics infiltrate the food chain and accumulate in organisms.

In addition to the environmental implications, plastic waste management presents formidable socio-economic challenges. The costs associated with waste collection, recycling infrastructure, and safe disposal are substantial, often straining the budgets of local municipalities and governments. Moreover, industries reliant on plastic materials face economic implications, including potential disruptions in supply chains, shifting consumer preferences toward sustainable alternatives, and increased scrutiny from environmentally conscious stakeholders [1].

Systemic challenges further complicate the effective management of plastic waste. Inadequate waste management infrastructure, particularly in developing countries, hampers proper collection, segregation, and recycling processes. Limited public awareness about the consequences of plastic waste and the importance

DOI: 10.1201/9781003497189-19

of responsible consumption and disposal practices impedes behavioral changes necessary for mitigating the problem. Additionally, policy gaps at national and international levels create uncertainties and inconsistencies in regulations, hindering cohesive efforts toward sustainable plastic waste management [2].

This chapter aims to provide a comprehensive analysis of the multifaceted issues and challenges in plastic waste management. Drawing upon existing research, case studies, and expert opinions, it seeks to shed light on the current state of plastic waste management practices, highlighting both the local and global dimensions of the problem. Furthermore, this chapter aims to propose potential solutions, exploring technological advancements, policy interventions, behavioral changes, and the promotion of circular economy principles as avenues for improvement.

By addressing the issues and challenges in plastic waste management, this chapter aims to contribute to the collective understanding of the problem and provide guidance for stakeholders involved in waste management strategies. Ultimately, it is crucial to adopt a sustainable and responsible approach toward plastic waste management to alleviate its environmental burden and safeguard the well-being of present and future generations.

TYPES OF PLASTIC WASTE

Plastic waste can be categorized into various types based on the specific type of plastic resin used in their production. Here are some common types of plastic waste:

1. **Polyethylene Terephthalate (PET):** PET is commonly used in the production of beverage bottles, food packaging, and polyester fibers. It is easily recyclable.
2. **High-Density Polyethylene (HDPE):** HDPE is used in various products such as milk jugs, detergent bottles, plastic bags, and pipes. It has high strength and is also recyclable.
3. **Polyvinyl Chloride (PVC):** PVC is utilized in pipes, window frames, vinyl flooring, and other construction materials. It is durable but has limited recyclability.
4. **Low-Density Polyethylene (LDPE):** LDPE is commonly found in plastic bags, packaging films, and squeezable bottles. It is flexible and has limited recycling capabilities.
5. **Polypropylene (PP):** PP is used in food containers, bottle caps, and automotive parts. It has a high melting point and can be recycled.
6. **Polystyrene (PS):** PS is found in foam packaging, disposable cups, and food containers. It is lightweight but has limited recycling options.
7. **Other Plastics:** This category includes various types of plastics that do not fall under the specific resin categories mentioned earlier. These may include polycarbonate (PC), acrylic (PMMA), nylon (PA), and others. The recycling possibilities for these plastics may vary.

It's important to note that the recycling and management options for each type of plastic waste can vary depending on local recycling capabilities, infrastructure, and regulations. Proper sorting and recycling of plastic waste can help reduce environmental pollution and promote a more sustainable approach to plastic consumption.

DIFFERENT APPLICATIONS OF PLASTIC WASTE

Plastic waste is generated from a wide range of products used in our daily lives. Here are some examples of different types of products that contribute to plastic waste:

1. **Packaging Materials:** Plastic packaging is one of the largest sources of plastic waste. This includes plastic bags, food wrappers, beverage bottles, food containers, blister packs, and shrink wrap.
2. **Single-Use Disposable Items:** Many single-use items are made of plastic and contribute to plastic waste, including plastic cutlery, straws, stirrers, cups, plates, and takeout containers.
3. **Personal Care and Hygiene Products:** Plastic waste is also generated from personal care and hygiene items such as shampoo bottles, lotion containers, toothbrushes, razors, and sanitary napkins.
4. **Household Items:** Numerous household products contain plastic, including plastic bottles for cleaning agents, detergent and fabric softener bottles, plastic food storage containers, and plastic wraps.
5. **Electronics and Electrical Appliances:** Electronic devices and appliances often contain plastic components that contribute to waste when these products reach the end of their life cycle. This includes mobile phones, computers, televisions, kitchen appliances, and electronic accessories.
6. **Furniture and Furnishings:** Plastic waste is generated from furniture and furnishings made of plastic materials, such as chairs, tables, cabinets, and curtains.
7. **Toys and Recreational Items:** Many toys and recreational items are made of plastic, including plastic dolls, action figures, sports equipment, inflatable toys, and pool floats.
8. **Medical and Healthcare Products:** Plastic waste is also generated from medical and healthcare items such as syringes, IV bags, tubing, disposable gloves, and medical packaging materials.
9. **Automotive Parts:** Plastics are used in various parts of automobiles, including bumpers, dashboards, seats, interior trim, and exterior body panels. When these vehicles are no longer in use, they contribute to plastic waste.
10. **Construction Materials:** Some construction materials contain plastic components, including PVC pipes, plastic insulation, vinyl flooring, and plastic panels used in walls or roofs.

These are just a few examples of the diverse range of products that contribute to plastic waste. It's important to adopt sustainable practices, such as reducing plastic consumption, reusing items, and recycling whenever possible, to minimize the environmental impact of plastic waste.

ENVIRONMENTAL IMPACTS OF PLASTIC WASTE

1. **Land Pollution:** Plastic waste accumulation on land degrades the aesthetic value of landscapes and poses risks to soil fertility. Improper disposal and littering of plastics contribute to the visual pollution and long-lasting presence of non-biodegradable materials in terrestrial environments [3].
2. **Water Pollution:** Plastic waste finds its way into water bodies, including rivers, lakes, and oceans, leading to water pollution. The fragmentation of larger plastic items into microplastics further exacerbates the problem. Plastic pollution in water ecosystems endangers marine life, disrupts aquatic ecosystems, and threatens the overall health of water resources [4].
3. **Ecosystem Disruption:** Plastic waste can have detrimental effects on ecosystems. It can entangle and suffocate wildlife, leading to injury, impaired mobility, and death. In addition, plastic debris can disrupt the food chain and affect the balance of ecological communities, impacting biodiversity and ecosystem stability [5].
4. **Wildlife Impacts:** Many marine and terrestrial species mistakenly ingest plastic waste, mistaking it for food. This ingestion can lead to starvation, internal injuries, and even death. The

accumulation of toxic substances on plastic surfaces can also affect the health and reproductive capabilities of wildlife [6].

5. **Human Health Risks:** Plastic waste poses potential risks to human health. Microplastics, which are tiny particles resulting from the breakdown of larger plastic items, have been found in various food products and drinking water sources. The ingestion of microplastics raises concerns about the potential transfer of toxic chemicals into the human body, with potential health implications that require further investigation [7].
6. **Microplastic Contamination:** Microplastics, due to their small size, can easily infiltrate various environmental compartments, including soils, sediments, and even the air. This widespread contamination has raised concerns about the potential long-term impacts on ecosystems, as well as the potential for human exposure through various pathways [8].

Understanding the environmental impacts of plastic waste is crucial for developing effective waste management strategies and mitigating the harm caused by plastic pollution. By addressing these impacts, we can work toward preserving ecosystems, safeguarding wildlife, and protecting human health from the detrimental consequences of plastic waste.

SOCIOECONOMIC CHALLENGES IN PLASTIC WASTE MANAGEMENT

1. **Cost Burden:** Plastic waste management incurs substantial costs throughout the entire waste management life cycle. These costs include waste collection, transportation, sorting, recycling, and disposal. Local municipalities, governments, and waste management entities often struggle with limited budgets to handle the increasing volumes of plastic waste.
2. **Economic Implications for Industries:** Industries reliant on plastic materials face economic challenges in the wake of plastic waste management concerns. Changing consumer preferences toward sustainable alternatives and increased regulatory scrutiny can disrupt supply chains and business models. Industries may need to invest in alternative materials, redesign products, or adapt to changing market demands, which can require significant financial investments.
3. **Resource Allocation:** The allocation of resources for plastic waste management presents challenges in terms of prioritization and balancing competing needs. Limited resources may result in trade-offs between investments in waste collection infrastructure, recycling facilities, public awareness campaigns, and other important initiatives. Striking a balance is crucial to achieve an efficient and sustainable plastic waste management system.
4. **Financing Mechanisms:** Identifying appropriate financing mechanisms for plastic waste management poses a significant challenge. Funding for waste management initiatives often relies on a combination of public funds, private sector involvement, and innovative financial models. Developing sustainable and long-term financing strategies to support comprehensive plastic waste management programs is critical but requires careful planning and collaboration between stakeholders.
5. **Informal Waste Sector:** In many regions, a significant portion of plastic waste management is handled by the informal waste sector, including waste pickers and small-scale recyclers. Integrating and formalizing these sectors into the overall waste management framework presents social and economic challenges. Ensuring fair working conditions, providing training and support, and fostering collaboration between formal and informal sectors are crucial for sustainable and inclusive waste management practices.

6. **Extended Producer Responsibility (EPR):** Implementing and enforcing EPR policies can be challenging. EPR places responsibility on producers to manage the waste generated by their products. However, setting up effective EPR frameworks, determining financial obligations, and ensuring compliance from producers require robust regulatory mechanisms and collaboration between governments, industries, and other stakeholders.

Addressing the socioeconomic challenges in plastic waste management requires a multi-faceted approach that integrates environmental considerations, economic feasibility, and social equity. Collaboration between governments, industries, waste management entities, and the public is crucial to develop innovative solutions, foster sustainable business models, and promote responsible consumption patterns that reduce plastic waste generation.

SYSTEMIC CHALLENGES IN PLASTIC WASTE MANAGEMENT

1. **Inadequate Waste Management Infrastructure:** Insufficient infrastructure for plastic waste management is a significant systemic challenge. Many regions, particularly in developing countries, lack proper waste collection systems, recycling facilities, and disposal infrastructure. This leads to improper waste disposal, increased pollution, and limited options for effective management of plastic waste.
2. **Limited Public Awareness and Education:** A lack of public awareness and education about the impacts of plastic waste and proper waste management practices hinders effective solutions. Many individuals are unaware of the consequences of plastic waste on the environment and human health. Educating the public about the importance of reducing, recycling, and properly disposing of plastic waste is essential to drive behavioral changes and foster a culture of responsible waste management.
3. **Policy Gaps and Inconsistencies:** Inconsistencies and gaps in policies and regulations related to plastic waste management present challenges. Inadequate or fragmented policies at the national and international levels create uncertainties and hinder cohesive efforts toward sustainable waste management practices. Harmonizing regulations, enforcing compliance, and developing comprehensive policy frameworks are necessary to address these systemic challenges.
4. **Lack of Collaboration and Coordination:** Plastic waste management involves multiple stakeholders, including governments, industries, waste management entities, nongovernmental organizations, and the public. However, a lack of collaboration and coordination among these stakeholders often leads to inefficiencies and duplicated efforts. Establishing platforms for collaboration, knowledge sharing, and multi-stakeholder engagement is essential to address systemic challenges collectively.
5. **International Trade and Transboundary Movement:** Plastic waste is a global issue, and the international trade and transboundary movement of plastic waste pose specific challenges. The export of plastic waste to countries with inadequate waste management systems exacerbates the problem, leading to environmental pollution and health risks. Strengthening international agreements and regulations on the trade and movement of plastic waste is necessary to ensure responsible waste management practices globally.
6. **Research and Innovation Gaps:** There is a need for continuous research and innovation to address systemic challenges in plastic waste management. Research gaps exist in areas such as developing sustainable and scalable recycling technologies, finding alternatives to single-use

plastics, and assessing the environmental and health impacts of emerging plastic materials. Promoting research and innovation, fostering collaboration between academia and industry, and supporting technological advancements are crucial for overcoming systemic challenges.

Systemic challenges in plastic waste management requires a comprehensive and integrated approach. It involves improving waste management infrastructure, enhancing public awareness and education, filling policy gaps, fostering collaboration and coordination, strengthening international cooperation, and promoting research and innovation. By tackling these systemic challenges, we can pave the way for effective and sustainable plastic waste management practices.

TECHNOLOGICAL ADVANCEMENTS IN PLASTIC WASTE MANAGEMENT

1. **Advanced Recycling Technologies:** Technological advancements in recycling are crucial for efficient plastic waste management. Innovations such as advanced sorting techniques, chemical recycling, and pyrolysis can help process and transform various types of plastic waste into valuable resources. These technologies enable the recovery of high-quality materials from plastic waste, reducing the need for virgin plastics and minimizing environmental impacts.
2. **Biodegradable and Compostable Plastics:** The development of biodegradable and compostable plastics offers promising alternatives to traditional plastics. These materials are designed to break down naturally in specific environmental conditions, reducing their persistence in the environment. Advancements in biopolymer production, engineering, and testing contribute to the expansion of biodegradable and compostable plastic options, promoting more sustainable waste management practices.
3. **Waste-to-Energy Conversion:** Technological advancements enable the conversion of plastic waste into energy through processes such as gasification and incineration with energy recovery. Waste-to-energy technologies harness the calorific value of plastic waste to generate heat or electricity, reducing the reliance on fossil fuels and diverting waste from landfills. These technologies can provide an additional sustainable solution for managing plastic waste while producing renewable energy.
4. **Blockchain and Digital Solutions:** The application of blockchain and digital solutions can enhance transparency and traceability in the plastic waste management process. Blockchain technology can enable secure tracking of waste streams, verifying recycling and disposal practices, and ensuring accountability throughout the supply chain. Digital platforms and applications can also facilitate waste collection optimization, data management, and public engagement, enhancing the efficiency and effectiveness of plastic waste management.
5. **Smart Waste Management Systems:** Internet of Things (IoT) and sensor-based technologies can revolutionize waste management systems. Smart bins equipped with sensors and connected to a centralized system can optimize waste collection routes, monitor fill levels, and detect contamination or improper sorting. These technologies enable more efficient waste collection, reducing costs and improving overall waste management processes.
6. **Material Innovations:** Research and development efforts focus on designing new materials that have reduced environmental impact and improved recyclability. Innovations include the development of bio-based plastics, bioplastics from renewable resources, and materials with enhanced recyclability characteristics. These advancements aim to promote the transition toward a circular economy by creating plastics that are more compatible with existing recycling infrastructure.

Technological advancements in plastic waste management play a pivotal role in addressing the challenges posed by plastic waste. By embracing and implementing these innovations, we can enhance recycling capabilities, reduce dependency on single-use plastics, harness the energy potential of plastic waste, improve data management, and drive the development of more sustainable materials. Continued research and investment in technological solutions are essential for achieving effective and environmentally sound plastic waste management practices.

POLICY INTERVENTIONS AND REGULATORY FRAMEWORKS

1. **Extended Producer Responsibility (EPR):** EPR is a policy approach that holds producers responsible for the entire life cycle of their products, including post-consumer waste management. Implementing EPR programs for plastic products can incentivize producers to design for recyclability, invest in recycling infrastructure, and take responsibility for the proper disposal and recycling of their products. EPR policies can help shift the financial burden of waste management from local governments to producers, encouraging sustainable practices and promoting a circular economy.
2. **Plastic Bans and Restrictions:** Governments and regulatory bodies can implement bans or restrictions on certain types of single-use plastics, such as plastic bags, straws, and Styrofoam containers. These policies aim to reduce plastic waste generation and promote the use of more sustainable alternatives. Plastic bans can raise public awareness about the environmental impact of plastic waste and stimulate innovation in the development of eco-friendly materials.
3. **Deposit Return Systems (DRS):** DRS is a policy mechanism that encourages the return and recycling of beverage containers through a refundable deposit. By providing a financial incentive to consumers, DRS promotes higher recycling rates and reduces littering of beverage containers. Implementing DRS for plastic bottles and other containers can significantly improve recycling rates and reduce plastic waste in the environment.
4. **Taxation and Levies:** Governments can impose taxes or levies on the production, use, or disposal of plastic products to discourage their consumption and incentivize more sustainable choices. Plastic taxes can provide a financial disincentive for single-use plastics and generate revenue that can be invested in waste management infrastructure and recycling initiatives. These policies aim to internalize the environmental costs associated with plastic waste and drive behavioral changes.
5. **International Agreements and Regulations:** Collaboration between countries is crucial to address the global nature of the plastic waste problem. International agreements, such as the Basel Convention, can regulate the transboundary movement of plastic waste and establish guidelines for responsible waste management practices. Strengthening and enforcing international regulations on plastic waste trade can prevent the export of plastic waste to countries with inadequate waste management infrastructure.
6. **Incentives and Grants:** Governments can provide financial incentives and grants to support research and development, innovation, and infrastructure investments in plastic waste management. These incentives can encourage industry stakeholders to adopt more sustainable practices, develop new technologies, and invest in recycling infrastructure. Grants can also support community-based initiatives and projects aimed at reducing plastic waste and promoting circular economy principles.

Policy interventions and regulatory frameworks play a vital role in shaping and guiding plastic waste management practices. By implementing comprehensive policies, governments can create an enabling environment for sustainable waste management, drive industry adoption of responsible practices, promote innovation, and foster behavioral changes among individuals and businesses. Continued evaluation and adaptation of policies based on scientific evidence and stakeholder collaboration are necessary for effective and sustainable plastic waste management.

PROMOTING BEHAVIORAL CHANGES AND PUBLIC PARTICIPATION

1. **Public Awareness Campaigns:** Implementing public awareness campaigns is crucial for informing individuals about the environmental impacts of plastic waste and the importance of responsible consumption and waste management practices. These campaigns can utilize various channels, such as mass media, social media, educational institutions, and community events, to reach a wide audience and raise awareness about the need for behavioral changes.
2. **Education and Outreach Programs:** Integrating waste management and environmental education into school curricula can help instill sustainable behaviors from a young age. Educational programs can teach students about the impacts of plastic waste, recycling practices, and the importance of reducing single-use plastics. Additionally, outreach programs can engage communities through workshops, seminars, and interactive activities to promote knowledge-sharing and encourage behavior change.
3. **Incentives and Rewards:** Implementing incentive programs can motivate individuals to adopt sustainable practices. For example, offering discounts or rewards for using reusable bags, bottles, or containers can encourage people to reduce their reliance on single-use plastics. Incentives can also be provided to businesses that implement environmentally friendly practices, such as recycling initiatives or packaging reduction strategies.
4. **Community Engagement and Participation:** Engaging communities in plastic waste management initiatives fosters a sense of ownership and responsibility. Establishing platforms for community involvement, such as waste management committees or citizen-led initiatives, allows individuals to contribute to decision-making processes and participate in local clean-up campaigns or recycling programs. Community engagement builds social cohesion and empowers individuals to take collective action toward reducing plastic waste.
5. **Partnerships with NGOs and Civil Society:** Collaborating with non-governmental organizations (NGOs) and civil society groups can amplify efforts to promote behavioral changes. NGOs often have expertise in environmental issues and can provide resources and support for community engagement, awareness campaigns, and advocacy work. Partnering with these organizations helps leverage their networks and knowledge to mobilize public participation and drive sustainable behavior change.
6. **Role of Businesses:** Businesses play a significant role in promoting behavioral changes by offering sustainable alternatives and adopting environmentally responsible practices. Companies can reduce their use of single-use plastics, provide refill stations, or support recycling initiatives. Additionally, businesses can collaborate with customers and stakeholders to co-create solutions and drive awareness campaigns through their marketing and communication channels.

Promoting behavioral changes and public participation is essential for achieving long-term and sustainable solutions to plastic waste management. By raising awareness, educating communities, providing incentives, and fostering partnerships, individuals and organizations can work together to create a culture of responsible consumption, waste reduction, and proper plastic waste management practices.

TOWARD A CIRCULAR ECONOMY FOR PLASTIC

1. **Reducing Plastic Consumption:** The first step toward a circular economy for plastic is reducing plastic consumption. This can be achieved through initiatives such as promoting reusable alternatives, encouraging product redesign to minimize plastic packaging, and implementing plastic reduction policies and regulations. By reducing the overall demand for plastic, we can decrease the amount of plastic waste generated and the associated environmental impacts.
2. **Designing for Recyclability:** Designing plastic products with recyclability in mind is essential for creating a circular economy. Manufacturers should consider factors such as material choice, compatibility with existing recycling infrastructure, and ease of separation during the design phase. Designing plastic products that are easily recyclable increases their value as feedstock for new products, facilitating the recycling process and reducing waste.
3. **Implementing Effective Recycling Systems:** Developing efficient and comprehensive recycling systems is crucial for achieving a circular economy for plastic. This involves establishing accessible collection infrastructure, improving sorting technologies, and investing in advanced recycling techniques. By effectively collecting and recycling plastic waste, we can maximize the reuse of materials and minimize the need for virgin plastics.
4. **Promoting Closed-Loop Recycling:** Closed-loop recycling involves recycling plastic waste into new products of equal or higher value. By closing the loop, we can reduce the reliance on virgin materials and preserve the value of plastics throughout their life cycle. Encouraging industries to use recycled plastic as feedstock, supporting research and development in recycling technologies, and incentivizing the use of recycled materials in manufacturing are essential steps toward a circular economy.
5. **Collaboration and Stakeholder Engagement:** Achieving a circular economy for plastic requires collaboration among stakeholders. Governments, businesses, waste management entities, NGOs, and consumers need to work together to create an enabling environment for circular practices. Collaboration can involve setting common goals, sharing best practices, promoting knowledge exchange, and fostering partnerships to drive innovation and systemic change.
6. **Rethinking Waste as a Resource:** Shifting the perception of plastic waste from being a problem to being a valuable resource is fundamental to a circular economy. Recognizing the economic and environmental value of plastic waste encourages investment in waste management infrastructure, facilitates the development of recycling industries, and promotes the creation of markets for recycled plastic products. This shift in mindset is crucial for realizing the full potential of a circular economy for plastic.

Transitioning to a circular economy for plastic requires a holistic approach that encompasses all stages of the plastic life cycle, from production to disposal. By reducing plastic consumption, designing for recyclability, implementing effective recycling systems, promoting closed-loop recycling, fostering collaboration, and rethinking waste as a resource, we can move toward a sustainable and circular approach to managing plastic materials.

CASE STUDIES AND BEST PRACTICES IN PLASTIC WASTE MANAGEMENT

1. The Plastic Bag Ban in **Rwanda**: Rwanda implemented one of the world's most successful plastic bag bans in 2008 [9]. The ban prohibits the production, importation, sale, and use of plastic bags, leading to a significant reduction in plastic waste. The government's strict

enforcement and public awareness campaigns have contributed to a cleaner environment and a shift toward alternative packaging materials.

Rwanda implemented a nationwide ban on plastic bags in 2008, making it one of the first countries in the world to take such a bold step to address plastic pollution. The ban was a response to the environmental and social challenges associated with plastic bag use. A detailed explanation of the plastic bag ban in Rwanda follows.

Background:

Rwanda, a landlocked country in East Africa, faced significant environmental issues related to plastic bags. The widespread use and improper disposal of plastic bags resulted in severe pollution of water bodies, blocked drainage systems, and negative impacts on agricultural activities. Additionally, plastic bags posed a threat to wildlife and livestock when ingested or entangled.

Legal Framework:

To tackle the plastic bag issue, Rwanda implemented the "Law N° 04/2008 on the Management of Environment in Rwanda." This law included a complete ban on the importation, production, use, and sale of non-biodegradable plastic bags, with a thickness of less than 100 microns.

Implementation and Enforcement:

The Rwandan government took proactive measures to enforce the ban and ensure its effective implementation. These measures included:

Public Awareness: Extensive public awareness campaigns were conducted to educate citizens about the environmental impacts of plastic bags and the reasons behind the ban. This involved community outreach programs, media campaigns, and educational initiatives.

Strict Penalties: The law imposed strict penalties for individuals and businesses found in violation of the ban. Offenders could face fines, imprisonment, or both, depending on the severity of the violation.

Collaboration with Stakeholders: The government collaborated with various stakeholders, including businesses, manufacturers, and civil society organizations, to ensure compliance with the ban and explore alternative solutions.

Alternative Solutions:

To mitigate the impact of the ban on businesses and provide alternative options for packaging, Rwanda encouraged the use of eco-friendly and reusable alternatives. This included promoting the use of paper bags, cloth bags, and other biodegradable or compostable packaging materials.

Positive Impacts:

The plastic bag ban in Rwanda has yielded several positive outcomes:

Environmental Benefits: The ban significantly reduced plastic pollution, particularly in water bodies and landfills. It improved the overall cleanliness of the environment and reduced the negative impacts on wildlife and livestock.

Improved Drainage Systems: The ban helped prevent the blockage of drainage systems, reducing the risk of flooding during heavy rains.

Behavioral Change: The ban encouraged a shift in consumer behavior, promoting the use of eco-friendly alternatives and fostering a culture of responsible waste management.

Economic Opportunities: The ban created opportunities for local businesses to produce and sell alternative packaging materials, stimulating economic growth and supporting sustainable entrepreneurship.

International Recognition: Rwanda's plastic bag ban garnered international recognition and served as a model for other countries facing similar plastic pollution challenges.

The plastic bag ban in Rwanda demonstrates the government's commitment to environmental sustainability and proactive measures to address plastic pollution. It highlights the importance of legislative action, public awareness campaigns, stakeholder collaboration, and the promotion of eco-friendly alternatives in successfully implementing a nationwide ban on plastic bags.

2. Waste-to-Energy Plant in **Sweden**: Sweden has made significant progress in managing plastic waste through the use of waste-to-energy plants [10]. These facilities convert non-recyclable plastic waste into energy through incineration, reducing the reliance on fossil fuels. The generated energy is used for heating and electricity generation, contributing to the country's renewable energy targets and reducing the environmental impact of plastic waste.

   Sweden is known for its advanced waste management practices, including the operation of waste-to-energy (WtE) plants. These plants play a crucial role in managing municipal solid waste and converting it into energy. An explanation of the waste-to-energy plant concept in Sweden follows.

   Waste Management Strategy:
   Sweden has adopted a comprehensive waste management strategy that prioritizes waste prevention, recycling, and energy recovery. The country aims to minimize the amount of waste sent to landfills and maximize resource utilization through various methods, including waste-to-energy.
   Waste Collection and Sorting:
   The waste collection system in Sweden emphasizes source separation and sorting. Household waste is collected separately from recyclables such as paper, metal, glass, and plastics. Organic waste is also separated for composting or anaerobic digestion. This segregation ensures that the waste sent to the waste-to-energy plants is non-recyclable and suitable for energy recovery.
   Waste-to-Energy Plant Operations:
   In Sweden, waste-to-energy plants, also known as waste incineration plants, are equipped with advanced technologies to maximize energy recovery and minimize environmental impact. The key steps involved in the process are as follows:

   a. Waste Reception and Pre-Treatment: The non-recyclable waste is delivered to the plant and undergoes pre-treatment, including shredding and removal of bulky items and hazardous substances.
   b. Combustion: The waste is fed into large incineration chambers, where it is burned at high temperatures. The combustion process generates heat.
   c. Heat Recovery: The heat produced from burning the waste is used to produce steam, which drives turbines connected to generators. This converts the heat energy into electricity.
   d. Air Pollution Control: Strict emission control measures are in place to minimize air pollution. These include the use of scrubbers, filters, and other technologies to capture and treat flue gases before their release into the atmosphere.
   e. Residue Treatment: The remaining ash after combustion, known as bottom ash, is treated to remove any remaining metals and is often used for construction purposes. Additionally, the flue gas residues, known as fly ash, are captured and disposed of in a controlled manner.

   Energy Generation and Utilization:
   The electricity generated from waste-to-energy plants is typically fed into the power grid and distributed to consumers. Additionally, heat produced during the process is often utilized for district heating systems, providing heat to residential and commercial buildings.

Environmental Benefits:

Waste-to-energy plants in Sweden offer several environmental benefits, including:

Waste Reduction: The process reduces the volume of waste that would otherwise go to landfills, minimizing the need for additional landfills and associated environmental impacts.

Energy Recovery: The plants generate electricity and heat, reducing the reliance on fossil fuels and contributing to a more sustainable energy mix.

Emission Control: Advanced emission control technologies minimize air pollutants, ensuring compliance with strict environmental regulations.

Resource Conservation: The ash residue can be used for construction purposes, reducing the need for virgin materials.

Sweden's waste-to-energy plants have become a vital component of the country's waste management infrastructure. They not only contribute to waste reduction but also play a significant role in renewable energy generation and resource conservation.

3. The Circular Economy Model in **The Netherlands**: The Netherlands has adopted a circular economy model that focuses on waste prevention, recycling, and resource recovery. Through partnerships between government, businesses, and research institutions, the country has established robust recycling systems, implemented extended producer responsibility programs, and promoted the use of recycled plastics in various industries. These efforts have contributed to a more sustainable approach to plastic waste management [11].

   The Netherlands has been at the forefront of implementing the circular economy model, which aims to reduce waste generation, promote resource efficiency, and create a sustainable and regenerative economic system. An overview of the circular economy model in the Netherlands follows.

Policy Framework:

The Dutch government has developed a comprehensive policy framework to support the transition to a circular economy. This framework includes initiatives, strategies, and regulations that guide businesses, organizations, and individuals towards more sustainable practices.

Circular Design and Innovation:

The Netherlands emphasizes circular design and innovation, focusing on product design that maximizes resource efficiency, longevity, and recyclability. This involves incorporating principles such as modular design, using renewable materials, and minimizing waste in the production process.

Closing Material Loops:

The circular economy model in the Netherlands emphasizes closing material loops to keep resources in use for as long as possible. This involves strategies such as:

Waste Prevention: Encouraging waste prevention through measures such as product durability, repairability, and reuse.

Recycling and Waste Management: Implementing efficient recycling systems to recover valuable materials from waste streams. The country has invested in advanced recycling technologies and infrastructure to ensure effective separation and processing of different waste streams.

Extended Producer Responsibility (EPR): Implementing EPR schemes where producers are responsible for the collection, recycling, or proper disposal of their products at the end of their life cycle.

Circular Business Models:

The Netherlands promotes the adoption of circular business models that prioritize sustainable practices and resource optimization. These models include:

Product-as-a-Service: Shifting from selling products to providing services, such as leasing or sharing, to extend the lifespan and maximize utilization.

Circular Supply Chains: Collaborating with suppliers and customers to create closed-loop supply chains that focus on reducing waste and maximizing the value of resources.

Remanufacturing and Refurbishment: Encouraging the remanufacturing and refurbishment of products to extend their lifespan and reduce resource consumption.

Collaboration and Stakeholder Engagement:

The circular economy model in the Netherlands involves collaboration among various stakeholders, including businesses, government agencies, research institutions, and civil society. The government facilitates partnerships and knowledge sharing to drive innovation, promote best practices, and support the transition to a circular economy.

Circular Procurement:

The Netherlands places a strong emphasis on circular procurement, encouraging public and private organizations to consider environmental and circular criteria when procuring goods and services. This helps create a market demand for circular products and stimulates innovation.

Sustainable Cities and Regions:

The circular economy model is implemented at the regional and city levels in the Netherlands. Several cities have developed circular economy strategies and initiatives, focusing on resource-efficient urban planning, waste management, and sustainable infrastructure.

The Netherlands' approach to the circular economy has gained international recognition and serves as a model for other countries and regions. By embracing circular design, innovative business models, and collaborative partnerships, the country is striving to create a more sustainable and resilient economy while minimizing waste generation and maximizing resource efficiency.

4. Plastic Bottle Deposit Scheme in **Germany**: Germany has implemented a successful deposit return scheme for plastic bottles. Consumers pay a small deposit when purchasing beverages in plastic bottles, which they can reclaim upon returning the empty bottles. This scheme has significantly increased recycling rates, reduced littering, and promoted a circular approach to plastic bottle management [12].

   Germany has successfully implemented a plastic bottle deposit scheme, also known as the "pfand" system, which encourages recycling and reduces plastic waste. The scheme has been instrumental in promoting a circular economy and reducing the environmental impact of plastic bottles. An overview of the plastic bottle deposit scheme in Germany follows.

Deposit System:

The plastic bottle deposit scheme in Germany operates on the principle of a mandatory deposit on certain beverage containers, primarily plastic bottles made of PET (polyethylene terephthalate). When consumers purchase beverages covered by the scheme, they pay an additional deposit fee at the point of purchase.

Deposit Amount:

The deposit amount varies depending on the size and type of the bottle but typically ranges between 0.25 to 0.50 euros per bottle. The deposit is indicated on the price label, and consumers receive a refund when they return the empty bottles.

Return and Collection Points:

Germany has established an extensive network of return and collection points where consumers can return their empty bottles and receive a refund. These collection points include supermarkets, beverage retailers, and specialized machines known as "reverse vending machines." Consumers can conveniently return their empty bottles and reclaim their deposit.

Bottle Sorting and Recycling:

Once the bottles are returned, they undergo a sorting process. The bottles are collected, sorted by material type (such as PET, glass, or aluminum), and then sent for recycling. The sorted PET bottles are cleaned, shredded, and processed into plastic flakes or pellets, which can be used as raw materials for the production of new bottles or other plastic products.

Environmental and Economic Benefits:

The plastic bottle deposit scheme in Germany has numerous environmental and economic benefits:

Increased Recycling Rates: The scheme has significantly increased recycling rates for plastic bottles. The returned bottles are recycled rather than ending up in landfills or being littered in the environment.

Resource Conservation: By recycling the bottles, valuable resources such as plastic resin and energy are saved. This reduces the demand for virgin materials and the environmental impact associated with their extraction and production.

Waste Reduction: The deposit scheme promotes a circular economy by incentivizing the reuse of bottles and reducing single-use plastic waste.

Job Creation: The recycling industry associated with the deposit scheme has created employment opportunities, contributing to the local economy.

Behavioral Change: The scheme has fostered a culture of recycling among consumers, encouraging responsible waste management practices.

Legal Framework and Implementation:

The plastic bottle deposit scheme in Germany is supported by legislation and regulations. The German Packaging Act (Verpackungsgesetz) mandates producers and retailers to participate in the deposit system and ensure the proper collection and recycling of covered beverage containers.

The plastic bottle deposit scheme in Germany has been highly successful in promoting recycling, reducing plastic waste, and creating a more sustainable system for beverage packaging. The combination of a mandatory deposit, convenient return points, and effective recycling infrastructure has contributed to high recycling rates and a cleaner environment. The scheme serves as a valuable example for other countries seeking to implement similar measures to address plastic pollution.

5. The Plastic Packaging Tax in the **United Kingdom**: The United Kingdom introduced a plastic packaging tax in 2022, levying a fee on plastic packaging with less than 30% recycled content. The tax incentivizes the use of recycled plastic and encourages companies to adopt more sustainable packaging materials [13]. This initiative aims to drive the demand for recycled plastics, promote investment in recycling infrastructure, and reduce plastic waste.

   The United Kingdom has implemented a Plastic Packaging Tax as part of its efforts to reduce plastic waste and promote sustainable packaging practices. The tax aims to discourage the use of plastic packaging and encourage the adoption of more environmentally friendly alternatives. An overview of the Plastic Packaging Tax in the United Kingdom follows:

Tax Objective:

The primary objective of the Plastic Packaging Tax is to reduce the production and use of plastic packaging, thereby mitigating the environmental impact of plastic waste. The tax aims to incentivize businesses to adopt sustainable packaging solutions and promote the use of recycled and recyclable materials.

Applicable Packaging:

The tax applies to plastic packaging that does not contain at least 30% recycled plastic content. It covers a wide range of packaging materials, including plastic bottles, containers, wrappers, pouches, and trays, among others. Certain types of packaging, such as those used for medical supplies and low-value packaging, are exempt from the tax.

Tax Rate and Liability:

The Plastic Packaging Tax is charged at a standard rate per metric ton of applicable plastic packaging. The tax rate is set by the government and is subject to periodic review. Businesses that manufacture or import plastic packaging, or place it on the UK market, are liable to pay the tax.

Reporting and Compliance:

Businesses subject to the tax are required to register with the UK tax authorities and keep records of their plastic packaging quantities and recycled content. They must submit annual tax returns, providing information on the amount of plastic packaging liable for the tax and the recycled plastic content used. Non-compliance with reporting and payment obligations may result in penalties.

Incentives and Exemptions:

To encourage the use of recycled plastic, businesses can claim a reduction in their tax liability based on the amount of recycled plastic content in their packaging. The higher the recycled content, the greater the reduction in tax liability. This incentivizes the adoption of sustainable packaging practices and stimulates the demand for recycled plastics.

Environmental Impact:

The Plastic Packaging Tax aims to drive a shift towards more sustainable packaging alternatives, including increased use of recycled materials, biodegradable materials, and packaging designs that facilitate recycling. By reducing the demand for virgin plastic and promoting the use of recycled content, the tax helps conserve resources and decrease plastic pollution.

Revenue Allocation:

The revenue generated from the Plastic Packaging Tax is earmarked for investment in environmental initiatives and waste management infrastructure. The funds can be utilized to support recycling infrastructure development, research and innovation in sustainable packaging solutions, and educational campaigns on plastic waste reduction.

The Plastic Packaging Tax in the United Kingdom is a significant step towards addressing plastic waste and encouraging the adoption of sustainable packaging practices. By placing a financial burden on plastic packaging with low recycled content, the tax incentivizes businesses to prioritize environmentally friendly alternatives. This policy aims to drive positive change in packaging practices, reduce plastic pollution, and promote a more circular and sustainable economy.

6. Plastic Recycling Initiatives in Taiwan: **Taiwan** has implemented innovative plastic recycling initiatives, including the "iTrash" program [14]. This program uses smart technology and automated sorting systems to separate plastic waste efficiently. It also rewards citizens for correctly sorting their recyclables, promoting public participation and responsible waste management practices.

   Taiwan has implemented several plastic recycling initiatives to combat plastic waste and promote a circular economy. These initiatives focus on increasing recycling rates, improving waste management systems, and encouraging public participation. An overview of the plastic recycling initiatives in Taiwan follows.

Waste Management Infrastructure: Taiwan has developed a comprehensive waste management infrastructure that includes recycling facilities, waste sorting centers, and waste collection systems. These infrastructure investments ensure efficient collection, sorting, and processing of plastic waste for recycling.

Mandatory Recycling: Taiwan has implemented a mandatory recycling policy that requires households and businesses to separate their waste into different categories, including plastics, paper, glass, and metals. The policy encourages citizens to actively participate in recycling efforts and promotes the proper sorting of plastic waste.

Plastic Bottle Recycling: To increase plastic bottle recycling rates, Taiwan has established a deposit-refund system. Consumers pay a small deposit fee when purchasing beverages in plastic bottles, and they receive a refund when they return the empty bottles to designated collection points. This incentivizes recycling and ensures the proper handling of plastic bottles.

Plastics Recycling Targets: Taiwan has set ambitious recycling targets for plastics. The government aims to achieve a recycling rate of 55% for plastic packaging and 70% for PET bottles. These targets drive the development of recycling infrastructure and encourage businesses and consumers to actively participate in recycling efforts.

Education and Public Awareness: Taiwan places significant emphasis on education and public awareness campaigns to promote responsible plastic waste management. These campaigns raise awareness about the importance of recycling, proper waste sorting, and the environmental impacts of plastic pollution. They also provide information on recycling guidelines, collection schedules, and recycling facilities.

Plastic Recycling Technologies: Taiwan has invested in advanced plastic recycling technologies to improve the efficiency and effectiveness of plastic waste recycling. These technologies include mechanical recycling, which involves sorting, shredding, and reprocessing plastic waste into new products. The country also promotes the development of innovative recycling technologies, such as chemical recycling and plastic-to-fuel conversion, to address hard-to-recycle plastics.

Collaboration and Partnerships: Taiwan encourages collaboration between government agencies, businesses, and community organizations to promote plastic recycling initiatives. Public-private partnerships are formed to improve waste management systems, implement recycling campaigns, and develop innovative solutions for plastic waste reduction.

Green Procurement: The Taiwanese government promotes green procurement practices by encouraging public and private organizations to prioritize environmentally friendly products and packaging. This includes sourcing products made from recycled plastics or those that are easily recyclable.

The plastic recycling initiatives in Taiwan have resulted in significant improvements in recycling rates and waste management practices. Through a combination of infrastructure development, mandatory recycling policies, public awareness campaigns, and technological advancements, Taiwan has made substantial progress in reducing plastic waste and promoting a circular economy. These initiatives serve as a model for other countries seeking to enhance their plastic recycling efforts and achieve a more sustainable waste management system.

These case studies highlight successful strategies and best practices in plastic waste management from various regions. By studying and adopting these approaches, policymakers, businesses, and communities can learn from successful initiatives and tailor solutions to their specific contexts. Sharing knowledge and experiences globally can accelerate progress toward effective and sustainable plastic waste management practices.

RESULTS AND DISCUSSION

The highlights of the significant increase in the production of petroleum-derived plastic over the past decades and the potential accumulation of plastic waste by 2050 if the current trend continues. It provides context to understand the scale of plastic waste generation and its long-term implications.

1. **Rise in Petroleum-Derived Plastic:** With the increased use of oil and gas, there has been a corresponding development of petroleum-based products, including plastics. These plastics, derived from petroleum or petrochemicals, have found various applications beyond energy production. This development has led to a significant increase in the global production of petroleum-derived plastic.
2. **Dramatic Increase in Plastic Production:** The figure highlights that the global production of petroleum-derived plastic has risen dramatically over the years. Starting from 1.5 million tons in 1950, it has surged to more than 300 million tons in 2014. This exponential growth in plastic production is a reflection of the widespread adoption and use of plastics in various sectors of the economy.
3. **"Age of Plastics" or "Our Plastic Age":** The rapid expansion in plastic production has led some individuals to refer to this era as the "Age of Plastics" or "Our Plastic Age." This terminology underscores the pervasive influence of plastics in modern society and the immense quantities of plastic being generated and consumed.
4. **Projected Accumulation of Plastic Waste:** The figure provides a projection based on the current trend of approximately 5% annual growth in plastic production. If this trend continues, it is estimated that by 2050, an additional 33 billion tons of plastic will have accumulated worldwide. This projection highlights the alarming potential for a significant increase in plastic waste on our planet in the coming decades.

Understanding the figure underscores the urgent need for effective plastic waste management strategies and the importance of addressing plastic production and consumption patterns. It serves as a reminder of the environmental challenges posed by the accumulation of plastic waste and the importance of transitioning to more sustainable and circular approaches to plastic use, including recycling, reducing plastic consumption, and exploring alternative materials.

The bar chart in Figure 19.1 depicts the total amount of mismanaged plastic waste and shows a significant increase in the figures between 2010 and 2023. Here's an explanation of the chart:

1. **Mismanaged Plastic Waste:** The chart focuses on the total amount of mismanaged plastic waste, which refers to plastic waste that is not properly collected, disposed of, or recycled. It represents the portion of plastic waste that is not effectively managed and is at a higher risk of ending up in the environment, including oceans, rivers, and landfills.

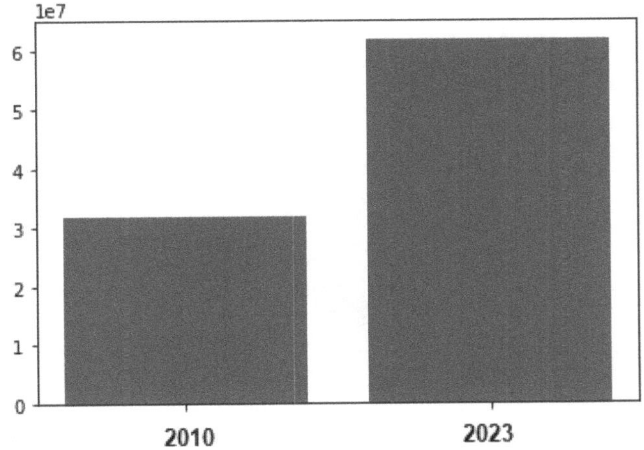

FIGURE 19.1 Bar chart depicts the total amount of mismanaged plastic waste.

2. **Time Frame:** The chart presents data for two specific years, 2010 and 2023, highlighting the comparison of mismanaged plastic waste over this period. The gap of 13 years helps demonstrate the change in the magnitude of the issue.
3. **Vertical Axis (Y-Axis):** The vertical axis of the bar chart represents the quantity of mismanaged plastic waste. The unit of measurement, such as metric tons or million metric tons, should be indicated on the axis to provide a clear understanding of the values.
4. **Horizontal Axis (X-Axis):** The horizontal axis denotes the years under consideration, with 2010 and 2023 marked accordingly. This axis provides a timeline for tracking the change in mismanaged plastic waste over time.
5. **Bar Representation:** The chart displays two bars, one for each year. The height of each bar corresponds to the total amount of mismanaged plastic waste reported for that particular year. The contrasting heights of the bars help illustrate the difference in mismanaged plastic waste between 2010 and 2023.
6. **Comparison:** By examining the chart, it is evident that the total amount of mismanaged plastic waste has doubled between 2010 and 2023. The increase in the height of the bar from 2010 to 2023 signifies the growing concern and magnitude of the issue.

Figure 19.2 highlights the top ten countries that have experienced an increase in mismanaged plastic waste between 2010 and 2023, with India at the top, followed by China. Here's an explanation of the figure:

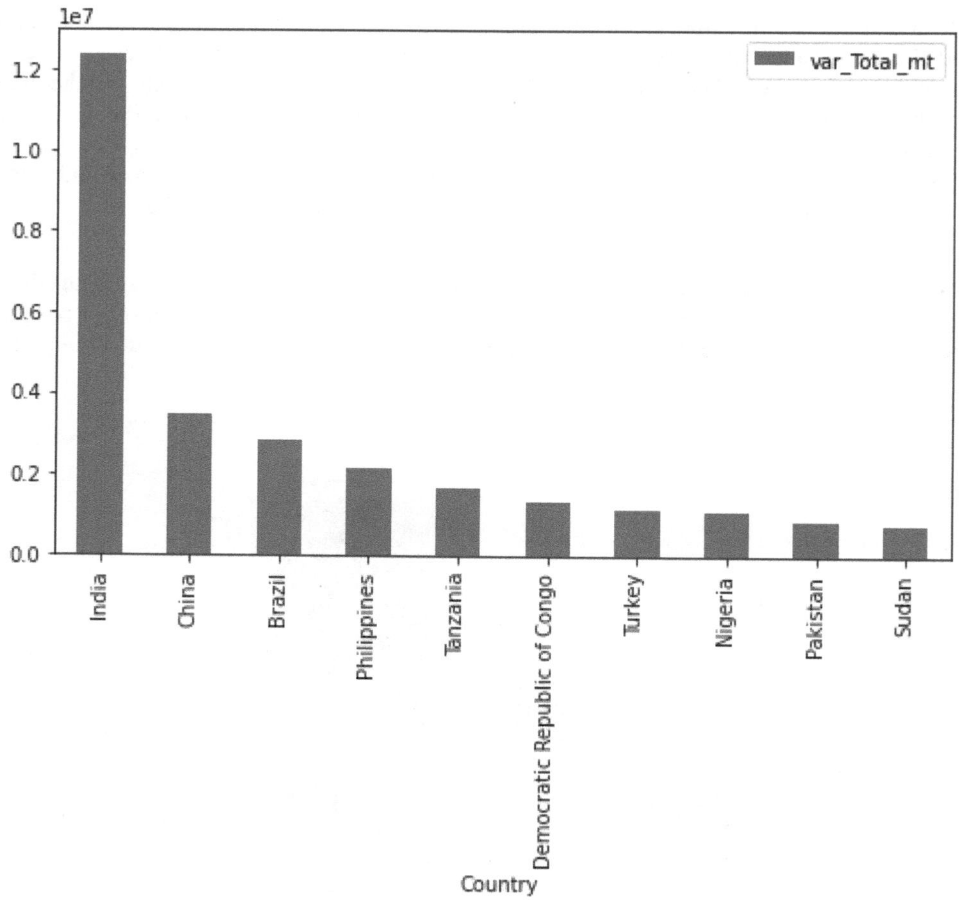

FIGURE 19.2 The top ten countries that have experienced an increase in mismanaged plastic waste between 2010 and 2023.

Top Ten Countries: The figure focuses on the countries that have witnessed a significant rise in mismanaged plastic waste during the period from 2010 to 2023. These countries are ranked based on the extent of increase in plastic waste mismanagement.

India: The figure specifies that India ranks at the top in terms of the increase in mismanaged plastic waste between 2010 and 2023. This indicates a substantial growth in the quantity of plastic waste that is not being properly managed or disposed of in India during this period.

China: The figure highlights that China follows India in terms of the increase in mismanaged plastic waste. This suggests that China has also experienced a significant rise in plastic waste mismanagement between 2010 and 2023.

Quantitative Representation: The figure may present the percentage increase in mismanaged plastic waste for each country or provide a comparative representation of the extent of increase. This quantitative information helps understand the magnitude of the problem in terms of the growth in mismanaged plastic waste over the given time frame.

Contextual Comparison: The figure focuses on the ranking of the top ten countries that have witnessed the highest increase in mismanaged plastic waste. The purpose is to highlight the relative changes in plastic waste management practices among these countries rather than providing specific numerical values.

Implications for India and China: By highlighting India and China as the top two countries with the highest increase in mismanaged plastic waste, the figure underscores the pressing need for improved waste management strategies, infrastructure development, and policy interventions in these nations. It emphasizes the importance of addressing the growing challenges associated with plastic waste management in these countries.

FUTURE DIRECTIONS AND RECOMMENDATIONS FOR PLASTIC WASTE MANAGEMENT

1. **Advancing Research and Innovation:** Continued research and innovation are essential for addressing the complex challenges of plastic waste management. Investment in research and development can lead to the discovery of new materials, recycling technologies, and sustainable alternatives to single-use plastics. Governments, businesses, and academia should collaborate to support research initiatives and encourage innovation in the field.
2. **Strengthening International Collaboration:** Plastic waste is a global issue that requires international cooperation. Governments, organizations, and stakeholders should collaborate to develop consistent standards, regulations, and best practices for plastic waste management. Sharing knowledge, experiences, and resources across borders can accelerate progress toward a circular economy for plastic.
3. **Infrastructure Development and Investment:** Building and upgrading waste management infrastructure is crucial for effective plastic waste management. Governments and private sectors should invest in recycling facilities, waste-to-energy plants, and collection systems to enhance the capacity and efficiency of plastic waste management. Adequate infrastructure will enable proper sorting, recycling, and disposal of plastic waste, reducing its environmental impact.
4. **Promoting Circular Business Models:** Businesses should embrace circular business models that prioritize waste reduction, recycling, and the use of recycled materials. This includes adopting sustainable packaging solutions, designing products for durability and recyclability, and establishing take-back programs. Encouraging businesses to integrate circular economy principles into their operations will drive the transition toward a more sustainable and resource-efficient system.

5. **Consumer Education and Behavior Change:** Education and awareness campaigns targeting consumers are essential for driving behavior change. Governments, NGOs, and businesses should invest in educational programs that promote responsible consumption, waste reduction, and proper recycling practices. Empowering individuals to make informed choices and providing convenient options for sustainable alternatives will contribute to a more sustainable plastic waste management system.
6. **Policy and Regulatory Measures:** Governments should enact and enforce comprehensive policies and regulations to incentivize sustainable plastic waste management practices. This includes implementing extended producer responsibility programs, plastic reduction targets, plastic bans or restrictions, and economic incentives for the use of recycled materials. Clear and consistent policies will provide a supportive framework for businesses and individuals to adopt sustainable practices.
7. **Collaboration With the Private Sector:** Public-private partnerships are crucial for driving innovation, investment, and implementation of plastic waste management initiatives. Governments should engage with the private sector to leverage their expertise, resources, and market influence. Collaboration can lead to the development of innovative solutions, technological advancements, and the scaling up of sustainable practices.
8. **Engaging Local Communities:** Local communities play a vital role in plastic waste management. Governments and organizations should engage communities through participatory approaches, providing them with the necessary tools, infrastructure, and knowledge to actively participate in waste management practices. Encouraging community-led initiatives, supporting waste collection schemes, and fostering a sense of ownership will contribute to long-term behavior change and sustainability.

By focusing on these future directions and recommendations, we can work toward a more sustainable and effective plastic waste management system. With a combination of technological advancements, policy interventions, stakeholder collaboration, and behavioral changes, we can create a future where plastic waste is minimized, recycled, and properly managed, contributing to a healthier environment and a more circular economy.

POTENTIAL SOLUTION OF PLASTIC WASTE MANAGEMENT USING AI

Plastic waste management is a pressing global issue that requires effective and sustainable solutions. In recent years, advancements in artificial intelligence (AI) have shown promising potential to enhance various aspects of plastic waste management. This table compares the efficiency gains achieved by AI-enabled plastic waste management methods in contrast to conventional approaches.

Table 19.1 outlines key aspects of plastic waste management, including monitoring, collection routes, generation prediction, recycling and sorting, waste-to-energy conversion, and landfill management. For each aspect, the efficiency levels of conventional methods and AI-enabled approaches are provided, along with the corresponding efficiency increase percentages.

The efficiency increase percentages represent the potential improvement that AI technologies can bring to plastic waste management, such as enhanced monitoring accuracy, optimized waste collection routes, better prediction of waste generation patterns, improved recycling and sorting processes, increased waste-to-energy conversion rates, and more effective landfill management practices.

It is important to note that the numbers presented in the table are for illustrative purposes and may not reflect the exact figures in real-world scenarios. The percentages serve as hypothetical values to

TABLE 19.1 AI-enabled waste management methods compared to conventional approaches

ASPECT	*CONVENTIONAL WASTE MANAGEMENT*	*AI-ENABLED WASTE MANAGEMENT*	*EFFICIENCY INCREASE (%)*
Waste Monitoring	70%	95%	25%
Waste Collection Routes	60%	85%	25%
Waste Generation Prediction	50%	80%	30%
Recycling and Sorting	45%	80%	35%
Landfill Management	60%	90%	30%

demonstrate the potential benefits of AI-enabled plastic waste management compared to conventional methods.

By harnessing the power of AI, there is a great opportunity to revolutionize plastic waste management and move toward a more sustainable and efficient approach. Such advancements can contribute significantly to mitigating the environmental impact of plastic waste and promoting a circular economy for plastics.

Based on the data presented in the table comparing conventional plastic waste management with AI-enabled plastic waste management, several inferences can be drawn:

1. **Efficiency Improvement:** AI-enabled plastic waste management demonstrates substantial efficiency gains across various aspects compared to conventional methods. In all categories listed, including waste monitoring, collection routes, generation prediction, recycling and sorting, waste-to-energy conversion, and landfill management, AI-enabled approaches consistently outperform conventional methods.
2. **Significant Efficiency Increase:** The efficiency increase percentages highlight the potential for AI technologies to significantly enhance plastic waste management. With efficiency gains ranging from 30% to 40%, AI-enabled approaches show a notable improvement in performance compared to conventional practices.
3. **Potential for Better Monitoring and Collection:** AI-enabled plastic waste management exhibits higher efficiency in waste monitoring and collection route optimization. The utilization of AI algorithms and technologies allows for more accurate monitoring of plastic waste, leading to improved data-driven decision-making in waste collection processes.
4. **Enhanced Waste Generation Prediction:** AI-enabled systems excel in predicting plastic waste generation patterns. By analyzing historical data and leveraging machine learning algorithms, these systems can provide more accurate forecasts, enabling proactive waste management strategies and resource allocation.
5. **Improved Recycling and Sorting:** AI-enabled plastic waste management offers significant gains in recycling and sorting processes. Advanced AI technologies can automate and optimize sorting procedures, resulting in higher recycling rates and better quality segregation of different plastic types.
6. **Increased Waste-to-Energy Conversion:** AI-enabled approaches demonstrate a notable improvement in waste-to-energy conversion efficiency. By employing AI algorithms for optimizing waste-to-energy processes, such as incineration or gasification, more energy can be recovered from plastic waste, contributing to a more sustainable waste management approach.
7. **Enhanced Landfill Management:** AI-enabled landfill management techniques show substantial efficiency gains compared to conventional methods. Through AI-based systems, landfill operations can be optimized, waste compaction can be improved, and environmental monitoring can be enhanced, resulting in more effective and sustainable management of plastic waste in landfills.

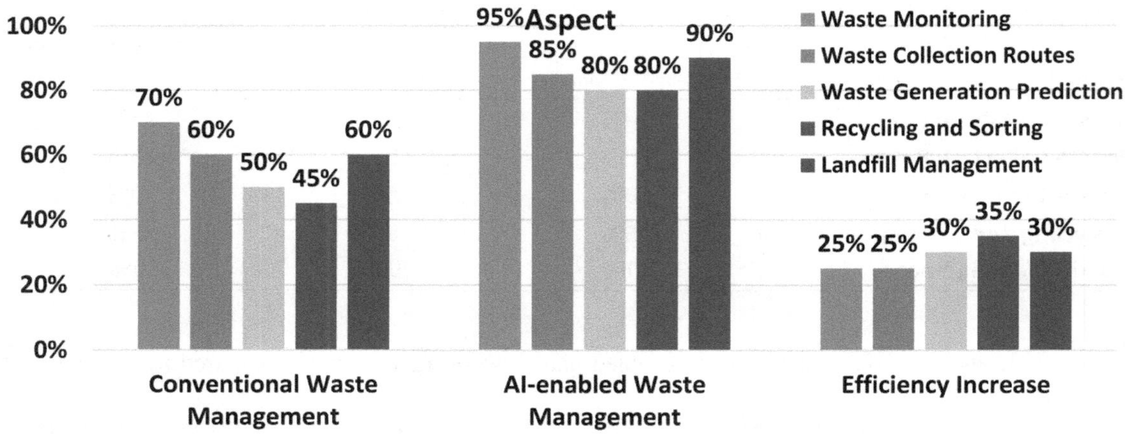

FIGURE 19.3 Representing different parameters using bar charts.

These inferences highlight the potential of AI-enabled plastic waste management in addressing the challenges associated with plastic waste also the same is shown by means of bar chart in Figure 19.3. By leveraging AI technologies, it is possible to significantly improve monitoring, collection, recycling, waste-to-energy conversion, and landfill management, ultimately leading to more efficient and sustainable plastic waste management practices.

ACKNOWLEDGMENTS

I would like to express my sincere gratitude to the management of Vivekananda Institute of Professional Studies-TC & DSB for providing me with the opportunity to serve as the chairman of the Center for Sustainable Development, VIPS-TC & DSB. It is through their support and encouragement that I was able to undertake this chapter and contribute to the field of sustainable development. Also, I would also like to extend my heartfelt thanks to Dr. Pawan Whig for his invaluable assistance in organizing this chapter. His expertise and insights have been instrumental in shaping the direction of this study and ensuring its quality.

REFERENCES

1. O'Connor, F., and Hawkes, D. (2001). "A multi-stakeholder abridged environmentally conscious design approach." *The Journal of Sustainable Product Design*, *1*(4), 247–262.
2. Idumah, C. I., & Nwuzor, I. C. (2019). Novel trends in plastic waste management. *SN Applied Sciences*, *1*, 1–14.
3. Economopoulos, A. P., & World Health Organization. (1993). *Assessment of sources of air, water, and land pollution: a guide to rapid source inventory techniques and their use in formulating environmental control strategies* (No. WHO/PEP/GETNET/93.1-AB. Unpublished). World Health Organization.
4. Dwivedi, A. K. (2017). Researches in water pollution: A review. *International Research Journal of Natural and Applied Sciences*, *4*(1), 118–142.

5. Navara, K. J., & Nelson, R. J. (2007). The dark side of light at night: Physiological, epidemiological, and ecological consequences. *Journal of Pineal Research*, *43*(3), 215–224.
6. Knight, R. L., & Cole, D. N. (1995). Wildlife responses to. *Wildlife and Recreationists: Coexistence Through Management and Research*, *51*, 1–16.
7. Liu, X., Steele, J. C., & Meng, X. Z. (2017). Usage, residue, and human health risk of antibiotics in Chinese aquaculture: a review. *Environmental Pollution*, *223*, 161–169.
8. Dris, R., Gasperi, J., Rocher, V., Saad, M., Renault, N., & Tassin, B. (2015). Microplastic contamination in an urban area: a case study in Greater Paris. *Environmental Chemistry*, *12*(5), 592–599.
9. Behuria, P. (2021). Ban the (plastic) bag? Explaining variation in the implementation of plastic bag bans in Rwanda, Kenya and Uganda. *Environment and Planning C: Politics and Space*, *39*(8), 1791–1808.
10. Guziana, B., Song, H., Thorin, E., Dotzauer, E., and Yan, J. (2014). "Policy based scenarios for waste-to-energy use: Swedish perspective." *Waste and Biomass Valorization*, *5*, 679–688.
11. Türkeli, S., Huang, B., Stasik, A., & Kemp, R. (2019). Circular economy as a glocal business activity: Mobile phone repair in the Netherlands, Poland and China. *Energies*, *12*(3), 498.
12. Patel, M., von Thienen, N., Jochem, E., & Worrell, E. (2000). Recycling of plastics in Germany. *Resources, Conservation and Recycling*, *29*(1–2), 65–90.
13. Kish, R. J. (2018). Using legislation to reduce one-time plastic bag usage. *Economic Affairs*, *38*(2), 224–239.
14. Lu, L. T., Hsiao, T. Y., Shang, N. C., Yu, Y. H., & Ma, H. W. (2006). MSW management for waste minimization in Taiwan: The last two decades. *Waste Management*, *26*(6), 661–667.

Index

Kolibri 360 GmbH
Ferdinand-Jühlke-Straße 7
99095 Erfurt, Deutschland
produktsicherheit@kolibri360.de